MULTIBODY DYNAMICS

Computational Methods in Applied Sciences

Volume 4

Series Editor

E. Oñate

Multibody Dynamics

Computational Methods and Applications

Edited by

JUAN CARLOS GARCÍA ORDEN

Universidad Politecnica de Madrid,
Madrid, Spain

JOSÉ M. GOICOLEA

Universidad Politecnica de Madrid,
Madrid, Spain

and

JAVIER CUADRADO

University of La Coruña
Ferrol, Spain

A C.I.P. Catalogue record for this book is available from the Library of Congress.

ISBN-10 1-4020-5683-4 (HB)
ISBN-13 978-1-4020-5683-3 (HB)
ISBN-10 1-4020-5684-2 (e-book)
ISBN-13 978-1-4020-5684-0 (e-book)

Published by Springer,
P.O. Box 17, 3300 AA Dordrecht, The Netherlands.

www.springer.com

Printed on acid-free paper

Table of Contents

Preface

Multibody Dynamics represents a challenging area of mechanical engineering both from the point of view of mathematical and computer models as well as from that of the applications. During the last years there have been significant contributions in the field, due to which increased maturity has been reached in theoretical models as well as in the availability of efficient commercial software. Furthermore, important improvements have been achieved in the coupling with Computer Aided Design (CAD) software, as well as with Finite Element Methods (FEM). Nevertheless, important contributions remain to be performed by the scientific and engineering community. Within the topics related to *methods and models*, open development areas include contact and impact, control and mechatronics, real-time simulation, optimization, flexible multibody systems, time integration schemes and software development. As regards *applications*, current areas of interest include robotics and walking machines, vehicle dynamics, aerospace, biomechanics, multidisciplinary applications and, finally, education.

The ECCOMAS Thematic Conference *Multibody Dynamics 2005* was held in Madrid from the 21st to the 24th of June, representing the second edition of a series which began in Lisbon 2003. The conference was held under the auspices of the European Community on Computational Methods in Applied Sciences ECCOMAS (`www.eccomas.org`). This conference provided a forum for discussion of several topics, through the participation of 161 researchers from 29 countries, mostly Europeans but also from Africa, America and Asia.

This book contains the revised and extended versions of selected conference communications, representing the state-of-the-art in the advances on computational multibody models, from the most abstract mathematical developments to practical engineering applications. We feel that this book will be highly valuable for those experienced researchers that want to keep updated on the details of the latest driving ideas in this field, but also to researchers approaching the field for the first time, since it provides a useful overview of the most active areas and the efforts devoted by many prominent research groups worldwide.

Juan Carlos García-Orden, José María Goicolea and Javier Cuadrado
June 2006

Modelling Multi-Body Systems Using the Master–Slave Approach

Gordan Jelenić[1] and José J. Muñoz[2]

[1]*Department of Civil Engineering, University of Rijeka, Rijeka, Republic of Croatia; E-mail: gordan@gradri.hr*
[2]*Department of Applied Mathematics III, Polytechnic University of Catalonia, Barcelona, Spain; E-mail: j.munoz@upc.edu*

Abstract. Many engineering structures involve some type of kinematic joints. Modelling such structures therefore represents a static or dynamic multi-body problem, be it deployment and retraction of roofs, domes, satellite aerials or rocket wings [25, 27, 28, 30, 41, 49, 55, 63], actuation of serial robot arms and parallel platforms and dynamics of rotating machinery [8, 20, 42, 43, 47, 61, 64, 65], operation of flexible mechanisms including suspension systems, steering mechanisms, railway collecting pantographs [23, 39, 40, 46, 54], or some other engineering multi-body problem.

Traditionally, these problems are analysed using specialist rigid-body mechanism codes, which often import flexibilities from finite element codes, but nowadays it is becoming exceedingly popular to employ non-linear finite element packages and enhance them so that they can handle this additional level of kinematic complexity. In this work, a particular technique within this approach has been applied to model a variety of simple and more complex joint types. The same method can be also employed to model general contact problems of elastic bodies, as a numerical example included at the end of this chapter demonstrates.

1 Introduction

Within the finite-element method, modelling the kinematics of joints by using the method of *Lagrange multipliers* [10, 21, 22, 26, 44, 62] is very popular. This technique generates certain difficulties, however, such as constraint violation [3] and the Lagrange multipliers pose as extra degrees of freedom, different in their physical character as well as their order of magnitude from the existing (kinematical) degrees of freedom. The resulting matrix of coefficients for such a mixed system is not necessarily well-posed and positive definite even in a stable region, which makes the occurrence of divergence-type instabilities less obvious. The extra degrees of freedom are avoided if the joints are simulated using a *penalty-type approach* [12, 15, 16, 42], but such a procedure is necessarily approximate, depending on the magnitude of the penalty parameters and is prone to numerical ill-conditioning. The two techniques

J.C. García Orden et al. (eds), Multibody Dynamics. Computational Methods and Applications, 1–22.

can be advantageously combined in the so-called *augmented Lagrange method* (e.g. [2]), which eliminates the approximate character of the penalty method while, for a suitably chosen penalty parameter, it retains the advantage of providing a positive definite Hessian at a stable equilibrium. Additional improvement follows by applying another constant scaling in order to eliminate the problem of different orders of magnitude for the unknown degrees of freedom. The augmented Lagrangian method thus eliminates most of the above problems at the expense of introducing two sets of problem-dependent constant parameters.

If, on the other hand, we introduce the joint kinematics at a node prior to the assembly of the finite element mesh, we arrive at the so-called *minimum set method*, also called the *master–slave method* [33, 34, 45], and the *parent–child method* [48]. This technique has many similarities with the *projection methods* in [1], the *joint coordinates* in [38], or the *constraint elimination* in [6]. In the terminology of Rosenberg [4], in this approach the constraints are *embedded* within the equilibrium equations, whereas with the Lagrange multipliers the constraints are *adjoined* to the equilibrium equations. A method originally employing the Lagrange multipliers, which are subsequently eliminated from the equilibrium equations by means of the basis matrix of the null space of the constraint Jacobian, called the *null-space method* [17], the *orthogonal complementary method* [31] or the *projection method* [19] leads to equivalent results.

Let us note that the term "master–slave" is also extensively used in the context of contact mechanics for denoting the surfaces in contact. Here, however, this term should be understood as the technique of embedding the joint constraints into the equilibrium equations.

In addition to the complexities associated with the non-linear character of joints, standard implicit time-integration strategies (which are generally considered as suitable for the long-term analysis in structural dynamics) cannot always be considered as reliable in the analysis of flexible mechanisms. Earlier work [24, 37, 59] has shown that the algorithms from the well-known Newmark family [53] and its Hilber–Hughes–Taylor generalisation [32], which are often used in commercial non-linear finite element codes, are not unconditionally stable when applied to problems of non-linear dynamics. In order to address this difficulty, new formulations have been proposed during the past decade, which are designed to preserve the energy and momenta for an ideal Hamiltonian problem with symmetries.

This work has been based on the use of standard three-dimensional beam elements which necessarily include rotations as variables. For the finite-element solution procedure to be successful when applied to problems of flexible multi-body mechanisms, a rigorous kinematic description of not only the joints, but also the beam model, is a must.

In the first part of this work, the method is used to formulate a mechanical model for a variety of joint types, including revolute and prismatic joints (joints with a single released degree of freedom); cylindrical, universal and spherical joints (joints with multiple independent degrees of freedom); and screw, rack-and-pinion and cam joints (joints with dependent degrees of freedom). These joints are defined at fixed points, which are taken to coincide with the finite-element nodes, and such a formulation is therefore named the *node-to-node* approach. The formulation presented in this part can be applied to any beam model with rotational degrees of freedom.

In the second part of the work, a specific problem of modelling a sliding joint on a deformable surface, named the *node-to-element* approach, is investigated. This problem introduces a range of complexities in comparison to the problem of modelling a simple prismatic or cylindrical joint moving along a rigid axis [13, 14], which result in the expansion of the null-space matrix [7] and require specially designed procedures to handle the transition of a contact point between elements [51]. In order to address the key ingredients of the approach, we will describe the method on a reduced model with no rotational degrees of freedom. Formulations for geometrically exact 3D beams, which include these degrees of freedom, can be found in [51, 52]. The first of these references shows that this approach is in fact applicable to the problems of modelling bilateral point-on-surface contact in 3D continua, and the numerical example included here illustrates this.

2 Node-to-Node Approach

2.1 Standard variational approach in analysis of flexible multi-body problems

The essence of the master–slave technique lies in the specific method in which inter-element kinematics is introduced into the finite-element process: the technique forms a module, where the kinematic releases that may exist at an inter-element connection are accounted for by defining the specific kinematic relationship between the elemental degrees of freedom at the connection. If the joint were not present at the connection, the degrees of freedom for each element at the connection would be simply made equal to each other. In this way, the mesh assembly – a standard operation within the finite-element methodology – is conveniently used to eliminate the algebraic constraint conditions. When a joint exists at a point different than a finite-element nodal point, for example in the case of sliding contact conditions, the mesh assembly is less straightforward, but it nonetheless follows the same basic principles as explained in Section 3 (see also [51]).

2.1.1 Kinematics of joints

Figure 1(a) shows a so-called *prismatic joint*, where a released variable s is the *amount of translation*, the direction of which is constrained to a continually moving axis, rigidly attached to the end of one of the elements. Let us name this end the *master* node, and the element to which it belongs the *master* element. A unit vector along this axis is one of the orthogonal vectors rigidly attached to the master node, which form an orthonormal *master* triad $\boldsymbol{\Lambda}_m \in SO(3) \mid \boldsymbol{\Lambda}_m^{-1} = \boldsymbol{\Lambda}_m^T$ and $\det \boldsymbol{\Lambda}_m = +1$ with an associated master rotation vector $\boldsymbol{\theta}_m \in \mathcal{R}^3 \mid \exp \widehat{\boldsymbol{\theta}}_m = \boldsymbol{\Lambda}_m$, where $\forall\, \boldsymbol{v}, \boldsymbol{w} \in \mathcal{R}^3\ \exists\, \widehat{\boldsymbol{v}}, \widehat{\boldsymbol{w}} \in so(3) \mid \boldsymbol{v} \times \boldsymbol{w} = \widehat{\boldsymbol{v}}\boldsymbol{w} = -\widehat{\boldsymbol{w}}\boldsymbol{v}$ [11, 57]. The orientation of the master node is defined by $\boldsymbol{\Lambda}_m$ (or $\boldsymbol{\theta}_m$) and its position is defined by the position vector $\boldsymbol{r}_m \in \mathcal{R}^3$ as shown in Figure 1(c).

For a *revolute joint* in Figure 1(b), the released joint variable is the rotation θ about a continuously moving axis of the master triad. Consequently, in an imaginary general case of a joint with all the degrees of freedom released, these *released degrees of freedom* will include the released displacement vector $\boldsymbol{r}_r \in \mathcal{R}^3$ and the released rotation vector $\boldsymbol{\theta}_r \in \mathcal{R}^3 \mid \exp \widehat{\boldsymbol{\theta}}_r = \boldsymbol{\Lambda}_r \in SO(3)$ as shown in Figure 1(c). The position and rotation of the node attached to the end of the other element in Figure 1(c) are denoted as $\boldsymbol{r}_s \in \mathcal{R}^3$ and $\boldsymbol{\theta}_s \in \mathcal{R}^3 \mid \exp \widehat{\boldsymbol{\theta}}_s = \boldsymbol{\Lambda}_r \in SO(3)$, and are named the *slave* position and rotation, and their node the *slave* node.

The position and rotation of the master and slave nodes are initially shared by two (or more) elements but, since they are not fully connected to each other, in the deformed configuration they are no longer completely shared and from Figure 1(c) we establish the relations

$$\boldsymbol{r}_s = \boldsymbol{r}_m + \boldsymbol{r}_r \tag{1}$$

$$\boldsymbol{\Lambda}_s = \boldsymbol{\Lambda}_r \boldsymbol{\Lambda}_m \Longleftrightarrow \exp \widehat{\boldsymbol{\theta}}_s = \exp \widehat{\boldsymbol{\theta}}_r \exp \widehat{\boldsymbol{\theta}}_m . \tag{2}$$

When modelling real joints, the master variables $\boldsymbol{r}_m$ and $\boldsymbol{\theta}_m$ are not entirely independent of the slave variables $\boldsymbol{r}_s$ and $\boldsymbol{\theta}_s$. Depending on the type of joint, some of the components of the position vectors $\boldsymbol{r}_m$ and $\boldsymbol{r}_s$, or the parameters defining the rotation vectors $\boldsymbol{\theta}_m$ and $\boldsymbol{\theta}_s$, can be the same. As shown in Figure 1, different types of joints are defined by releasing displacements and rotations with respect to the relevant axes. Since these axes rotate with the structure, the components of the released displacement or rotation need to be defined with respect to these continually rotating axes and for this reason it is necessary to introduce the *material* forms of the released degrees of freedom defined as

$$\boldsymbol{r}_R = \boldsymbol{\Lambda}_m^T \boldsymbol{r}_r \tag{3}$$

$$\boldsymbol{\theta}_R = \boldsymbol{\Lambda}_m^T \boldsymbol{\theta}_r , \tag{4}$$

where the material vector of released displacements $\boldsymbol{r}_R$, and the material released rotation vector $\boldsymbol{\theta}_R$ have zero components in non-released directions.

Following on from this discussion, (1) and (2) may be rewritten as

$$\boldsymbol{r}_s = \boldsymbol{r}_m + \boldsymbol{\Lambda}_m \boldsymbol{r}_R \tag{5}$$

$$\boldsymbol{\Lambda}_s = \boldsymbol{\Lambda}_m \exp \widehat{\boldsymbol{\theta}}_R \Longleftrightarrow \exp \widehat{\boldsymbol{\theta}}_s = \exp \widehat{\boldsymbol{\theta}}_m \exp \widehat{\boldsymbol{\theta}}_R . \tag{6}$$

More detail on the kinematics of the master–slave method can be found in [34, 37].

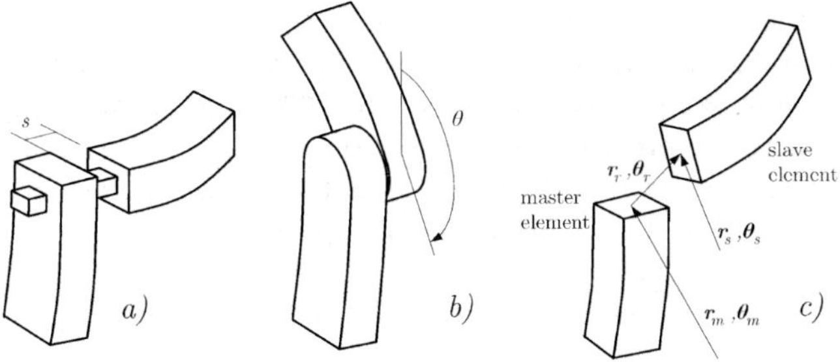

Fig. 1. Kinematics of joints: (a) prismatic joint, (b) revolute joint, (c) degrees of freedom.

2.1.2 Variational approach

For a static analysis, the principle of virtual work may be used to obtain the weak form of the structural equilibrium. Central to this is the application of the variational calculus in order to obtain the *virtual displacements*, in particular the virtual master, slave and released degrees of freedom which, as we know now, are not independent from one another. In particular, we aim to express $\delta\boldsymbol{\beta}_s \in \mathcal{R}^3$, associated with the variation of $\boldsymbol{\Lambda}_s$, in terms of $\delta\boldsymbol{\beta}_m \in \mathcal{R}^3$, associated with the variation of $\boldsymbol{\Lambda}_m$, and $\delta\boldsymbol{\beta}_R \in \mathcal{R}^3$, associated with the variation of $\exp\widehat{\boldsymbol{\theta}}_R$. In order to alleviate the notation, from now on we will drop the index "s" from the description of the slave variables.

The variation of the slave rotation matrix $\boldsymbol{\Lambda}$ can be expressed in terms of the variation $\delta\boldsymbol{\beta}$ as [57]

$$\delta\boldsymbol{\Lambda} = \widehat{\delta\boldsymbol{\beta}}\boldsymbol{\Lambda}, \tag{7}$$

and, in conjunction with (6), this results in [34]

$$\delta\boldsymbol{\beta} = \delta\boldsymbol{\beta}_m + \boldsymbol{\Lambda}_m\delta\boldsymbol{\beta}_R, \tag{8}$$

where $\delta\boldsymbol{\beta}$, $\delta\boldsymbol{\beta}_m$, $\delta\boldsymbol{\beta}_R$ are often referred to as the *spin* slave, master and released variables (in contrast to $\delta\boldsymbol{\theta}$, $\delta\boldsymbol{\theta}_m$, $\delta\boldsymbol{\theta}_R$, referred to as the additive slave, master and released rotational variations). In the same way, (5) may be varied to obtain

$$\delta\boldsymbol{r} = \delta\boldsymbol{r}_m + \boldsymbol{\Lambda}_m\delta\boldsymbol{r}_R - \widehat{\boldsymbol{\Lambda}_m\boldsymbol{r}_R}\delta\boldsymbol{\beta}_m, \tag{9}$$

The relation between the master and the released variations on the right-hand side and the slave variations on the left-hand side can be now written in a matrix form as

$$\begin{Bmatrix} \delta\boldsymbol{r} \\ \delta\boldsymbol{\beta} \end{Bmatrix} = \begin{bmatrix} \boldsymbol{\Lambda}_m & \mathbf{0} & \mathbf{I} & -\widehat{\boldsymbol{\Lambda}_m\boldsymbol{r}_R} \\ \mathbf{0} & \boldsymbol{\Lambda}_m & \mathbf{0} & \mathbf{I} \end{bmatrix} \begin{Bmatrix} \delta\boldsymbol{r}_R \\ \delta\boldsymbol{\beta}_R \\ \delta\boldsymbol{r}_m \\ \delta\boldsymbol{\beta}_m \end{Bmatrix}. \tag{10}$$

In practice, not all of the "releasable" variables will be really released: some of them will be instead set to zero, so that most often there will be only one non-zero component in $\delta \boldsymbol{r}_R$ or $\delta \boldsymbol{\beta}_R$. For each node i on a beam finite element, (10) can be rewritten in a compact form as

$$\delta \boldsymbol{p}_i = \mathbf{N}_i \delta \boldsymbol{p}_{Rm,i}, \tag{11}$$

where $\delta \boldsymbol{p}_i^T = \{\delta \boldsymbol{r}_i^T \ \delta \boldsymbol{\beta}_i^T\}$, $\delta \boldsymbol{p}_{Rm,i}^T = \{\delta \boldsymbol{r}_{R,i}^T \ \delta \boldsymbol{\beta}_{R,i}^T \ \delta \boldsymbol{r}_{m,i}^T \ \delta \boldsymbol{\beta}_{m,i}^T\}$ and

$$\mathbf{N}_i = \begin{bmatrix} \boldsymbol{\Lambda}_{m,i} & \mathbf{0} & \mathbf{I} & -\widehat{\boldsymbol{\Lambda}_{m,i} \boldsymbol{r}_{R,i}} \\ \mathbf{0} & \boldsymbol{\Lambda}_{m,i} & \mathbf{0} & \mathbf{I} \end{bmatrix}. \tag{12}$$

Application of the principle of virtual work to a general N-noded beam element leads to the relationship

$$\sum_{i=1}^{N} \delta \boldsymbol{p}_i \cdot (\boldsymbol{q}^i - \boldsymbol{f}^i) = 0, \tag{13}$$

where $\boldsymbol{q}^i$ and $\boldsymbol{f}^i$ are respectively the internal and the external load vectors at node i and $\boldsymbol{q}^i - \boldsymbol{f}^i$ is the static residual vector at this node.

Inserting (11) into (13) now leads to the virtual work equation

$$\sum_{i=1}^{N} \delta \boldsymbol{p}_{Rm,i} \cdot \mathbf{N}_i^T (\boldsymbol{q}^i - \boldsymbol{f}^i) = 0, \tag{14}$$

and, as this equation must hold for any virtual variables, $\delta \boldsymbol{p}_{Rm,i}$ $(i = 1, \ldots, N)$, we obtain the vector equilibrium at every node i as

$$\mathbf{N}_i^T (\boldsymbol{q}^i - \boldsymbol{f}^i) = \boldsymbol{0}. \tag{15}$$

A different approach in obtaining this result is given in [17], where it comes as a last step in the process of modelling the joints as the kinematic constraints introduced into the equilibrium equation through the Lagrange multipliers, and subsequently eliminating them through the equivalent premultiplication of the result by the null-space matrix $\mathbf{N}_i^T$.

2.1.3 Application to end-point dynamics

Extending the method to cater for dynamic problems is remarkably straightforward: the transpose of the linear operator (the null-space matrix) between the variations of the kinematic variables in joints, should now act on the vector of *dynamic*, rather than the static residuals. In this way, the standard end-point approaches like Newmark's [53] and Hilber, Hughes and Taylor's α-method [32] can be immediately applied.

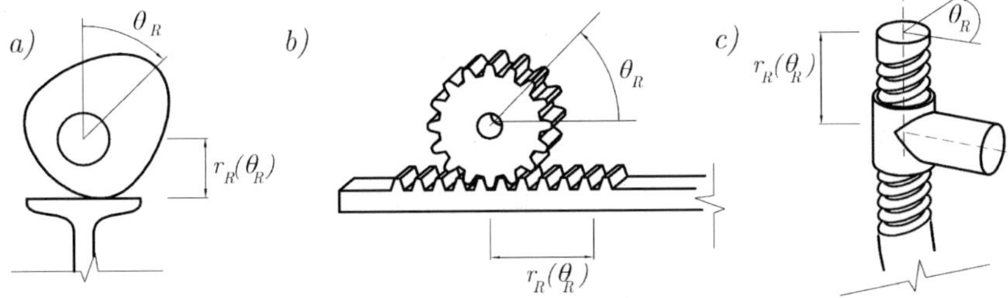

Fig. 2. Joints with dependent degrees of freedom.

The formulation has been here applied only to a particular type of rotational variations – the spin variables $\delta\boldsymbol{\beta}$ – and, therefore, any nodal dynamic residual can be processed within the present master–slave methodology provided it is work conjugate to $\delta\boldsymbol{p}_i^T = \{\delta\boldsymbol{r}_i^T\ \delta\boldsymbol{\beta}_i^T\}$. In the case of isoparametric beams based on the Reissner–Simo theory [56], the dynamic residuals given in [35, 36, 58, 60], among others, may all be used. Subject to the above condition on the variations of the slave degrees of freedom, co-rotational formulations, e.g. [24], are equally applicable. In all of these elements the integration in time has been performed by employing the Newmark trapezoidal rule [53] or its α-generalisation [32], but other time-stepping schemes are also applicable.

2.2 Joints with dependent degrees of freedom

To consider practical problems of rotating machinery, steering mechanisms and suspension systems, more complex joint models than those given earlier are needed. Examples include the eccentric cam joint depicted in Figure 2(a), the rack-and-pinion joint depicted in Figure 2(b) and the screw joint depicted in Figure 2(c).

In these joints the released displacement and rotation vectors are not independent of one another. For the cam joint in Figure 2(a) with a given cam-lobe profile function, the released displacement r_R, for example, depends on an *independent* rotation θ_R. For the rack-and-pinion joint in Figure 2(b) and the screw joint in Figure 2(c), the released displacement r_R is again dependent, even though this time linearly, on the independent released rotation θ_R. All of them, therefore, have *only one independent degree of freedom*: rotation θ_R. A similar approach can also be applied to worm gears, bevel gears, and helical gears [5].

Incorporating the joints with several mutually dependent released degrees of freedom into the presented master–slave method comes as very natural: the relationship between the released degrees of freedom simply serves to reduce the dimension of the transpose of the linear operator (the null-space matrix) between the kinematic variables in joints [50]. As a special case, a rigid segment also falls into the category of joints with dependent degrees of freedom [33].

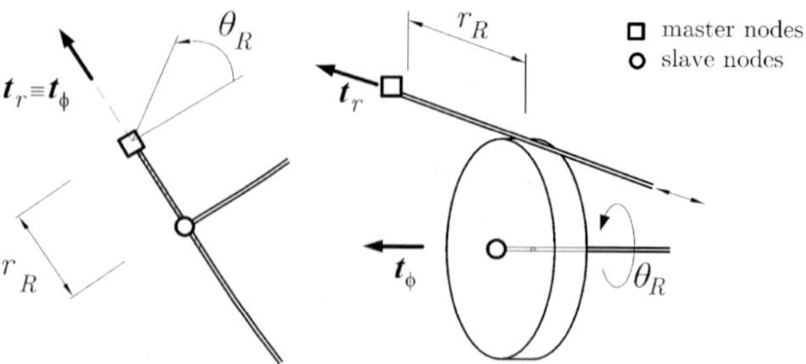

Fig. 3. Scheme of the screw joint and the rack-and-pinion joint.

2.2.1 Kinematics of joints with dependent degrees of freedom

Setting the released displacement $\boldsymbol{r}_R$ as the dependent variable and the released rotation $\boldsymbol{\theta}_R$ as the independent variable, the relationship between the two of them may be written through a function $\boldsymbol{r}_R = \boldsymbol{f}(\boldsymbol{\theta}_R)$.

In the joints commonly considered, only one component of the released rotation will be non-zero, in which case $\boldsymbol{\theta}_R$ and $\delta\boldsymbol{\beta}_R$ are coaxial. Therefore $\delta\boldsymbol{\theta}_R = \delta\boldsymbol{\beta}_R$, and the variation of $\boldsymbol{r}_R = \boldsymbol{f}(\boldsymbol{\theta}_R)$ reads $\delta\boldsymbol{r}_R = \frac{\partial \boldsymbol{f}(\boldsymbol{\theta}_R)}{\partial \boldsymbol{\theta}_R}\delta\boldsymbol{\beta}_R$. Replacing $\delta\boldsymbol{r}_R$ into (9) the matrix $\mathbf{N}$ turns into

$$\mathbf{N} = \begin{bmatrix} \mathbf{0} & \boldsymbol{\Lambda}_m \frac{\partial \boldsymbol{f}(\boldsymbol{\theta}_R)}{\partial \boldsymbol{\theta}_R} & \mathbf{I} & -\widehat{\boldsymbol{\Lambda}_m \boldsymbol{r}_R} \\ \mathbf{0} & \boldsymbol{\Lambda}_m & \mathbf{0} & \mathbf{I} \end{bmatrix}. \tag{16}$$

2.2.2 Screw joint and rack-and-pinion joint

These joints have important common features that enable defining them in a unique manner. In both cases there exists a released rotation around axis $\boldsymbol{t}_\phi$ (see Figure 3), and a released displacement with components along axes $\boldsymbol{t}_r$ (attached to the master node) and $\boldsymbol{t}_r \times \boldsymbol{t}_\phi$. For the screw joint (sc) axes $\boldsymbol{t}_r$ and $\boldsymbol{t}_\phi$ are coincident, while in the rack-and-pinion joint (rp) they are perpendicular to each other. The latter also has a non zero initial displacement (the radius of the pinion) in a direction perpendicular to $\boldsymbol{t}_r$ and $\boldsymbol{t}_\phi$.

The particular form of $\boldsymbol{r}_R = \boldsymbol{f}(\boldsymbol{\theta}_R)$ for these joints can be written using the scalar parameters b and c as

$$\boldsymbol{r}_R = \boldsymbol{f}(\boldsymbol{\theta}_R)_{sc} = \boldsymbol{f}(\boldsymbol{\theta}_R)_{rp} = b(\boldsymbol{t}_r \times \boldsymbol{t}_\phi) + c(\boldsymbol{\theta}_R \cdot \boldsymbol{t}_\phi)\boldsymbol{t}_r = b\widehat{\boldsymbol{t}}_r \boldsymbol{t}_\phi + c(\boldsymbol{t}_r \otimes \boldsymbol{t}_\phi)\boldsymbol{\theta}_R, \tag{17}$$

where b is the radius of the pinion and c has a different meaning for the two joints. In the rack-and-pinion joint $c = b$ and in the screw joint it corresponds to the pitch of the thread. Note that since $\boldsymbol{t}_r$ and $\boldsymbol{t}_\phi$ have the same direction in the screw joint, the constant term $b\widehat{t_r t_\phi}$ in this case vanishes. Differentiating the kinematic relation (17) gives rise to the terms

$$\left.\frac{\partial \boldsymbol{f}(\boldsymbol{\theta}_R)}{\partial \boldsymbol{\theta}_R}\right|_{rp} = \left.\frac{\partial \boldsymbol{f}(\boldsymbol{\theta}_R)}{\partial \boldsymbol{\theta}_R}\right|_{sc} = c\boldsymbol{t}_r \otimes \boldsymbol{t}_\phi \tag{18}$$

needed in (16) and subsequently in (14) and (15).

2.2.3 Rigid segment

Within the master–slave approach, a rigid segment (rs) can be formulated using the definition of the rack-and-pinion joint in which the rack is fixed to the pinion. This is equivalent to setting $c = 0$ in (17), in which case the master and the slave node remain at a fixed distance b throughout the motion with the relative orientation of one with respect to the other also fixed. In this case $\left.\frac{\partial \boldsymbol{f}(\boldsymbol{\theta}_R)}{\partial \boldsymbol{\theta}_R}\right|_{rs} = \boldsymbol{0}$, which inserted in (16) defines the linear operator $\mathbf{N}$ at the node.

2.2.4 Cam joint

Let us consider a cam joint with a simple eccentric cam-lobe profile as shown in Figure 4. By setting the upper node of the rotating element B as the slave node and the left-end node of the element A as the master node, the relation between the released translational displacement $\boldsymbol{r}_R$ and the released rotation $\boldsymbol{\theta}_R$ may be written as

$$\boldsymbol{r}_R = \boldsymbol{f}(\boldsymbol{\theta}_R)_{cam} = (R\cos|\boldsymbol{\theta}_R| - R - a)\boldsymbol{t}_r, \tag{19}$$

where a and $2R + a$ are the minimum and maximum released displacements of the arm A and $\boldsymbol{t}_r$ is the unit vector attached to the master node (direction in which the relative translation takes place). The term $\frac{\partial \boldsymbol{f}(\boldsymbol{\theta}_R)}{\partial \boldsymbol{\theta}_R}$ for the cam joint is computed as

$$\left.\frac{\partial \boldsymbol{f}(\boldsymbol{\theta}_R)}{\partial \boldsymbol{\theta}_R}\right|_{cam} = -R\sin|\boldsymbol{\theta}_R|\boldsymbol{t}_r \otimes \boldsymbol{t}_\phi, \tag{20}$$

which is needed in (16) in order to define the linear operator $\mathbf{N}$ at the node.

2.3 Numerical stability and conservative integration

In order to avoid the problems with numerical stability in the standard end-point numerical integration of equations of motion, a different, *conservative*, class of methods for numerical integration can be applied to the problem. For a conservative force-free class of dynamical problems, this procedure introduces an energy- and momentum-conserving transformation which relates the *finite changes of the slave and the master/released variables over a time step* rather than the *variations* of the slave and

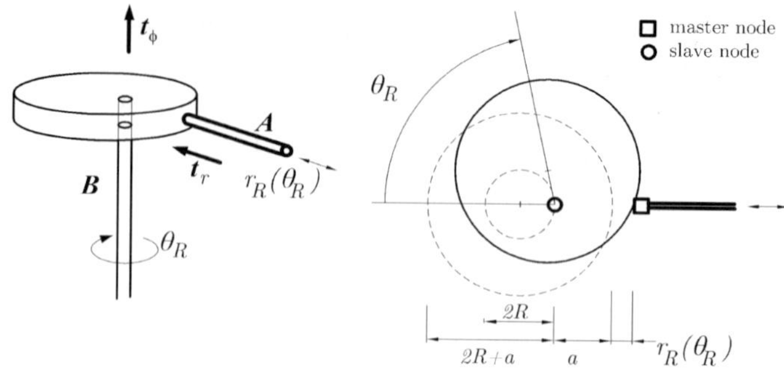

Fig. 4. Scheme of the cam joint.

the master/released variables, utilised in the original formulation [33, 37]. The theory is quite involved and, rather than presenting it in a general case, a simple model problem will be studied in the next section, where it will be also shown how an energy-momentum conserving linear operator (the null-space matrix) can be derived for a sliding joint.

3 Node-to-Element Approach

The master–slave approach will be now extended to enable the modelling of sliding joints. The theory is relevant for the analysis of mechanisms [48], but also for the modelling of more general node-on-segment contact [51, 52].

In order to emphasise the key ingredients of the method, we will describe the node-to-element approach with the help of reduced models depicted in Figure 5. Although the additional complexities of the rotational degrees of freedom are circumvented, the example contains the essence of the master–slave approach for this kind of joints. References [51, 52] contain a more complete description of sliding joints with rotational variables.

On this example we will also demonstrate how the master–slave technique can be applied to obtain the necessary matrices used in the null-space or the orthogonal complementary method [17, 31]. In order to point out the similarities and differences between these methods and the master–slave method, we will give a brief description of the Lagrange multipliers technique first.

3.1 Variational approach

In order to derive the equilibrium equations of the model depicted in Figure 5(b), let us first express the total energy of the system. Mass 2 is connected to mass 1 via a massless spring of initial length ℓ_0. Mass 1 is forced to satisfy the constraint equation $\boldsymbol{r}_1 = \bar{\boldsymbol{r}}$, where $\bar{\boldsymbol{r}} \in \mathcal{R}^{ndim}$ is a smooth C^2 curve, and $ndim$ is the number of dimensions of the problem (2 or 3). If no external loads are applied, the total energy is given as $E = T + V$, where the kinetic and elastic energy, T and V, are given by

$$T = \frac{1}{2} m_1 \dot{\boldsymbol{r}}_1 \cdot \dot{\boldsymbol{r}}_1 + \frac{1}{2} m_2 \dot{\boldsymbol{r}}_2 \cdot \dot{\boldsymbol{r}}_2,$$

$$V = \frac{1}{2} k(|\boldsymbol{r}_2 - \boldsymbol{r}_1| - \ell_0)^2. \tag{21}$$

We will assume throughout this section that $|\boldsymbol{r}_2 - \boldsymbol{r}_1| \neq 0$.

3.1.1 Equations of the unconstrained model

The equilibrium equations can be derived by resorting to the principle of conservation of energy and setting $\dot{E} = 0$, which leads to

$$\dot{\boldsymbol{r}}_1 \cdot \left(m_1 \ddot{\boldsymbol{r}}_1 + \bar{k}(\boldsymbol{r}_1 - \boldsymbol{r}_2)\right) + \dot{\boldsymbol{r}}_2 \cdot \left(m_2 \ddot{\boldsymbol{r}}_2 + \bar{k}(\boldsymbol{r}_2 - \boldsymbol{r}_1)\right) = 0, \tag{22}$$

with $\bar{k} = \frac{k(|\boldsymbol{r}_2 - \boldsymbol{r}_1| - \ell_0)}{|\boldsymbol{r}_2 - \boldsymbol{r}_1|}$. In case of an *unconstrained system* (model in Figure 5(a)), the velocities $\dot{\boldsymbol{r}}_i$ are independent from one another, and thus the following dynamic equilibrium equations are obtained:

$$m_1 \ddot{\boldsymbol{r}}_1 + \bar{k}(\boldsymbol{r}_1 - \boldsymbol{r}_2) = \boldsymbol{0},$$

$$m_2 \ddot{\boldsymbol{r}}_2 + \bar{k}(\boldsymbol{r}_2 - \boldsymbol{r}_1) = \boldsymbol{0}. \tag{23}$$

By defining the matrices $\mathbf{M}_{12} = \begin{bmatrix} m_1 \mathbf{I} & \mathbf{0} \\ \mathbf{0} & m_2 \mathbf{I} \end{bmatrix}$ and $\mathbf{K}_{12} = \bar{k} \begin{bmatrix} \mathbf{I} & -\mathbf{I} \\ -\mathbf{I} & \mathbf{I} \end{bmatrix}$, with $\mathbf{I}$ the $ndim \times ndim$ unit matrix, we can rewrite $\dot{E}$ in (22) as

$$\dot{E} = \dot{\boldsymbol{r}} \cdot (\mathbf{M}_{12} \ddot{\boldsymbol{r}} + \mathbf{K}_{12} \boldsymbol{r}) = 0, \tag{24}$$

where $\boldsymbol{r}^T = \{\boldsymbol{r}_1^T \; \boldsymbol{r}_2^T\}$ is the vector of global displacements. It has $ndof$ components, and in our case $ndof = 2 \times ndim$. The dynamic equilibrium equations in (23) can then be written for short in the following standard form:

$$\mathbf{M}_{12} \ddot{\boldsymbol{r}} + \mathbf{K}_{12} \boldsymbol{r} = \boldsymbol{0}, \tag{25}$$

where, owing to the presence of $\bar{k}$, $\mathbf{K}_{12}$ is non-constant and must not be mistaken for a tangent stiffness matrix.

3.1.2 Equations of the constrained model

By defining the constraint function $\boldsymbol{\Phi} := \boldsymbol{r}_1 - \bar{\boldsymbol{r}}$, the equilibrium of the constrained system can be obtained by simultaneously solving equations

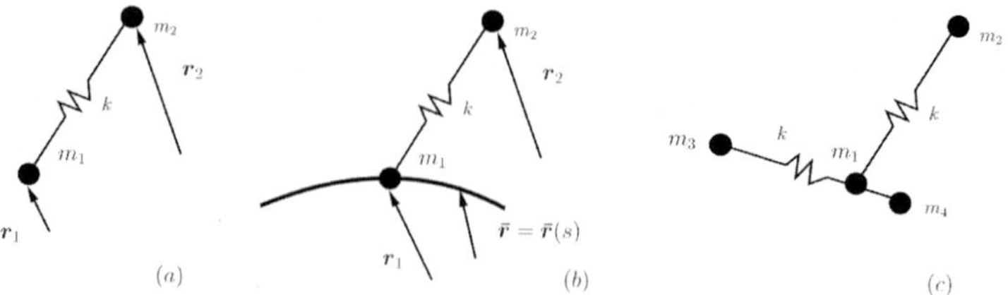

Fig. 5. Reduced (a) unconstrained model, (b) single-element constrained model and (c) two-element constrained model.

$$\mathbf{M}_{12}\ddot{\boldsymbol{r}} + \mathbf{K}_{12}\boldsymbol{r} = \boldsymbol{0}, \tag{26a}$$

$$\boldsymbol{\Phi} = \boldsymbol{0}, \tag{26b}$$

i.e. a system of differential and algebraic equations (DAE). It can be interpreted as the restriction of the solution of (25) to the constraint $\boldsymbol{\Phi} = \boldsymbol{0}$. The constrained system can be basically solved by two groups of techniques: by adding the constraint equation (26b) to the solution of (26a), or by rewriting the differential equation (26a) as a different differential equation, the solution of which remains by construction such that $\boldsymbol{\Phi} = \boldsymbol{0}$. The first group is equivalent to the use of Lagrange multipliers, penalty methods or augmented Lagrangian technique [2, 6]. The second group is representative of the master–slave approach which, as mentioned in the introduction, has similarities with other methods (embedding constraint technique, coordinate partitioning, joint or relative coordinates and the null-space method). Although we aim to explore the latter group of methods, it is helpful to write the equilibrium equations using Lagrange multipliers first.

3.1.3 Lagrange multipliers

Let us assume that in a general case we have c constraint equations, which implies that $\boldsymbol{\Phi} \in \mathcal{R}^c$ (in our reduced model, $c = ndim$). The total energy of the system is now extended by using a vector of Lagrange multipliers $\boldsymbol{\lambda} \in \mathcal{R}^c$ as

$$E = T + V + \boldsymbol{\lambda} \cdot \boldsymbol{\Phi}.$$

The new dynamic equilibrium equations now follow as

$$\begin{aligned} \mathbf{M}_{12}\ddot{\boldsymbol{r}} + \mathbf{K}_{12}\boldsymbol{r} + \mathbf{C}^T\boldsymbol{\lambda} &= \boldsymbol{0}, \\ \boldsymbol{\Phi} &= \boldsymbol{0}, \end{aligned} \tag{27}$$

where the matrix $\mathbf{C} = \frac{\partial \boldsymbol{\Phi}}{\partial \boldsymbol{r}} = \nabla \boldsymbol{\Phi}$ is for the reduced model given by

$$\mathbf{C} = \left[\frac{\partial \boldsymbol{\Phi}}{\partial \boldsymbol{r}_1} \; \mathbf{0}_{c \times ndim} \right]_{c \times ndof}.$$

The rows of $\mathbf{C}$ define a basis $\boldsymbol{c}_i^T = \{\frac{\partial \Phi_i}{\partial \boldsymbol{r}_1}\ \mathbf{0}_{1\times ndim}\} \in \mathcal{R}^{ndof}$, $i = 1, \ldots, c$, of the space where the constraint forces act. The Lagrange multipliers are the components of these constraint forces in this basis, in our case acting on mass m_1. Note that the algebraic equation $\boldsymbol{\Phi} = \boldsymbol{0}$ remains in the system, and thus the method must still solve a system of differential and algebraic equations.

3.1.4 Master–slave and null-space methods

Null-space methods are based on the projection of governing equations (26) onto a space orthogonal to vectors $\boldsymbol{c}_i$. Such an orthogonal space is spanned by a set of vectors $\mathbf{p}_i \in \mathcal{R}^{ndof}$, which can be collected within a matrix $\mathbf{P}^T = \left[\mathbf{p}_1 \ldots \mathbf{p}_{ndof-c}\right]$. This space is in the literature also called the orthogonal complement or the null-space of $\mathbf{C}$, i.e. $range(\mathbf{P}) = null(\mathbf{C})$. It then follows that $\mathbf{P}^T\mathbf{C}^T = \mathbf{0}$ and $\mathbf{CP} = \mathbf{0}$ and, premultiplying equations (27) by $\mathbf{P}^T$, we obtain the following system of equations:

$$
\begin{aligned}
\mathbf{P}^T\left(\mathbf{M}_{12}\ddot{\boldsymbol{r}} + \mathbf{K}_{12}\boldsymbol{r}\right) &= \boldsymbol{0}, \\
\boldsymbol{\Phi} &= \boldsymbol{0}.
\end{aligned}
$$

Several methods have been proposed to obtain a suitable expression of matrix $\mathbf{P}$ [17, 18]. The master–slave method provides a direct way to construct such a matrix, and it is based in the fact that the slideline, given in the implicit form as $\boldsymbol{\Phi} = \mathbf{0}$, can be also written in a parametric form as $\boldsymbol{r}_1 = \bar{\boldsymbol{r}}(s)$. The variation of $\boldsymbol{r}_1$ in time can be then obtained as

$$
\dot{\boldsymbol{r}}_1 = \frac{\partial \bar{\boldsymbol{r}}(s)}{\partial s}\dot{s}. \tag{28}
$$

The parameter s determines the displacement of mass m_1 along the slideline. In the master–slave context, this displacement is referred to as the *released displacement*. We introduce the global vector of degrees of freedom $\boldsymbol{r}_{Rm}^T = \{s\ \boldsymbol{r}_2^T\}$, which contains the minimum set of independent degrees of freedom of the system. From equation (28), we can derive a transformation matrix that relates $\delta\boldsymbol{r}$ to $\delta\boldsymbol{r}_{Rm}$ as

$$
\dot{\boldsymbol{r}} = \mathbf{N}\dot{\boldsymbol{r}}_{Rm}; \qquad \mathbf{N} = \begin{bmatrix} \frac{\partial \bar{\boldsymbol{r}}}{\partial s} & \mathbf{0} \\ \mathbf{0} & \mathbf{I} \end{bmatrix}. \tag{29}
$$

By noting that $\frac{\partial \bar{\boldsymbol{r}}(s)}{\partial s}$ is tangent to the slideline and thus $\frac{\partial \boldsymbol{\Phi}}{\partial \boldsymbol{r}} \cdot \frac{\partial \bar{\boldsymbol{r}}(s)}{\partial s} = 0$, it follows that $\mathbf{N}^T\mathbf{C}^T = \mathbf{0}$ and $\mathbf{CN} = \mathbf{0}$, and therefore matrix $\mathbf{N}$ is a suitable choice for the null-space matrix $\mathbf{P}$ mentioned above and the new set of equations can be written using the minimum set of degrees of freedom as

$$
\mathbf{N}^T\left(\mathbf{M}_{12}\ddot{\boldsymbol{r}} + \mathbf{K}_{12}\boldsymbol{r}\right) = \boldsymbol{0}. \tag{30}
$$

It is obvious that this equation has the same form as the equilibrium equation derived for the node-to-node master–slave approach in statics in (15). Note that the condition $\boldsymbol{\Phi} = \boldsymbol{0}$ has been removed because the slave displacements are computed using the parameter s as $\boldsymbol{r}_1 = \bar{\boldsymbol{r}}(s)$, and thus $\boldsymbol{\Phi} = \boldsymbol{0}$ is satisfied by construction. Alternatively, the previous equation can be obtained directly by inserting (29) into $\dot{E}$ in (24), which gives rise to

$$\dot{E} = \dot{\boldsymbol{r}}_{Rm} \cdot \mathbf{N}^T \left(\mathbf{M}_{12}\ddot{\boldsymbol{r}} + \mathbf{K}_{12}\boldsymbol{r}\right) = 0. \tag{31}$$

It is worth noting that while in the unconstrained system the global displacements in $\boldsymbol{r}$ are arbitrary, in the constrained system they are not due to $\boldsymbol{r}_1 = \bar{\boldsymbol{r}}(s)$. In other words, the components of $\dot{\boldsymbol{r}}_1$ must satisfy $\frac{\partial \boldsymbol{\Phi}}{\partial \boldsymbol{r}}\dot{\boldsymbol{r}}_1 = \boldsymbol{0}$. However, no relation between the components of $\dot{\boldsymbol{r}}_{Rm}$ exists in (31), and therefore (30) is implied by (31).

3.2 Incremental approach: Conserving time-integration algorithms

The system of differential equations in (30) can be solved using a number of methods described in the literature, the previously mentioned Newmark and generalised HHT-α method being among them. Owing to their improved numerical stability, specially attractive are the so-called conserving schemes, which algorithmically preserve some of the constants of motion, such as the total energy (in conservative systems) and the linear and angular momenta (in force-free systems). The design of such algorithms involves a different system of equations, which in fact parallel the steps described in the previous section. Instead of setting the dynamic equilibrium from the conservation law defined over an infinitesimal period of time $\dot{E} = 0$, we will apply this law over a finite period of time, $\Delta E = 0$ where, for a quantity $(\bullet)$, $\Delta(\bullet)$ denotes a change of this quantity in time, i.e. $\Delta(\bullet) = (\bullet)_{n+1} - (\bullet)_n$.

In what follows we will use the conservative constraint model shown in Figure 5(c) with springs of equal stiffnesses. Mass 1 is constrained to slide along the line defined by the position of masses m_3 and m_4. In the master–slave context, nodes 3 and 4 are the *master nodes* belonging to the *master element*, and node 1 is the *slave node*, whose coordinates depend on the positions of the master nodes and a released displacement.

3.2.1 Equations for the unconstrained model

From the definitions of the kinetic and elastic energy in (21), the increments of these energies between the times t_n and t_{n+1} can be written as

$$\begin{aligned} \Delta T &= \sum_{i=1}^{4} m_i \boldsymbol{v}_{i,n+\frac{1}{2}} \cdot \Delta \boldsymbol{v}_i, \\ \Delta V &= \xi_{12}(\Delta \boldsymbol{r}_2 - \Delta \boldsymbol{r}_1) \cdot (\boldsymbol{r}_{2,n+\frac{1}{2}} - \boldsymbol{r}_{1,n+\frac{1}{2}}) \\ &\quad + \xi_{34}(\Delta \boldsymbol{r}_4 - \Delta \boldsymbol{r}_3) \cdot (\boldsymbol{r}_{4,n+\frac{1}{2}} - \boldsymbol{r}_{3,n+\frac{1}{2}}), \end{aligned} \tag{32}$$

with

$$(\bullet)_{n+\frac{1}{2}} = \frac{1}{2}\left((\bullet)_{n+1} + (\bullet)_n\right),$$

$$\xi_{12} = \frac{2\Delta V_{12}}{\ell^2_{12,n+1} - \ell^2_{12,n}}, \qquad \xi_{34} = \frac{2\Delta V_{34}}{\ell^2_{34,n+1} - \ell^2_{34,n}},$$

$$\ell_{12} = |\boldsymbol{r}_2 - \boldsymbol{r}_1|, \quad \ell_{34} = |\boldsymbol{r}_4 - \boldsymbol{r}_3|,$$

$$V_{12} = \frac{k}{2}(\ell_{12} - \ell_0)^2, \quad V_{34} = \frac{k}{2}(\ell_{34} - \ell_0)^2$$

and $\boldsymbol{v}_i$ the algorithmic velocity vector at node i. By defining the algorithmic mid-point velocity as $\boldsymbol{v}_{n+\frac{1}{2}} = \frac{\Delta \boldsymbol{r}}{\Delta t}$, and defining the vector of global positions as $\boldsymbol{r}^T = \{\boldsymbol{r}_1^T \; \boldsymbol{r}_2^T \; \boldsymbol{r}_3^T \; \boldsymbol{r}_4^T\}$, the total energy change over a time step $\Delta E = \Delta T + \Delta V$ can be written as

$$\Delta E = \Delta \boldsymbol{r} \cdot \left(\mathbf{M}\frac{\Delta \boldsymbol{v}}{\Delta t} + \mathbf{K}\boldsymbol{r}_{n+\frac{1}{2}}\right), \tag{33}$$

with $\mathbf{M}$ and $\mathbf{K}$ now defined as

$$\mathbf{M} = \begin{bmatrix} \mathbf{M}_{12} & \mathbf{0} \\ \mathbf{0} & \mathbf{M}_{34} \end{bmatrix} \text{ and } \mathbf{K} = \begin{bmatrix} \mathbf{X}_{12} & \mathbf{0} \\ \mathbf{0} & \mathbf{X}_{34} \end{bmatrix}, \tag{34}$$

where

$$\mathbf{M}_{12} = \begin{bmatrix} m_1\mathbf{I} & \mathbf{0} \\ \mathbf{0} & m_2\mathbf{I} \end{bmatrix}, \quad \mathbf{M}_{34} = \begin{bmatrix} m_3\mathbf{I} & \mathbf{0} \\ \mathbf{0} & m_4\mathbf{I} \end{bmatrix},$$

$$\mathbf{X}_{12} = \xi_{12}\begin{bmatrix} \mathbf{I} & -\mathbf{I} \\ -\mathbf{I} & \mathbf{I} \end{bmatrix} \quad \text{and} \quad \mathbf{X}_{34} = \xi_{34}\begin{bmatrix} \mathbf{I} & -\mathbf{I} \\ -\mathbf{I} & \mathbf{I} \end{bmatrix}.$$

If the algorithmic energy is to be conserved at any discrete point in time, (33) needs to be zero for arbitrary $\Delta \boldsymbol{r}$, which requires

$$\mathbf{M}\frac{\boldsymbol{v}_{n+1} - \boldsymbol{v}_n}{\Delta t} + \mathbf{K}\boldsymbol{r}_{n+\frac{1}{2}} = \boldsymbol{0}. \tag{35}$$

3.2.2 Lagrange multipliers

The incremental version of (27) is given by [17],

$$\mathbf{M}\frac{\boldsymbol{v}_{n+1} - \boldsymbol{v}_n}{\Delta t} + \mathbf{K}\boldsymbol{r}_{n+\frac{1}{2}} + \bar{\mathbf{C}}^T\boldsymbol{\lambda} = \boldsymbol{0}, \tag{36a}$$

$$\boldsymbol{\Phi}_{n+1} = \boldsymbol{0}, \tag{36b}$$

where matrix $\bar{\mathbf{C}}$ is a discrete version of $\mathbf{C} = \frac{\partial \boldsymbol{\Phi}}{\partial \boldsymbol{r}}$. It can be expressed as $\bar{\mathbf{C}} = \bar{\nabla}\boldsymbol{\Phi}$, where $\bar{\nabla}$ is a discrete derivative [29], defined in such way that it satisfies the property $\bar{\nabla}\boldsymbol{\Phi} \cdot \Delta \boldsymbol{r} = \boldsymbol{\Phi}_{n+1} - \boldsymbol{\Phi}_n$, where in this case the right-hand side vanishes due to (36b). However, the system of equations in (36) is still a system of differential and algebraic equations.

3.2.3 Master–slave and null-space methods

As before, within the null-space method it is necessary to construct a matrix $\bar{\mathbf{P}}$ as an orthogonal complement of $\bar{\mathbf{C}}$, i.e. such that $\bar{\mathbf{P}}^T\bar{\mathbf{C}}^T = \mathbf{0}$. This route is investigated in [17] for non-sliding joints.

Again, the master–slave technique allows us to construct a matrix equivalent to $\bar{\mathbf{P}}$ directly: we define the vector of independent (i.e. master and released) degrees of freedom $\boldsymbol{r}_{Rm}^T = \{s\ \boldsymbol{r}_2^T\ \boldsymbol{r}_3^T\ \boldsymbol{r}_4^T\}$, and we derive a matrix $\bar{\mathbf{N}}$ such that

$$\Delta\boldsymbol{r} = \bar{\mathbf{N}}\Delta\boldsymbol{r}_{Rm}. \tag{37}$$

Inserting this relationship into the increments of energy in (33), and noting that the incremental master and released displacements $\Delta\boldsymbol{r}_{Rm}$ are independent and arbitrary, the conservation of energy is implied by the following equation:

$$\bar{\mathbf{N}}^T\left(\mathbf{M}\frac{\Delta\boldsymbol{v}}{\Delta t} + \mathbf{K}\boldsymbol{r}_{n+\frac{1}{2}}\right) = \boldsymbol{0}. \tag{38}$$

We will now presume that $\Delta\boldsymbol{r}_1$ can be expressed as:

$$\Delta\boldsymbol{r}_1 = \tilde{I}^3\Delta\boldsymbol{r}_3 + \tilde{I}^4\Delta\boldsymbol{r}_4 + \mathbf{A}\Delta s, \tag{39}$$

where $\tilde{I}^3$, $\tilde{I}^4$ is a pair of shape functions and $\mathbf{A}$ is an $ndim$ vector to be determined in the following paragraphs. The previous expression allows us to write the following general form for matrix $\bar{\mathbf{N}}$ as

$$\bar{\mathbf{N}} = \begin{bmatrix} \mathbf{A} & \mathbf{0} & \tilde{I}^3\mathbf{I} & \tilde{I}^4\mathbf{I} \\ \mathbf{0} & \mathbf{I} & \mathbf{0} & \mathbf{0} \\ \mathbf{0} & \mathbf{0} & \mathbf{I} & \mathbf{0} \\ \mathbf{0} & \mathbf{0} & \mathbf{0} & \mathbf{I} \end{bmatrix}. \tag{40}$$

We will next describe two definitions of $\tilde{I}^3$, $\tilde{I}^4$ and $\mathbf{A}$ that have been implemented and tested for geometrically exact 3D beams in bilateral sliding contact [52].

Contact condition consistent with finite element interpolation

Using the standard finite element nodal interpolation functions, the points on the slideline between the master nodes 3 and 4 are obtained as $I^\alpha\boldsymbol{r}_\alpha$, $\alpha = 3, 4$ (summation convention implied). Therefore, the discrete form of the contact condition is expressed as $\boldsymbol{r}_{1,n} = I_n^\alpha\boldsymbol{r}_{\alpha,n}$. The slave displacement increments $\Delta\boldsymbol{r}_1$ can be then interpolated via a parameter γ in the following way:

$$\Delta\boldsymbol{r}_1 = I_\gamma^\alpha\Delta\boldsymbol{r}_\alpha + \Delta I^\alpha\boldsymbol{r}_{\alpha,1-\gamma}, \tag{41}$$

where $(\bullet)_\gamma = \gamma(\bullet)_n + (1-\gamma)(\bullet)_{n+1}$. The shape functions $\tilde{I}^\alpha$ and matrix $\mathbf{A}$ in (39) are then given by:

$$\tilde{I}^{\alpha} = I_{\gamma}^{\alpha} \tag{42a}$$

$$\mathbf{A} = \frac{\Delta I^{\alpha}}{\Delta s} \boldsymbol{r}_{\alpha,1-\gamma}. \tag{42b}$$

It is demonstrated in Appendix A that for the conservation of the angular momentum, the shape functions $\tilde{I}^{\alpha}$ must satisfy equation (48), which according to the current expressions in (42a) turns into

$$\boldsymbol{r}_{1,n+\frac{1}{2}} = I_{\gamma}^{3} \boldsymbol{r}_{3,n+\frac{1}{2}} + I_{\gamma}^{4} \boldsymbol{r}_{4,n+\frac{1}{2}}. \tag{43}$$

This condition clearly violates the finite element interpolation of the slideline, and thus, the algorithm does not conserve the angular momentum. However, a good approximation of the angular momentum condition is obtained for $\gamma = \frac{1}{2}$ [9].

Contact condition not consistent with finite element interpolation

An algorithm that conserves the angular momentum can be built by respecting condition (48) in Appendix A with $\tilde{I}^{\alpha} = I_{\gamma}^{\alpha}$, which is equivalent to using (43). The coordinates of the slave node $\boldsymbol{r}_{1,n+1}$ are then given by

$$\boldsymbol{r}_{1,n+1} = 2 I_{\gamma}^{3} \boldsymbol{r}_{3,n+\frac{1}{2}} + 2 I_{\gamma}^{4} \boldsymbol{r}_{4,n+\frac{1}{2}} - \boldsymbol{r}_{1,n}. \tag{44}$$

An expression with the general form in (39) can be now written as

$$\Delta \boldsymbol{r}_{1} = I_{\gamma}^{3} \Delta \boldsymbol{r}_{3} + I_{\gamma}^{4} \Delta \boldsymbol{r}_{4} + 2 \frac{I_{\gamma}^{\alpha} \boldsymbol{r}_{\alpha,n} - \boldsymbol{r}_{1,n}}{\Delta s} \Delta s, \tag{45}$$

which gives rise to the following definitions of the components of $\bar{\mathbf{N}}$:

$$\begin{aligned} \tilde{I}^{\alpha} &= I_{\gamma}^{\alpha} \\ \mathbf{A} &= 2 \frac{(\gamma \Delta I^{\alpha} + I_{n+1}^{\alpha}) \boldsymbol{r}_{\alpha,n} - \boldsymbol{r}_{1,n}}{\Delta s}. \end{aligned} \tag{46}$$

Note that when $\Delta s \rightarrow 0$ we have $\Delta \boldsymbol{r}_{1} = \gamma I^{\alpha\prime} \boldsymbol{r}_{\alpha,n} + \lim_{\Delta s \rightarrow 0} \frac{I_{n}^{\alpha} \boldsymbol{r}_{\alpha,n} - \boldsymbol{r}_{1,n}}{\Delta s}$, which may become singular. Alternatively, we can compute $\boldsymbol{r}_{n+1}^{1}$ according to (43) and keep expression for $\mathbf{A}$ in (42) which is not singular when $\Delta s \rightarrow 0$. In this later case though, energy conservation is lost at the expense of conserving the angular momentum. Nonetheless, numerical experiments with the geometrically exact beams show that the energy increments remain relatively small [9, 52].

4 Numerical Example

The node-to-element master–slave approach has been applied to the static analysis of a semi-spherical shell being pushed against a rigid surface, which has been modelled

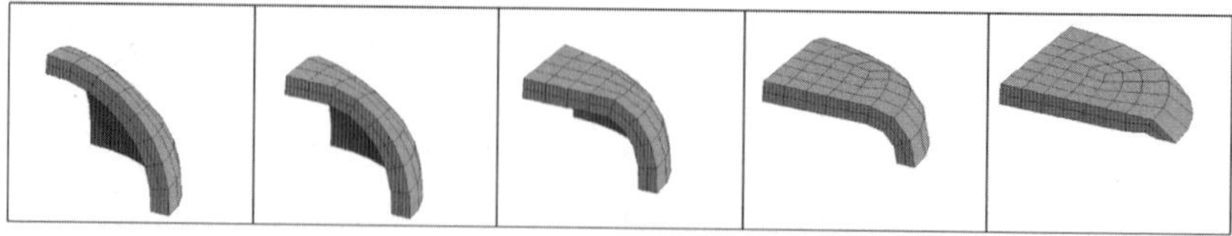

Fig. 6. Deformed configurations of a semi-sphere approaching a rigid plane.

as a single rigid element. The nodes at the surface parallel to the rigid plane have their coordinate perpendicular to the rigid plane prescribed. The shell has been made of as an isotropic hyperelastic material.

The model requires the definition of *coupling elements*, which include the displacements of the master element and the slave element (see [51] for details). Although details have been omitted here due to space limitations, we mention that in this problem this definition is configuration dependent. A kinematic based contact detection algorithm has been implemented, which updates the coupling elements. Figure 6 shows a sequence of deformed configurations of one quarter of the shell.

A Conservation of Angular Momentum

Let us for each node i define the vectors $\boldsymbol{q}^i_m$ and $\boldsymbol{q}^i_k$ as the kinetic and elastic (internal) part of the dynamic nodal residual vector, respectively, as $\boldsymbol{q}^i_m = m_i \frac{\Delta \boldsymbol{v}_i}{\Delta t}$ and $\boldsymbol{q}^i_k = (\mathbf{K}\boldsymbol{r}_{n+\frac{1}{2}})^i$.

With these definitions and the general form of $\bar{\mathbf{N}}$ in (40) at hand, we can rewrite equilibrium equation (38) as

$$\begin{aligned} \mathbf{A}^T(\boldsymbol{q}^1_m + \boldsymbol{q}^1_k) &= 0 \\ \boldsymbol{q}^2_m + \boldsymbol{q}^2_k &= \boldsymbol{0} \\ \boldsymbol{q}^3_m + \boldsymbol{q}^3_k + \tilde{I}^3(\boldsymbol{q}^1_m + \boldsymbol{q}^1_k) &= \boldsymbol{0} \\ \boldsymbol{q}^4_m + \boldsymbol{q}^4_k + \tilde{I}^4(\boldsymbol{q}^1_m + \boldsymbol{q}^1_k) &= \boldsymbol{0}. \end{aligned} \tag{47}$$

On the other hand, the angular momentum for the current model is defined as $\boldsymbol{J} = \sum_i \widehat{\boldsymbol{r}}_i m_i \boldsymbol{v}_i$. The increment of J per time step then reads $\frac{\Delta \boldsymbol{J}}{\Delta t} = \sum_i \widehat{\boldsymbol{r}}_{i,n+\frac{1}{2}} m_i \frac{\Delta \boldsymbol{v}_i}{\Delta t} = \sum_i \widehat{\boldsymbol{r}}_{i,n+\frac{1}{2}} \boldsymbol{q}^i_m$, where the first identity follows from the algorithmic mid-point velocity $\boldsymbol{v}_{n+\frac{1}{2}} = \frac{\Delta \boldsymbol{r}}{\Delta t}$. Using the equilibrium equations in (47), and noting that $\widehat{\boldsymbol{r}}_{1,n+\frac{1}{2}} \boldsymbol{q}^1_k + \widehat{\boldsymbol{r}}_{2,n+\frac{1}{2}} \boldsymbol{q}^2_k = \boldsymbol{0}$ and $\widehat{\boldsymbol{r}}_{3,n+\frac{1}{2}} \boldsymbol{q}^3_k + \widehat{\boldsymbol{r}}_{4,n+\frac{1}{2}} \boldsymbol{q}^4_k = \boldsymbol{0}$, $\frac{\Delta \boldsymbol{J}}{\Delta t}$ can be rewritten as

$$\frac{\Delta \boldsymbol{J}}{\Delta t} = \widehat{\boldsymbol{r}}_{1,n+\frac{1}{2}} \boldsymbol{q}_m^1 - \widehat{\boldsymbol{r}}_{2,n+\frac{1}{2}} \boldsymbol{q}_k^2 - (\tilde{I}^3 \widehat{\boldsymbol{r}}_{3,n+\frac{1}{2}} + \tilde{I}^4 \widehat{\boldsymbol{r}}_{4,n+\frac{1}{2}})(\boldsymbol{q}_m^1 + \boldsymbol{q}_k^1)$$
$$= \left(\widehat{\boldsymbol{r}}_{1,n+\frac{1}{2}} - (\tilde{I}^3 \widehat{\boldsymbol{r}}_{3,n+\frac{1}{2}} + \tilde{I}^4 \widehat{\boldsymbol{r}}_{4,n+\frac{1}{2}})\right)(\boldsymbol{q}_m^1 + \boldsymbol{q}_k^1).$$

It then follows that the angular momentum is conserved if the following condition is satisfied:

$$\boldsymbol{r}^1_{n+\frac{1}{2}} = \tilde{I}^3 \boldsymbol{r}_{3,n+\frac{1}{2}} + \tilde{I}^4 \boldsymbol{r}_{4,n+\frac{1}{2}} \tag{48}$$

Acknowledgements

This work was supported by Engineering and Physical Sciences Research Council of Great Britain under grants GR 04171/01 and AF/1000089.

References

1. García de Jalón J, Bayo E (1994) Kinematic and Dynamic Simulation of Multibody Systems – The Real-Time Challenge. Springler-Verlag, New York
2. Géradin MA, Cardona A (2001) Flexible Multibody Dynamics. A Finite Element Approach. John Wiley & Sons, New York
3. Hairer E, Wanner G (1996) Solving Ordinary Differential Equations II. Stiff and Differential-Algebraic Problems. Springer-Verlag, Berlin
4. Rosenberg RM (1977) Analytical Dynamics of Discrete Systems. Plenum Press
5. Shigley JE, Uicker JJ (1995) Theroy of Machines and Mechanisms. McGraw-Hill, New York
6. Wriggers P (2002) Computational Contact Mechanics. John Wiley & Sons, New York
7. Jelenić G, Crisfield MA (2002) Frictionless bilateral contact using minimum set method: Application in beams with sliding joints, In: Bonet J et al. (eds) Extended Abstracts for the 10th Annual Conference of ACME. University of Swansea, Swansea
8. Midha A (1993) Elastic Mechanisms. In: Erdman A (ed) Modern Kinematics: Development in the Last Forty Years. Wiley Series in Design Engineering. John Wiley & Sons, New York
9. Munoz JJ (2004) Finite-element analysis of flexible mechanisms using the master–slave approach with emphasis on the modelling of joints. PhD Thesis, London University, London
10. Anantharaman M, Hiller M (1991) Numerical simulation of mechanical systems using methods for differential-algebraic equations. Int J Num Meth Eng 32:1531–1542
11. Argyris J (1982) An excursion into large rotations. Comp Meth Appl Mech Eng 32:85–155
12. Avello A, Garcia de Jalón J, Bayo E (1991) Dynamics of flexible multibody systems using Cartesian co-ordinates and large displacement theory. Int J Num Meth Eng 32:1543–1563

13. Bauchau O (2000) On the modeling of prismatic joints in flexible multy-body systems. Comp Meth Appl Mech Eng 181:87–105
14. Bauchau O, Bottasso, CL (2001) Contact conditions for cylindrical, prismatic, and screw joints in flexible multibody systems. Multibody Syst Dyn 5:251–278
15. Bayo E, Garcia de Jalón J, Serna MA (1988) A modified Lagrangian formulation for the dynamic analysis of constrained mechanical systems. Comp Meth Appl Mech Eng 71:183–195
16. Bayo E, Ledesma R (1996) Augmented Lagrangian and mass-orthogonal projection methods for constrained multibody dynamics. Nonlinear Dynamics 9:113–130
17. Betsch P (2005) The discrete null space method for the energy consistent integration of constrained mechanical systems, Part I: Holonomic constraints. Comp Meth Appl Mech Eng 194:5159–5190
18. Blajer W (1995) An orthonormal tangent space method for constrained multibody systems. Comp Meth Appl Mech Eng 121:45–57
19. Blajer W (2001) A Geometrical interpretation and unifrm matrix formulation of multibody system dynamics. J Appl Math Mech (ZAMM) 81(4):247–259
20. Boyer F, Khalil W (1998) An efficient calculation of flexible manipulator inverse dynamics. Int J Robotics Research 17:282–293
21. Cardona A, Géradin M (1989) Time integration of the equations of motion in mechanism analysis. Comput Struct 33:801–820
22. Cardona A, Huespe A (1998) Continuation methods for tracing the equilibrium path in flexible mechanism analysis. Eng Computat 15:190–220
23. Chu S-C, Pan KC (1975) Dynamic response of a high-speed slider-crank mechanism with and elastic connecting rod. ASME J Eng Industry 542–550
24. Crisfield MA, Galvanetto U, Jelenić G (1997) Dynamics of 3-D co-rotational beams. Computat Mech 20:507–519
25. Downer JD, Park KC (1993) Formulation and solution of inverse spaghetti problem: application to beam deployment dynamics. AIAA Journal 31:339–347
26. Downer JD, Park KC, Chiou JC (1992) Dynamics of flexible beams for multibody systems: A computational procedure. Comp Meth Appl Mech Eng 96:373–408
27. Gantes C, Connor JJ, Logcher RD (1991) Combining numerical analysis and engineering judgement to design deployable structures. Comput Struct 40:431–440
28. Gantes C, Giakoumakis A, Vousvounis P (1997) Symbolic manipulation as a tool for design of deployable domes. Comput Struct 64:865–878
29. Gonzalez O (1996) Time integration and discrete Hamiltonian systems. J Nonlin Sci 8:449–467
30. Guest SD, Pellegrino S (1994) The folding of triangulated cylinder, Part I: Geometric considerations. ASME J Appl Mech 61:773–777
31. Hemami H, Weimer FC (1981) Modeling of nonholonomic dynamic systems with applications. J Appl Mech 48:177–182
32. Hilber H M, Hughes TJR, Taylor RL (1977) Improved numerical dissipation for time integration algorithms in structural dynamics. Earthquake Eng Struct Dyn 5:283–292
33. Ibrahimbegović A, Mamouri S (2000) On rigid components and joint constraints in nonlinear dynamics of flexible multibody systems employing 3D geometrically exact beam model. Comp Meth Appl Mech Eng 188:805–831
34. Jelenić G, Crisfield MA (1996) Non-linear “master–slave" relationships for joints in 3-D beams with large rotations. Comp Meth Appl Mech Eng 135:211–228

35. Jelenić G, Crisfield MA (1998) Interpolation of rotational variables in nonlinear dynamics of 3D beams. Int J Num Meth Eng 43:1193–1222
36. Jelenić G, Crisfield MA (1999) Geometrically exact 3D beam theory: Implementation of a strain-invariant finite element for static and dynamics. Comp Meth Appl Mech Eng 171:141–171
37. Jelenić G, Crisfield MA (2001) Dynamic analysis of 3D beams with joints in presence of large rotations. Comp Meth Appl Mech Eng 190:4195–4230
38. Jerkovsky W (1978) The structure of multibody dynamics equations. J Guidance Control 1:173–182
39. Kang HY, Suh CH (1994) Synthesis and analysis of spherical-cylindrical (SC) link in the McPherson strut suspension mechanism. ASME J Mech Design 116:599–606
40. Koppens WP, Sauren AAHJ, Veldpaus FE, van Campen DH (1988) The dynamics of a deformable body experiencing large displacements. ASME J Appl Mech 55:676–680
41. Kroyer R (1999) Wing mechanism analysis. Comput Struct 72:253–265
42. Ledesma R, Bayo E (1994) A Lagrangian approach to the non-causal inverse dynamics of flexible multibody systems: The three-dimensional case. Int J Num Meth Eng 37:3343–3361
43. Lee I-PJ, Bagci C (1975) The RCRRC five-link space mechanism – Displacement analysis, force and torque analysis and its transmission criteria. ASME J Eng Industry 581–594
44. Lin ST, Hong M C (1998) Stabilization method for numerical integration of multibody mechanical systems. ASME J Mech Design 120:565–572
45. Marjamäki H, Mäkinen J (2003) Modelling telescopic boom – The plane case: Part I. Comput Struct 81:1597–1609
46. Mayo J, Dominguez J, Shabana AA (1995) Geometrically nonlinear formulations of beams in flexible multibody dynamics. ASME J Vibration Acoustics 117:501–509
47. Meirovitch L, Chen Y (1995) Trajectory and control optimization for flexible space robots. J Guidance Cont Dyn 18:493–502
48. Mitsugi J(1997) Direct strain measure for large displacement analysis on hinge connected beam structures. Comput Struct 64:509–517
49. Modi V J, Suleman A, Ng AC (1991) An approach to dynamics and control of orbiting flexible structures. Int J Num Meth Eng 32:1727–1748
50. Muñoz JJ, Jelenić G, Crisfield MA (2003) Master–slave approach for the modelling of joints with dependent degrees of freedom in flexible mechanisms. Comm Num Meth Eng 19:689–702
51. Muñoz JJ, Jelenić G (2004) Sliding contact conditions using the master–slave approach with application on geometrically non-linear beams. Int J Solids Struct 41:6963–6992
52. Muñoz JJ, Jelenić G (2005) Sliding joints in 3D beams: conserving algorithms using the master–slave approach. Submitted for publication
53. Newmark NM (1959) A method of computation for structural dynamics. ASCE J Eng Mech Div EM3 85:67–94
54. Okamura H, Shinno A, Yamanaka T, Suzuki A, Sogabe K (1995) Simple modeling and analysis for crankshaft three-dimensional vibrations, part 1: background and application to free vibrations. ASME J Vibration Acoustics 117:70–79
55. Pellegrino S (1995) Large retractable appendages in spacecraft. J Spacecraft Rockets 32:1006–1014

56. Simo J (1985) A finite strain beam formulation: The three-dimensional dynamic problem. Part I. Comp Meth Appl Mech Eng 49:55–70
57. Simo J, Vu-Quoc L (1986) A three-dimensional finite strain rod model. Part II: Computational aspects. Comp Meth Appl Mech Eng 58:79–116
58. Simo J, Vu-Quoc L (1988) On the dynamics in space of rods undergoing large motions – A geometrically exact approach. Comp Meth Appl Mech Eng 66:125–161
59. Simo J, Tarnow N (1992) The discrete energy-momentum method. Conserving algorithms for nonlinear elastodynamics. J Appl Math Phys 43:757–792
60. Simo J, Tarnow N, Doblare M (1995) Nonlinear dynamics of three-dimensional rods: exact energy and momentum conserving algorithms. Int J Num Meth Eng 38:1431–1473
61. Wang H-L, Eischen JW, Silverberg LM (1998) On control and optimization of elastic multilink mechanisms. Comput Struct 67:483–502
62. Wang JT, Huston RL (1987) Kane's equations with undetermined multipliers – Application to constrained multibody systems. ASME J Appl Mech 54:424–429
63. You Z, Pellegrino S (1997) Foldable bar structures. Int J Solids Struct 34:1825–1847
64. Zhang C, Song, S-M (1994) Forward position analysis of nearly general Stewart platforms. ASME J Mech Design 116:54–60
65. Zhiming J (1994) Dynamics decomposition for Stewart platforms. ASME J Mech Design 116:67–69

Aspects of Contact Problems in Computational Multibody Dynamics

Saeed Ebrahimi and Peter Eberhard

Institute of Engineering and Computational Mechanics, University of Stuttgart, Pfaffenwaldring 9, 70569 Stuttgart, Germany;
E-mail: {ebrahimi, eberhard}@itm.uni-stuttgart.de

Abstract. This paper, first of all, explains briefly some of the well-known and frequently used methods for contact treatment through finite element methods and multibody systems dynamics and gives a short description of their computational aspects in applications dealing with contact problems. Then, as the core of this paper, a formulation for considering unilateral contact of constrained and non-constrained planar deformable bodies in a multibody system leading to Linear Complementarity Problems (LCPs) is presented. In doing so, kinematic relationships governing the behavior of contact are formulated in a way that they consider the effect of deformations. As a general approach for flexible multibody systems, the moving frame of reference approach and modal coordinates are used to describe deformable bodies. In this formulation the effects of deformation are taken into account starting from the relative velocity of contact points in the normal and tangential direction and then the procedure is followed by introducing the relative acceleration of contact points which includes all the necessary terms needed for considering deformations. Then, the complementarity relations are reformulated following the same procedure as for rigid bodies contact. Therefore, the main difference of this algorithm compared to the rigid body case is in the formulation of the kinematics of contact. This formulation just considers the continual contact case of deformable bodies and for impact calculation this formulation has to be extended.

1 Introduction

There are many applications dealing with contact problems which are simulated by multibody systems in industry and engineering and make their modeling an essential and very demanding topic in multibody dynamics. Therefore, a lot of research focuses on this topic and many theoretical and mathematical methods are developed for considering contact.

With increasing availability of fast computers, special attention was paid to the numerical investigation of contact problems and their application. Meanwhile, more and more approaches have been developed and taken into service. These approaches

J.C. García Orden et al. (eds), Multibody Dynamics. Computational Methods and Applications, 23–47.

are based on the physics of contact and take into account some assumptions in order to model the contact problem physically correct.

The numerical approaches which are being used widely in contact analysis can be divided into two main groups: first Finite Element Methods (FEM) [4, 36–38] and second approaches based on Multi-Body System dynamics (MBS) [4, 7, 12, 16, 24]. The FEM is without doubt the most powerful numerical method in the field of contact modeling. Although it is well suited for particularly high accuracy requirements, its very high computational effort for contact treatment causes some practical difficulties such as very long simulation times for computations over a long period of time. In such situations, MBS can often already model the contact with acceptable accuracy and considerably less computational effort compared to the FEM.

In general the problem of contact modeling in multibody systems consists of two major parts, see [6, 15, 16]:

1. Search for contact between moving bodies which is often named collision detection.
2. Computation of the contact forces and/or impulses which are the results of contact between bodies.

Collision detection is an important aspect of contact modeling of moving bodies, i.e., to find when, where and which bodies are in contact and therefore, plays an essential role. It is the main task of collision detection to check if potential contact bodies are in contact or not but not to take any action to prevent penetration. In general, contact objects have complicated geometries and, therefore, the requirement for efficient and fast methods of collision detection is inevitable. The accuracy of the results will depend on the accuracy and smoothness of the geometry definition of the contacting bodies.

After detecting the region of contact on the contacting bodies, contact forces and/or impulses have to be determined based on the area of this region, the geometric and material properties and the relative velocity of the bodies. From the modeling methodology point of view, several different methods have been introduced in the literature during the past two decades for the modeling of contact. In the next section, among all these methods the most well-known and frequently used ones are explained briefly and then, some application features of contact modeling in granular media and geared systems are mentioned. Then, in Section 3 as the core of our work in this paper, the continual contact modeling of rigid bodies formulated as a linear complementarity problem in [12, 24] is extended for planar deformable bodies, see also [5, 6]. The paper ends with a conclusion and a list of references.

2 Contact Treatment in Multibody Systems

In this section some well-known and frequently used formulations in finite element methods and multibody system dynamics which are applied to incorporate the contact constraints into the governing equations of the system are introduced briefly. Among all possible and implemented procedures, herein the penalty technique, the Lagrange multiplier method, their combination which is known as augmented Lagrange multipliers method and finally procedures yielding Linear Complementarity Problems (LCPs) are explained. Furthermore, some special formulations used in multibody system dynamics such as the impulse-based approach, the Polygonal Contact Model (PCM) and time-stepping methods are also introduced shortly. Each approach has some advantages and some disadvantages and depending on the system in which the contact problem occurs, an appropriate procedure has to be chosen.

2.1 The penalty approach

Since contact events yield a dependency between coordinates of the contacting bodies, their corresponding constraints have to be imposed in the system. The formulation of this procedure starts with considering the weak form of equilibrium equations for contacting bodies and inserting contact contributions. In doing so, from the equilibrium equations using a weighted residual approach its corresponding weak form which is equivalent to the principle of virtual work has to be derived. Then, one has to follow a discretization process to reach a discretized form of equations. These obtained equations contain nonlinearities arising from constitutive equations, geometrical nonlinearities and unilateral contact conditions. Consequently, their linearization is required in order to be solvable numerically. Contact problems may usually be interpreted as solving an optimization problem in which the total potential energy $P(\mathbf{u})$ of the system subjected to unilateral contact constraints $\mathbf{g}_N(\mathbf{u}) \geq \mathbf{0}$ has to be minimized. This condition states that the normal gap between the possible contact regions on the contacting bodies as a function of displacement fields vector $\mathbf{u}$ must not become negative. In other words, no penetration in the normal direction is allowed.

Let us first consider the formulation of contact in normal direction and without friction. Then, the total potential energy from the penalty approach with considering contact contributions as the result of the virtual work principle may be considered as

$$P^{Pen}(\mathbf{u}) = P(\mathbf{u}) + \int_{\Gamma_C} \frac{1}{2}\epsilon g_N^2(\mathbf{u})\, dA, \tag{1}$$

see [4, 36, 37], in which the potential energy corresponding to the penalty force is considered in the integral and $P(\mathbf{u})$ corresponds to the potential energy without considering contact forces. The integration is taken over the contact surface Γ_C. In fact, the term $\epsilon g_N(\mathbf{u})$ can be interpreted as the force of a spring element with stiffness ϵ and deformation $g_N(\mathbf{u})$. The variational form of Equation (1) as our objective function for minimization yields

$$\delta P^{Pen}(\mathbf{u}) = \delta P(\mathbf{u}) + \int_{\Gamma_C} \epsilon g_N(\mathbf{u})\delta g_N(\mathbf{u})\, dA = 0. \tag{2}$$

This continuous form has to be transformed to a discretized FE form and then must be linearized. Therefore, instead of the continuous field variable $\delta\mathbf{u}$ one deals with the variations of the nodal displacements $\delta\mathbf{U}$

$$\delta\Pi + \delta\mathbf{U}\cdot\epsilon(\mathbf{C}^T\cdot\mathbf{g}_N) = 0, \tag{3}$$

where $\delta\Pi$ arises from the discretization of $\delta P(\mathbf{u})$ and $\mathbf{C} = \partial\mathbf{g}_N/\partial\mathbf{U}$. Then, one can write the linearized form of the above equation

$$\delta\mathbf{U}\cdot(\mathbf{K}_T\cdot\Delta\mathbf{U} - \mathbf{r}) + \delta\mathbf{U}\cdot\epsilon\left(\mathbf{C}^T\cdot\mathbf{g}_N + \frac{\partial(\mathbf{C}^T\cdot\mathbf{g}_N)}{\partial\mathbf{U}}\cdot\Delta\mathbf{U}\right) = 0. \tag{4}$$

Here the tangential stiffness matrix $\mathbf{K}_T$ and the residuum $\mathbf{r}$ are terms arising from the linearization of $\delta\Pi$, see [4]. After introducing $\mathbf{K}_{TNP} = \partial(\mathbf{C}^T\cdot\mathbf{g}_N)/\partial\mathbf{U}$ and rearrangement of this equation and due to the independence of $\delta\mathbf{U}$ our penalty iteration may be formulated as

$$(\mathbf{K}_T + \epsilon\mathbf{K}_{TNP})\cdot\Delta\mathbf{U} = \mathbf{r} - \epsilon\mathbf{C}^T\cdot\mathbf{g}_N. \tag{5}$$

Now, utilizing the Newton–Raphson iteration this equation can be solved iteratively for the displacements $\mathbf{U}$. The iteration will start by choosing a penalty factor ϵ and assigning initial values of vector $\mathbf{U}$. Then, at each iteration step the active sets which are in contact have to be found. Solving Equation (5) and increasing the displacements leads the new values of $\mathbf{U}$ and then the norm of $\mathbf{r} - \epsilon\mathbf{C}^T\cdot\mathbf{g}_N$ has to be checked for convergence. Eventually the penalty parameter may be updated for better convergence.

In the case of frictional contact, two cases of either sliding or sticking may happen. The procedure will be the same as discussed for normal contact but the contact contributions coming from tangential contact have to be considered, too. One can find a detailed description of this formulation in [4, 36].

The penalty formulation prevents any penetration when the penalty parameter approaches infinity, $\epsilon\to\infty$, and consequently fulfills exactly the unilateral contact constraints. However, this introduces some severe difficulties in the numerical solution process. Choosing large values for ϵ increases the condition number of the corresponding stiffness matrix and, therefore, leads to ill-conditioned problems. There are some proposed approaches from which a proper penalty factor for normal contact may be chosen. One possibility is to relate the penalty factor with the material properties of contacting bodies including Young's modulus, Poisson's ratio and a so-called elastic layer thickness, see [18]. Another approach presented in [23] approximates the penalty factor based on an error analysis considering round-off errors as well as penalty approach errors.

The penalty approach for contact in multibody systems may also be known as the *surface compliance method*. In fact, this method is a widely used method in contact modeling of multibody systems with multiple contacts between complex objects. In this approach each contact region is modeled with a spring-damper element. The magnitudes of stiffness and deflection of the spring-damper element are computed based on the penetration, material properties and surface geometries of the colliding bodies, see [16, 18]. The normal force calculation prevents penetration and is done by determination of the elastic and damping shares. The elastic share based on the elastic foundation model is described by

$$F_{ek} = c_l A_k u_{nk}, \tag{6}$$

where c_l is the combined layer stiffness, A_k is the area of the contact element and u_{nk} is the penetration. Here, the indices 'e' for elastic, 'l' for layer, 'n' for normal direction and 'k' for the k-th contact element are used. The damping share is determined by

$$F_{dk} = \begin{cases} d_l A_k v_{nk} & \text{for} \quad u_{nk} \geq u_d, \\ d_l A_k v_{nk} \dfrac{u_{nk}}{u_d} & \text{for} \quad u_{nk} < u_d. \end{cases} \tag{7}$$

In this equation d_l is the areal layer damping factor, v_{nk} is the normal component of the relative velocity of the contacting bodies at the contact element position and u_d is a given transition depth.

In contrast to kinematic constraint methods like the Lagrange multiplier method, see Section 2.2, in which the contact is considered by contact constraints, in the surface compliance method no explicit kinematic constraint is considered. Instead, the resulting contact forces impose this condition only approximately.

As another feature of this approach, it can be pointed out that friction forces are considered as well. In addition to these features, contact objects with all complicated geometries involved can be treated. However, this method requires precise contact detection in order to apply the contact forces to contact regions and subsequently it imposes costly numerical calculations due to collision detection. At each time step of the simulation, contact forces in the normal and tangential directions (which are defined based on the contact plane) are computed. Based on the magnitude and direction of the contact forces, the new positions and velocities of the contacting bodies are computed and the collision detection process for the current condition is implemented. This procedure is performed until the end of simulation.

2.2 The Lagrange multiplier approach

In order to impose the contact constraints in the system, instead of the penalty method one can use the Lagrange multiplier method in which contact forces are handled by introducing some additional quantities, see [4, 36, 38]. Utilizing this approach the unilateral contact constraints can be fulfilled exactly. The starting point of this formulation is the same as for the penalty method but instead, the total potential energy has to be defined as follows

$$P^{Lag}(\mathbf{u}) = P(\mathbf{u}) + \int_{\Gamma_C} \lambda_N g_N(\mathbf{u})\, dA, \tag{8}$$

where λ_N denotes the normal contact force and is an additional unknown. Similar to the penalty formulation, the formulation of normal contact is given here. For the formulaion of frictional contact one can refer to [4, 36]. Analogous to Equation (1) the variational form of $P^{Lag}(\mathbf{u})$ leads to

$$\delta P^{Lag}(\mathbf{u}) = \delta P(\mathbf{u}) + \int_{\Gamma_C} (\delta\lambda g_N(\mathbf{u}) + \lambda \delta g_N(\mathbf{u}))\, dA = 0, \tag{9}$$

and its discretized form can be written as

$$\delta\Pi + (\delta\boldsymbol{\lambda} \cdot \mathbf{g}_N(\mathbf{u}) + \delta\mathbf{U} \cdot (\mathbf{C}^T \cdot \boldsymbol{\lambda})) = 0. \tag{10}$$

Following the same procedure as used in [4] and due to the independence of variables $\delta\mathbf{U}$ and $\delta\boldsymbol{\lambda}$, its linearized form yields

$$\begin{aligned} (\mathbf{K}_T \cdot \Delta\mathbf{U} - \mathbf{r}) + (\mathbf{C}^T \cdot \boldsymbol{\lambda} + \mathbf{K}_{TNL} \cdot \Delta\mathbf{U} + \mathbf{C}^T \cdot \Delta\boldsymbol{\lambda}) &= \mathbf{0}, \\ \mathbf{g}_N + \mathbf{C} \cdot \Delta\mathbf{U} &= \mathbf{0}, \end{aligned} \tag{11}$$

in which $\Delta\mathbf{U}$ and $\Delta\boldsymbol{\lambda}$ arise from the linearization and $\mathbf{K}_{TNL} = \partial(\mathbf{C}^T \cdot \boldsymbol{\lambda})/\partial\mathbf{U}$ has been substituted. Finally, one may summarize these equations in a matrix form

$$\begin{bmatrix} \mathbf{K}_T + \mathbf{K}_{TNL} & \mathbf{C}^T \\ \mathbf{C} & \mathbf{0} \end{bmatrix} \cdot \begin{bmatrix} \Delta\mathbf{U} \\ \Delta\boldsymbol{\lambda} \end{bmatrix} = \begin{bmatrix} \mathbf{r} - \mathbf{C}^T \cdot \boldsymbol{\lambda} \\ -\mathbf{g}_N \end{bmatrix}. \tag{12}$$

The solution of that problem can be obtained by solving Equation (12) iteratively for the unknowns $\Delta\mathbf{U}$ and $\Delta\boldsymbol{\lambda}$ using the Newton–Raphson iteration. The iteration starts by assigning initial values of $\mathbf{U}$ and $\boldsymbol{\lambda}$. The iteration procedure is the same as described for the penalty approach.

The Lagrange multiplier method after linearization leads to a coefficient matrix whose diagonal components associated with the Lagrange multipliers are zero, as it can be seen from Equation (12), and this is not desirable numerically. Therefore, another special formulation, which is known as the perturbed Lagrange method, a combination of both the penalty and the Lagrange multiplier methods has been developed.

The formulation described above is sometimes used in finite element formulation of contact problems. However, in multibody systems a simpler formulation may be offered keeping the same idea. In this approach for all contact possibilities between bodies, kinematic constraint equations are derived and appended to the system of equations of motion in multibody systems, see e.g. [29]. There, situations of continuous contact are modeled well since contact constraints are already considered in the equations of motion and so they are fulfilled in each evaluation of the system dynamics. Examples of such situations are e.g. contact between a rolling ball and the ground, or contact between wheel and rail in railway vehicles.

Apart from the above situation, this method cannot be used easily in the case of contact modeling between bodies with complicated geometries since in this method body surfaces should be described in such a way that it would be possible to derive the contact constraints for the contact between these surfaces. Another situation that causes some difficulties in the implementation of this method is the case of non-continuous contact. Due to these reasons, kinematic constraint equations of contact are almost impossible to be formulated in many practical cases.

2.3 The augmented Lagrange multiplier approach

Both the penalty and the Lagrange multiplier methods have some advantages and disadvantages. In order to overcome the drawbacks of each approach and to gain their advantages, the augmented Lagrange method as an idea mixing both approaches has been constructed [36–38].

Then, following the same procedure as for penalty and Lagrange approaches, the final equations will be summarized in a matrix form which can be solved by Newton–Raphson iteration and using Uzawa's algorithm [36, 37]. This algorithm consists of two iteration loops where in the inner loop the penalty factor and the Lagrange multipliers are kept constant and the displacements vector is calculated iteratively till convergence. Then, in the outer loop they are updated and used for a new iteration in the inner nested loop. This procedure can considerably improve the convergence.

2.4 The impulse-based approach

The concept of an impulse-based approach to contact problems has been utilized e.g. by Mirtich [21]. The fundamental idea of this approach is to define a relationship between the relative velocities of contact points before and after contact through the coefficient of restitution. The main advantage of this approach is its capability to handle different types of contact under a common formulation. This approach can easily be implemented for contact between free contacting bodies with simple geometries. In addition, no contact coherence is required and it is computationally robust. The other advantages such as strict enforcement of non-penetration conditions, natural handling of colliding contact and decoupling of bodies during dynamics integration may also be mentioned. However, this approach has some drawbacks compared to constraint-based approaches like the Lagrange multiplier method. E.g., it cannot exactly handle situations where sticking contact appears. Poor handling of continuous and simultaneous contacts is another problem of using this approach.

Critical components of these approaches are the following steps:

1. Collision detection: For each pair of contacting bodies a maximum time step must be determined in such a way that no collision between them occurs. Then, at the end of the time interval usually only a pair of bodies which is known as critical pair must be checked for collision.
2. Dynamic evolution: The system of contacting bodies is simulated by integration of the equations of motion for each body.
3. Collision response: Detection of the collision between critical pair of bodies is done. If any collision is detected, then collision impulses are applied to them.

2.5 Polygonal Contact Model

The Polygonal Contact Model (PCM) is an algorithm for contact modeling between objects with complicated geometries in multibody dynamics based on a penalty method. The original implementation of PCM is described in [15, 16] where this algorithm is proposed as a general algorithm for contact modeling between rigid bodies with polygonal surfaces. In the course of our work, this algorithm was modified and extended, so that elastic bodies may be considered, too, see [7, 8].

PCM is based on some facts:

- The geometry of the contacting bodies is described by a polygonal approximation.
- An areal discretization of the contact patch results in contact elements.
- Contact forces are computed by the elastic foundation model.

A polygonal surface is defined by a set of polygons in three dimensional space. In order to define polygons, the vertex coordinates and the faces which contain these vertices have to be available. This information is given to PCM in the format of wavefront object files [35]. Polygonal surfaces that PCM uses must not contain duplicated vertices and their polygons have to be oriented consistently. The polygons have to be connected but no closed surface is required. This kind of representation is widely used in computer graphics, computational geometry and computer vision.

Contact analysis in PCM consists of three basic steps:

1. Collision detection based on a hierarchical bounding volumes (BVs) representation,
2. construction of the intersection areas and discretization of the corresponding contact patches, and
3. calculation of the contact forces of each contact element and their application to the contacting bodies.

From the multibody dynamics point of view, PCM behaves as a force element which is going to be used as a user-defined routine in the commercial MBS code SIMPACK [31] without requiring internal changes. The approach used in PCM for calculation of contact forces is based on the penalty approach and uses spring-damper elements for contact force computation, see Section 2.1.

PCM implements an exact and efficient algorithm based on BV hierarchies [39]. The collision detection process is done at each time step and checks if two surfaces intersect for a given relative position and orientation. The BV trees which are used in the collision detection are created only once for every surface during preprocessing. In doing so, PCM takes the polygonal surface of each body as input and creates the BV tree based on the axis aligned bounding box approach, see [14, 39]. PCM follows the top-down approach and starts to create the BV tree from the root element and splits it up into two sub-volumes. Then it follows this procedure recursively to the leaf elements.

Because of the technical importance of contact problems for elastic bodies, considerable effort has been made in the investigation of this topic. Therefore, the extension of PCM for elastic bodies is very desirable and has been investigated [7, 8].

The rigid body version of PCM cannot be used directly for elastic bodies but served as a reliable basis for further developments. In doing so, following steps have been performed to modify the original PCM for the consideration of elastic bodies, too:

1. Data object modification,
2. determination of the geometry of elastic bodies,
3. updating the bounding volumes trees,
4. recalculating normal vectors, areas and barycenter positions,
5. modification of the relative velocities of contact elements, and
6. modification of the implementation of contact forces.

2.6 Contact treatment yielding complementarity problems

There exist other contact modeling approaches which avoid penetration and yield a mathematical formulation based on the kinematics of contacting bodies in a complementarity form. For planar systems linear complementarity problems (LCPs) must be solved yielding the exact solution to the contact problem, see e.g. [24, 25].

Many researchers use this type of formulation for contact modeling of rigid bodies due to its capabilities in handling of frictional contact, including many slip/stick contact transitions. In this approach, two subproblems must be formulated, the continual contact and impact.

Mathematicians are interested to investigate the occurring LCPs from the mathematical point of view and treat aspects like necessary conditions for existence and uniqueness of the solutions. In some of their formulations, the basic theory of convolution complementarity problems is given in which it deals with impact problems for elastic bodies with Coulomb friction, see e.g. [33].

Some others are more interested to investigate the implementation of contact in LCP form and its applications in engineering. As a result, some algorithms were developed based on unilateral constraints of the kinematics of contact points. Initially, only planar relative contact kinematics was considered and then this algorithm was extended for the case of spatial contact yielding sometimes nonlinear complementarity problems, see [12].

LCPs give mathematically the exact solution to the contact problem. Other advantages of these methods are that they do not result in stiff equations of motion and they show a lower effort for time integration compared to applying a brute force Lagrange multiplier method. However, their implementation and formulation of contact specially in the presence of friction is not trivial and needs high efforts and skills. Other disadvantages that can be mentioned are the neglection of the contact patch deflection, extreme simplification of complex physical phenomena and open problems in frictional impact theory. In Section 3 a formulation for continual contact of planar deformable bodies as the core of this paper will be presented.

2.7 Time-stepping techniques

In the presence of impulsive forces which are exerted during a very short period of time and as well in the presence of friction, the problem of discontinuities may frequently occur and, therefore, many difficulties can arise during the simulation of frictional contact in multibody systems. Among all approaches to overcome these problems, also time-stepping methods [1, 34] may be mentioned by which the equations of motions are discretized and then reformulated in a linear complementarity formalism. In this formulation, a linear complementarity relationship holds between the system generalized velocities and the impulsive forces.

In this framework, the integration process is combined with the equations of motion in order to reach a discretized formulation of equations of motion. Unfortunately, most time-stepping approaches use the Euler integration method which normally requires choosing a very small time step size and additionally is not suitable for handling impact problems of stiff multibody systems and multibody systems including deformable bodies. Therefore, almost all time-stepping approaches are developed for the case of rigid bodies impact and there is only a very limited number of researches which focus on their implementation for stiff multibody systems [2] and deformable bodies, see e.g. [32].

2.8 Applications of contact modeling in multibody systems

Contact events can frequently happen in multibody systems and in many cases the functionality of mechanical systems is based on them. Contact between tire and road in automobiles, wheel and rail in railway vehicles, contact in robotics and grasping machines, power transmission systems like geared systems, cam and follower, chain and gear, contact in granular media, ... are common examples of systems in which contact events play an essential and inevitable role. Among all these systems, here just a very brief reference of contact in granular media and geared systems as applications the authors are dealing with is given.

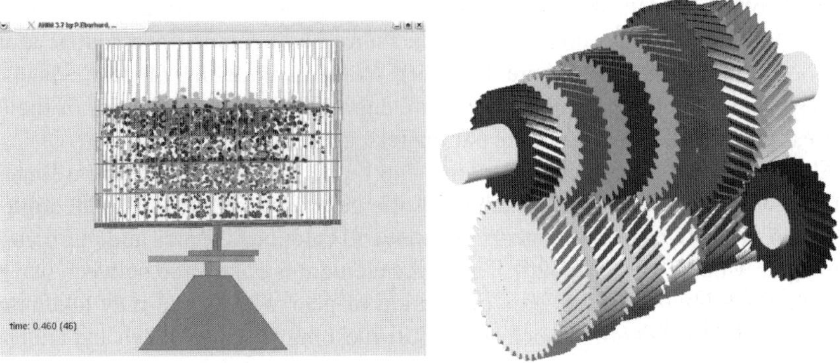

Fig. 1. Contact in granular media and geared systems.

An important feature of contact modeling deals with particles and granular media where the motion of many bodies is investigated by means of molecular dynamics, but usually no real consideration of flexibility is taken into account [22]. For efficient simulation of granular media, the techniques which are being used in molecular dynamics and multibody systems have to be combined. The particles are usually considered as being rigid and small overlaps are allowed. The contact formulation is based on simple models in order to keep the computational times within a feasible range. For an example, see Figure 1 (left).

Different approaches may be followed for finding neighboring objects of each body to reduce the high computational effort which is required for collision detection. Some of them divide the regions occupied by particles into small subregions and consider all particles within the same subregion as neighbors. In others, a bounding box will surround each particle and then the neighboring particles are found by detecting the collision between bounding boxes. For this purpose, in [22] three methods namely the Verlet-neighbor list (VL), Linked cell method (LC) and Linked linear list (LLL) are compared. The results presented in [22] show a desirable performance of VL approach for systems with only a small number of particles. However, for large systems the LC and LLL approaches should be used to handle such situations with better performance and efficiency.

After the neighboring pairs have been identified, they have to be checked for collision. Although this step may be done easily for spherical particles, dealing with polygonal-shaped particles is a sophisticated task which requires following some special algorithms in order to find the regions of contact [22]. Once the amount of penetration is found, normal contact forces arising from contact between particles are modeled utilizing spring-damper elements which corresponds to the penalty approach.

As another aspect, contact of meshing gear wheels may be mentioned. In many applications in mechanical engineering gear wheels are used to transmit power between rotating shafts and, therefore, the ability to incorporate them into multibody systems and to simulate contact between them, has become an essential topic in multibody dynamics, see Figure 1 (right). Contact modeling of gear wheels has some special difficulties which arise from, e.g., the nonlinear behavior of the tooth stiffness, backlash, gear geometric parameters, see e.g. [19].

Contact modeling of geared systems can be performed by defining a force element between each gear pair which takes the geometry, initial rotational angle and rotational velocity of meshing gears as input and calculates forces and moments acting on gear wheels as output [20]. This force element is connected between two body fixed markers located on the axis of rotation of gear wheels and may take into account some important points such as involute meshing teeth, backlash and addendum modification, tip relief factor, changes in axes distance, relative axial movement of gear pairs and parabolic behavior of the tooth stiffness. The calculation of contact forces is also performed here using the penalty method and uses spring-damper elements located on teeth surfaces in order to apply appropriate forces according to the amount of penetration and relative velocities in the normal and tangential directions. The algorithm and implementation of contact modeling of rigid gear wheels [20] is based on three main steps

1. calculation of basic parameters related to the gear geometry,
2. calculation of contact geometry,
3. determination of contact forces and resulting torques.

However, in advanced applications where the flexibility of gear bodies and meshing teeth cannot be neglected, rigid gear contact modeling cannot achieve realistic results and in some situations the simulation results based on this approach might have a big difference to real measurements. Therefore, as a compromise between the fully rigid modeling and the fully elastic FEM approaches, an extension of the rigid modeling approach of [20] by considering elastic elements between single teeth and the gear body for each gear wheel may be proposed [9]. In this approach, the tangentially movable teeth and the body of each gear wheel are still rigid but they are connected to each other by elastic elements. The required steps for doing these extensions are

1. introducing teeth coordinates as new force states into the equations of motion,
2. considering elastic elements between teeth and gear body,
3. utilizing a new search algorithm for finding contact situations and
4. modification of applying contact forces and resulting torques.

Simulation of several numerical examples shows a considerable improvement in the results of this approach which are closer to the accurate results of FEM compared to the fully rigid gear wheel. For getting such results, the new procedure will not need much more computational effort compared to the rigid modeling and can be used for simulation of contact of multibody geared systems over a long period of simulation time with several revolutions of the gears which is almost impossible using FEM.

3 LCP Formulation for Continual Contact of Planar Deformable Bodies

In applications where the flexibility of contacting bodies is not negligible, rigid body contact modeling cannot be used and the deformability of contacting bodies must be taken into account.

Therefore, it has been tried to develop an approach by reformulating the kinematic equations governing rigid body contact in order to be able to consider the deformation of contacting bodies too, see also [5, 6]. In doing so, the moving frame of reference approach is utilized in order to introduce the elastic coordinates into the equations of motion, see [28, 30]. It is important to emphasize that this paper just considers the continual contact of deformable bodies and for impact calculation this formulation has to be extended.

This procedure starts by reformulating the kinematics of contact for flexible bodies and introduces this in the complementarity form known from rigid body formulations. Then, a comparison between the presented formulation and the rigid body formulation is given. Finally, some results of a simulated example will be presented in order to show the feasibility of the described approach.

3.1 Contact formulation

Deformable bodies are modeled here using the well-known moving frame of reference approach [28, 30]. By introducing rigid and elastic coordinates, the movement of bodies is separated in two independent parts, the rigid body movement and the small elastic deformations. In this approach the deformation of bodies is assumed to be small compared to the dimension of bodies, but not negligible.

3.1.1 Kinematics of contact points

In Figure 2 two deformable bodies i and j are depicted, which are in contact in point k. In order to describe the position and orientation of these deformable bodies with respect to the global coordinate system, two sets of generalized coordinates $\mathbf{q}_i = (\mathbf{R}_i,\ \boldsymbol{\theta}_i,\ \mathbf{q}_{f_i})$ and $\mathbf{q}_j = (\mathbf{R}_j,\ \boldsymbol{\theta}_j,\ \mathbf{q}_{f_j})$ including the rigid $(\mathbf{R},\ \boldsymbol{\theta})$ and elastic $(\mathbf{q}_f)$ generalized coordinates are used. The rigid ones specify the position $\mathbf{R}$ and orientation $\boldsymbol{\theta}$ of the body reference coordinate systems with respect to the global coordinate system. In addition, the elastic coordinates are defined with respect to the body reference coordinate systems and are supposed to specify the position and orientation of any point on the bodies locally. Later all generalized coordinates are summarized in $\mathbf{q}$. However, $\mathbf{q}$ is not a vector of minimal coordinates since constraints from joints, ... must be considered by Lagrange multipliers additionally.

Considering the deformable body i, one can calculate the velocity of each arbitrary point p_i located on the body i from

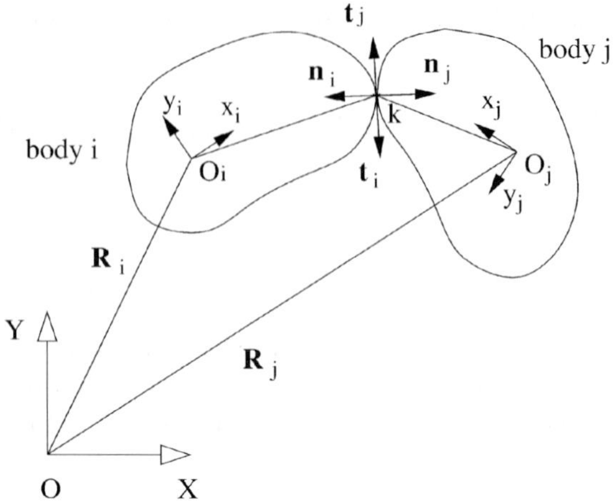

Fig. 2. Contact between two deformable bodies.

$$\mathbf{v}^{p_i} = \dot{\mathbf{R}}_i + \dot{\mathbf{A}}_i \cdot \bar{\mathbf{u}}_i + \mathbf{A}_i \cdot \dot{\bar{\mathbf{u}}}_i, \tag{13}$$

where $\mathbf{A}_i$ denotes the transformation matrix of body i and $\bar{\mathbf{u}}_i$ is the position vector of the point p_i in the reference coordinate system of body i in the deformed configuration, see [28]. By substituting the velocities $\dot{\bar{\mathbf{u}}}_i = \mathbf{S}_i \cdot \dot{\mathbf{q}}_f$, this equation can also be written as

$$\mathbf{v}^{p_i} = \dot{\mathbf{R}}_i + \dot{\mathbf{A}}_i \cdot \bar{\mathbf{u}}_i + \mathbf{A}_i \cdot \mathbf{S}_i \cdot \dot{\mathbf{q}}_f. \tag{14}$$

In this relation, $\mathbf{S}_i$ is the shape matrix of body i which is independent of time. One can write the second term of the right hand side of this equation in terms of the derivative of the orientation $\boldsymbol{\theta}_i$ with respect to time as

$$\dot{\mathbf{A}}_i \cdot \bar{\mathbf{u}}_i = \mathbf{B}_i \cdot \dot{\boldsymbol{\theta}}, \tag{15}$$

where $\mathbf{B}_i$ is a 2×1 matrix in the case of planar flexible bodies. It is defined in terms of the partial derivative of the transformation matrix $\mathbf{A}_i$ times the local position vector $\bar{\mathbf{u}}_i$ of the point p_i with respect to the rotational coordinates $\boldsymbol{\theta}_i$, see [28],

$$\mathbf{B}_i = \frac{\partial}{\partial \boldsymbol{\theta}_i}(\mathbf{A}_i \cdot \bar{\mathbf{u}}_i) = \frac{\partial}{\partial \boldsymbol{\theta}_i}(\mathbf{A}_i \cdot (\bar{\mathbf{u}}_{0i} + \mathbf{S}_i \cdot \mathbf{q}_{fi})). \tag{16}$$

In this relation, the vector $\bar{\mathbf{u}}_{0i}$ denotes the position of point p_i on the undeformed body i. Using Equation (15), Equation (14) can be written in terms of the generalized velocity vector $\dot{\mathbf{q}}_i$ as

$$\mathbf{v}^{p_i} = \mathbf{L}_i^{p_i} \cdot \dot{\mathbf{q}}_i. \tag{17}$$

In this relation, the vector $\dot{\mathbf{q}}_i$ includes rigid and elastic generalized velocities $\dot{\mathbf{q}}_i = (\dot{\mathbf{R}}_i,\ \dot{\boldsymbol{\theta}}_i,\ \dot{\mathbf{q}}_{f_i})$ and $\mathbf{L}_i$ is a matrix which projects the generalized velocity vector $\dot{\mathbf{q}}_i$ to the velocities in the global coordinate system

$$\mathbf{L}_i = \begin{bmatrix} \mathbf{I} & \mathbf{B}_i & \mathbf{A}_i \cdot \mathbf{S}_i \end{bmatrix}, \tag{18}$$

where $\mathbf{I}$ is the 2×2 identity matrix.

According to Figure 2 and noticing that at any contact point k the normal vectors $(\mathbf{n}_i, \mathbf{n}_j)$ and tangential vectors $(\mathbf{t}_i, \mathbf{t}_j)$ of body i and body j are aligned with each other, the relative velocities in normal and tangential directions for contact point k can be written as

$$\dot{g}^k_{N_{ij}} = \mathbf{n}^k_i \cdot (\mathbf{L}^k_i \cdot \dot{\mathbf{q}}_i - \mathbf{L}^k_j \cdot \dot{\mathbf{q}}_j), \qquad \dot{g}^k_{T_{ij}} = \mathbf{t}^k_i \cdot (\mathbf{L}^k_i \cdot \dot{\mathbf{q}}_i - \mathbf{L}^k_j \cdot \dot{\mathbf{q}}_j). \tag{19}$$

According to these relationships, for any contact point k between bodies i and j, the matrices $\mathbf{L}^k_i$ and $\mathbf{L}^k_j$ must be calculated since they depend on the positions.

The relative accelerations in the normal and tangential directions for contact point k are calculated by taking the derivative of the relative velocities

$$\ddot{g}^k_{N_{ij}} = (\mathbf{W}^k_N)^T_{ij} \cdot \ddot{\mathbf{q}}_{ij} + (\mathbf{w}^k_N)^T_{ij} \cdot \dot{\mathbf{q}}_{ij}, \qquad \ddot{g}^k_{T_{ij}} = (\mathbf{W}^k_T)^T_{ij} \cdot \ddot{\mathbf{q}}_{ij} + (\mathbf{w}^k_T)^T_{ij} \cdot \dot{\mathbf{q}}_{ij}, \tag{20}$$

where

$$\dot{\mathbf{q}}_{ij} = \begin{bmatrix} \dot{\mathbf{q}}_i \\ \dot{\mathbf{q}}_j \end{bmatrix}, \quad \ddot{\mathbf{q}}_{ij} = \begin{bmatrix} \ddot{\mathbf{q}}_i \\ \ddot{\mathbf{q}}_j \end{bmatrix}, \quad (\mathbf{W}^k_N)_{ij} = \begin{bmatrix} (\mathbf{n}^k_i \cdot \mathbf{L}^k_i)^T \\ -(\mathbf{n}^k_i \cdot \mathbf{L}^k_j)^T \end{bmatrix}, \quad (\mathbf{W}^k_T)_{ij} = \begin{bmatrix} (\mathbf{t}^k_i \cdot \mathbf{L}^k_i)^T \\ -(\mathbf{t}^k_i \cdot \mathbf{L}^k_j)^T \end{bmatrix},$$

$$(\mathbf{w}^k_N)_{ij} = \begin{bmatrix} (\dot{\mathbf{n}}^k_i \cdot \mathbf{L}^k_i + \mathbf{n}^k_i \cdot \dot{\mathbf{L}}^k_i)^T \\ -(\dot{\mathbf{n}}^k_i \cdot \mathbf{L}^k_j + \mathbf{n}^k_i \cdot \dot{\mathbf{L}}^k_j)^T \end{bmatrix}, \quad (\mathbf{w}^k_T)_{ij} = \begin{bmatrix} (\dot{\mathbf{t}}^k_i \cdot \mathbf{L}^k_i + \mathbf{t}^k_i \cdot \dot{\mathbf{L}}^k_i)^T \\ -(\dot{\mathbf{t}}^k_i \cdot \mathbf{L}^k_j + \mathbf{t}^k_i \cdot \dot{\mathbf{L}}^k_j)^T \end{bmatrix}.$$

It is clear that effects of deformations are introduced to the kinematic relation of the relative normal and tangential accelerations $\ddot{g}_N$ and $\ddot{g}_T$ of contact point k through the matrices $\mathbf{L}^k_i$ and $\mathbf{L}^k_j$. Although Equation (20) looks similar to the rigid body case, its computation is very different. This equation which holds for contact point k can be used to obtain the matrix form of the relative normal and tangential accelerations $\ddot{\mathbf{g}}_N$ and $\ddot{\mathbf{g}}_T$ for all n_c contact points between n_b bodies

$$\ddot{\mathbf{g}}_N = \mathbf{W}^T_N \cdot \ddot{\mathbf{q}} + \mathbf{w}^T_N \cdot \dot{\mathbf{q}}, \qquad \ddot{\mathbf{g}}_T = \mathbf{W}^T_T \cdot \ddot{\mathbf{q}} + \mathbf{w}^T_T \cdot \dot{\mathbf{q}}, \tag{21}$$

where $\mathbf{W}_N$, $\mathbf{W}_T$, $\mathbf{w}_N$ and $\mathbf{w}_T$ consist of submatrices which are calculated for the corresponding contact pairs from Equation (20) and the vectors $\dot{\mathbf{q}}$ and $\ddot{\mathbf{q}}$ are the generalized velocities and accelerations of the system.

Next, the relative accelerations $\ddot{\mathbf{g}}_N$ and $\ddot{\mathbf{g}}_T$ and the equations of motion of flexible bodies will be used together in order to form the complementarity relationships between the contact forces and the relative accelerations.

3.1.2 Deformable bodies in multibody systems

In this section, first the equations of motion of the flexible body i taking into account the contact forces in a system of interconnected rigid and flexible bodies are written and then the equations of motion of flexible multibody systems are given in matrix form.

Considering the flexible body i, one can write its nonlinear equations of motion as

$$\mathbf{M}_i \cdot \ddot{\mathbf{q}}_i + \mathbf{C}_i \cdot \dot{\mathbf{q}}_i + \mathbf{K}_i \cdot \mathbf{q}_i + (\frac{\partial \mathbf{c}}{\partial \mathbf{q}_i})^T \cdot \boldsymbol{\lambda}_c = \mathbf{F}_{ext_i} + \mathbf{F}_{v_i} + \mathbf{F}_{c_i}. \tag{22}$$

Here $\mathbf{M}_i$ is the mass matrix of body i without considering constraints, $\mathbf{C}_i$ and $\mathbf{K}_i$ are the damping and stiffness matrices arising from the elastic coordinates, $\partial\mathbf{c}/\partial\mathbf{q}_i$ is the Jacobian matrix containing the derivatives of all constraints $\mathbf{c} = \mathbf{0}$ except the contact constraints, $\boldsymbol{\lambda}_c$ are the Lagrange multipliers corresponding to the constraint forces, $\mathbf{F}_{ext_i}$ is the vector of generalized external forces, $\mathbf{F}_{v_i}$ is the vector of generalized Coriolis forces and $\mathbf{F}_{c_i}$ is the vector of generalized contact forces. These equations can be summarized for the whole system together with the second derivatives of the constraints in matrix form for flexible multibody systems including constrained and non-constrained bodies, see [28],

$$\underbrace{\begin{pmatrix} \mathbf{M} & \mathbf{C}_q^T \\ \mathbf{C}_q & \mathbf{0} \end{pmatrix}}_{\mathbf{M}_c} \cdot \underbrace{\begin{pmatrix} \ddot{\mathbf{q}} \\ \boldsymbol{\lambda}_c \end{pmatrix}}_{\ddot{\mathbf{q}}_c} = \underbrace{\begin{pmatrix} \mathbf{F}_{ext} + \mathbf{F}_v - \mathbf{C} \cdot \dot{\mathbf{q}} - \mathbf{K} \cdot \mathbf{q} \\ -\partial(\frac{\partial \mathbf{c}}{\partial \mathbf{q}} \cdot \dot{\mathbf{q}})/\partial \mathbf{q} \cdot \dot{\mathbf{q}} - 2\frac{\partial^2 \mathbf{c}}{\partial \mathbf{q} \partial t} \cdot \dot{\mathbf{q}} - \frac{\partial^2 \mathbf{c}}{\partial t \partial t} \end{pmatrix}}_{\mathbf{h}_c} + \begin{pmatrix} \mathbf{F}_c \\ \mathbf{0} \end{pmatrix}. \tag{23}$$

When two bodies come in contact, normal and tangential contact forces arise as the result of the collision. Therefore, the contact force $\mathbf{F}_c$ in Equation (23) for n_c contact points can be supposed to be the summation of normal and tangential forces which is written in terms of two different vectors $\boldsymbol{\lambda}_N$ and $\boldsymbol{\lambda}_T$, for details refer to [24],

$$\begin{aligned} \mathbf{F}_c &= \left(\mathbf{W}_N^1 \ldots \mathbf{W}_N^{n_c}\right) \cdot \begin{pmatrix} \boldsymbol{\lambda}_N^1 \\ \vdots \\ \boldsymbol{\lambda}_N^{n_c} \end{pmatrix} + \left(\mathbf{W}_T^1 \ldots \mathbf{W}_T^{n_c}\right) \cdot \begin{pmatrix} \boldsymbol{\lambda}_T^1 \\ \vdots \\ \boldsymbol{\lambda}_T^{n_c} \end{pmatrix} \\ &= \mathbf{W}_N \cdot \boldsymbol{\lambda}_N + \mathbf{W}_T \cdot \boldsymbol{\lambda}_T. \end{aligned} \tag{24}$$

By separating the tangential contact forces into the sliding and sticking contacts, one can rewrite Equation (24) to get

$$\begin{aligned} \mathbf{F}_c = &\underbrace{\left(\mathbf{W}_{N_1} + \mathbf{W}_{G_1} \cdot \boldsymbol{\mu}_{G_1} \ldots \mathbf{W}_{N_{nc}} + \mathbf{W}_{G_{nc}} \cdot \boldsymbol{\mu}_{G_{nc}}\right)}_{\mathbf{W}_N + \mathbf{W}_G \cdot \boldsymbol{\mu}_G} \cdot \boldsymbol{\lambda}_N + \\ &\underbrace{\left(\mathbf{W}_{H_1} \ldots \mathbf{W}_{H_{nc}}\right)}_{\mathbf{W}_H} \cdot \boldsymbol{\lambda}_H. \end{aligned} \tag{25}$$

In this equation $\mathbf{W}_G$ and $\mathbf{W}_H$ are matrices extracted from the matrix $\mathbf{W}_T$ which correspond to the sliding and sticking parts, respectively.

The equations of motion (23) can be reformulated by substituting the contact forces from Equation (25)

$$\mathbf{M}_c \cdot \ddot{\mathbf{q}}_c - \mathbf{h}_c - \underbrace{\begin{pmatrix} \mathbf{W}_N + \mathbf{W}_G \cdot \mu_G & \mathbf{W}_H \\ \mathbf{0} & \mathbf{0} \end{pmatrix}}_{\mathbf{W}_{NH}} \cdot \underbrace{\begin{pmatrix} \boldsymbol{\lambda}_N \\ \boldsymbol{\lambda}_H \end{pmatrix}}_{\boldsymbol{\lambda}} = \mathbf{0}. \tag{26}$$

Now the equations of motion for multibody systems including constrained and non-constrained rigid and flexible bodies in terms of contact forces (which are denoted by $\boldsymbol{\lambda}_N$ and $\boldsymbol{\lambda}_H$) have been obtained and this form of equations will be used in the next section to construct the complementarity form of the equations of motion with contact.

3.1.3 Construction of the complementarity form

In this section, it has been tried to summarize a similar procedure as described in [24] by constructing the complementarity equations for continual contact of deformable bodies utilizing the equations obtained in the previous sections. Maybe at the first view, these equations seem to be very similar to the rigid ones but as it was pointed out before, all effects of deformabilities are taken into account in the kinematic quantities and the calculation procedure of these quantities is totally different.

Starting from Equation (26), supposing that there is no redundant constraint in the system such that $\mathbf{M}_c$ is a regular matrix, we can find the acceleration $\ddot{\mathbf{q}}_c$ as a function of the Lagrange multipliers $\boldsymbol{\lambda}$

$$\ddot{\mathbf{q}}_c = \mathbf{M}_c^{-1} \cdot \mathbf{h}_c + \mathbf{M}_c^{-1} \cdot \mathbf{W}_{NH} \cdot \boldsymbol{\lambda}. \tag{27}$$

Then, Equation (21) is rewritten for the sliding and sticking contacts

$$\ddot{\mathbf{g}}_{NH} = \begin{bmatrix} \ddot{\mathbf{g}}_N \\ \ddot{\mathbf{g}}_H \end{bmatrix} = \begin{bmatrix} \mathbf{W}_N & \mathbf{W}_H \end{bmatrix}^T \cdot \ddot{\mathbf{q}} + \begin{bmatrix} \mathbf{w}_N & \mathbf{w}_H \end{bmatrix}^T \cdot \dot{\mathbf{q}}. \tag{28}$$

The vector $\mathbf{w}_H$ is the part of the vector $\mathbf{w}_T$ which corresponds to the sticking contact. By inserting the Lagrange multipliers $\boldsymbol{\lambda}_c$, which denote the constraint forces (except contact forces), into the unknown variables as we did for the generation of equations of motion in Equation (23), the above relation is rewritten as

$$\ddot{\mathbf{g}}_{NH} = \begin{bmatrix} \ddot{\mathbf{g}}_N \\ \ddot{\mathbf{g}}_H \end{bmatrix} = \underbrace{\begin{bmatrix} \mathbf{W}_N^T & \mathbf{0} \\ \mathbf{W}_H^T & \mathbf{0} \end{bmatrix}}_{\mathbf{W}^T} \cdot \underbrace{\begin{bmatrix} \ddot{\mathbf{q}} \\ \boldsymbol{\lambda}_c \end{bmatrix}}_{\ddot{\mathbf{q}}_c} + \underbrace{\begin{bmatrix} \mathbf{w}_N^T \cdot \dot{\mathbf{q}} \\ \mathbf{w}_H^T \cdot \dot{\mathbf{q}} \end{bmatrix}}_{\mathbf{w}}. \tag{29}$$

In the next step, the vector $\ddot{\mathbf{q}}_c$ from Equation (27) can be substituted in this equation

$$\ddot{\mathbf{g}}_{NH} = \mathbf{W}^T \cdot \mathbf{M}_c^{-1} \cdot \mathbf{h}_c + \mathbf{W}^T \cdot \mathbf{M}_c^{-1} \cdot \mathbf{W}_{NH} \cdot \boldsymbol{\lambda} + \mathbf{w}\,. \tag{30}$$

By following the same procedure as described in [24] and by using Equation (30) one can construct the complementarity form of the equations of motion and then simulate the continual contact problem of deformable bodies, too. In doing so and in order to handle the condition of switching between sliding and sticking contact, the tangential part of the contact forces must be decomposed into two different parts and, thereby, the frictional case of contact will be handled appropriately.

The final form of the complementarity equations based on our notation and according to the above mentioned points is

$$\begin{bmatrix} \ddot{\mathbf{g}} \\ \boldsymbol{\lambda}_{H_0} \end{bmatrix} = \begin{bmatrix} \mathbf{W}^T \cdot \mathbf{M}_c^{-1} \cdot \mathbf{W}_{NH} & \mathbf{I}^T \\ \mathbf{N}_H - \mathbf{I} & \mathbf{0} \end{bmatrix} \cdot \begin{bmatrix} \boldsymbol{\lambda} \\ \mathbf{z} \end{bmatrix} + \begin{bmatrix} \mathbf{W}^T \cdot \mathbf{M}_c^{-1} \cdot \mathbf{h}_c + \mathbf{w} \\ \mathbf{0} \end{bmatrix},$$

$$\begin{bmatrix} \ddot{\mathbf{g}} \\ \boldsymbol{\lambda}_{H_0} \end{bmatrix} \geq \mathbf{0}, \quad \begin{bmatrix} \boldsymbol{\lambda} \\ \mathbf{z} \end{bmatrix} \geq \mathbf{0}, \quad \begin{bmatrix} \ddot{\mathbf{g}} \\ \boldsymbol{\lambda}_{H_0} \end{bmatrix} \cdot \begin{bmatrix} \boldsymbol{\lambda} \\ \mathbf{z} \end{bmatrix} = 0. \tag{31}$$

The parameters $\boldsymbol{\lambda}_{H_0}$, $\mathbf{N}_H$ and $\mathbf{z}$ in this equation are chosen in the same way as in [24] and have the same meaning. These parameters are used in order to formulate the complementarity form of the equations in such a way to handle switching between sliding and sticking cases of contact.

3.2 Comparison between the elastic and the rigid formulation

The presented formulation in this paper is applicable for continual contact of all planar multibody systems including constrained and non-constrained rigid and flexible bodies supposing that the deformations are small. In the case of rigid bodies, this formulation leads to the same results as the formulation developed for rigid bodies in [24].

For a comparison between the obtained formulation for $\ddot{\mathbf{g}}$ for deformable bodies and the case of rigid bodies, one can start from the presented relationships and reach to the formulations in [24] by setting the elastic coordinates to zero coordinates which denotes that there is no deformation in the system.

Considering two rigid bodies i and j and referring to Equation (20), we have

$$\ddot{g}^k_{N_{ij}} = (\mathbf{W}^k_N)^T_{ij} \cdot \ddot{\mathbf{q}}_{ij} + (\mathbf{w}^k_N)^T_{ij} \cdot \dot{\mathbf{q}}_{ij}. \tag{32}$$

For simplicity and without loss of generality, we suppose that one of the bodies, for example body j, is the ground and therefore, its corresponding coordinates vanish from the above equation. With this assumption, the above equation can be expanded in terms of its parameters

$$\ddot{g}^k_{N_{ij}} = \mathbf{n}^k_i \cdot \mathbf{L}^k_i \cdot \ddot{\mathbf{q}}_i + (\mathbf{n}^k_i \cdot \dot{\mathbf{L}}^k_i + \dot{\mathbf{n}}^k_i \cdot \mathbf{L}^k_i) \cdot \dot{\mathbf{q}}_i. \tag{33}$$

Since the body i is assumed as being rigid for this comparison, all its elastic coordinates vanish and from Equation (16) it remains only $\mathbf{B}_i = \partial(\mathbf{A}_i \cdot \bar{\mathbf{u}}_{0i})/\partial\boldsymbol{\theta}_i$. In the case of rigid planar systems, the matrix $\mathbf{L}_i^k$ is a 2×3 matrix and $\boldsymbol{\theta}_i$ is replaced with the scalar variable θ_i. Therefore, the above equation is simplified to

$$\ddot{g}_{N_{ij}}^k = \underbrace{\mathbf{n}_i^k \cdot (\ddot{\mathbf{R}}_i + \ddot{\theta}_i \mathbf{B}_i^k + \dot{\theta}_i \dot{\mathbf{B}}_i^k)}_{\text{part 1}} + \underbrace{\dot{\mathbf{n}}_i^k \cdot (\dot{\mathbf{R}}_i + \dot{\theta}_i \mathbf{B}_i^k)}_{\text{part 2}} = \mathbf{n}_i^k \cdot \mathbf{a}_i^k + \dot{\mathbf{n}}_i^k \cdot \mathbf{v}_i^k. \quad (34)$$

From this equation it is clear that the relative normal acceleration $\ddot{\mathbf{g}}_N$ for any contact point k consists of two parts. The first part is due to the acceleration of that point in the direction of the normal vector $\mathbf{n}_i^k$ and the second part is due to the variation of the direction of the normal vector.

This procedure can be followed exactly in the same way for the relative tangential accelerations $\ddot{\mathbf{g}}_T^k$ of the contact point k and then extended for all contact points

$$\begin{aligned} \ddot{g}_T^k &= \mathbf{t}_i^k \cdot \mathbf{L}_i^k \cdot \ddot{\mathbf{q}}_i + (\mathbf{t}_i^k \cdot \dot{\mathbf{L}}_i^k + \dot{\mathbf{t}}_i^k \cdot \mathbf{L}_i^k) \cdot \dot{\mathbf{q}}_i \\ &= \mathbf{t}_i^k \cdot (\ddot{\mathbf{R}}_i + \ddot{\theta}_i \mathbf{B}_i^k + \dot{\theta}_i \dot{\mathbf{B}}_i^k) + \dot{\mathbf{t}}_i^k \cdot (\dot{\mathbf{R}}_i + \dot{\theta}_i \mathbf{B}_i^k) = \mathbf{t}_i^k \cdot \mathbf{a}_i^k + \dot{\mathbf{t}}_i^k \cdot \mathbf{v}_i^k. \end{aligned} \quad (35)$$

Therefore, one can easily verify that the presented relationships for relative accelerations result in the accelerations of rigid bodies when deformations of flexible bodies are set to zero. Compared to the continual contact of rigid bodies, these four matrices are the most important quantities which have to be reformulated in a way to lead to a correct formulation for the continual contact of deformable bodies. The results shown in the following section confirm the validity of this procedure as well.

3.3 Numerical example: sliding and sticking of an elastic rectangular block

The presented formulation is implemented and examined here, see also the results in [5]. With due attention to the numerical results, one can verify the validity and feasibility of the described approach. In this examples, the continual contact of an elastic rectangular block on a rigid foundation is investigated. The elastic rectangular block slides on the foundation and after a while due to the effect of the friction force sliding contact changes to sticking contact. The shape of the rigid foundation is chosen in such a way to activate all the important terms in the presented formulation. The elastic block and the rigid foundation are depicted in Figure 3. According to this figure, the foundation consists of three parts: two inclined straight parts and a curved part which is an arc of a circle with radius $r = 10$ m. The length of each inclined part is 10 m and the inclination angle is $\alpha = 30$ Deg. The simulation is done for $t_{end} = 10$ s and for friction coefficient $\mu = 0.2$. Some results of this simulation are illustrated in Figure 5.

In Figure 4 the elastic rectangular block in an arbitrary position and orientation with respect to the global coordinate system O is shown. The coordinate system O_i is attached rigidly to node 4 of the elastic block and is considered to be the block reference coordinate system. The x-axis is aligned to edge (4, 1). Thus, the rigid coordinates $\mathbf{R}$ and θ of the block are calculated from the position and orientation of this

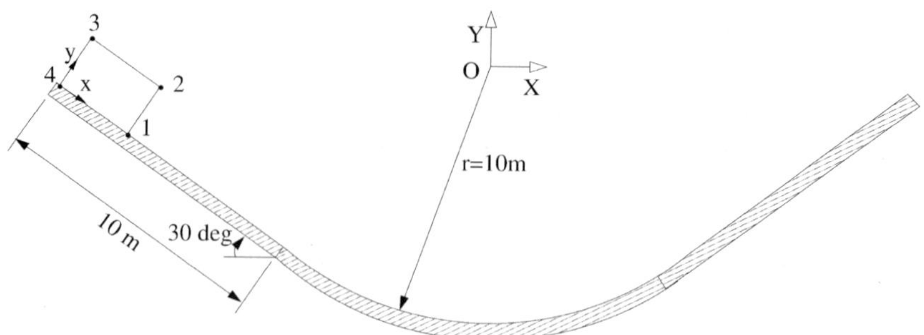

Fig. 3. Sliding and sticking contact of a non-constrained elastic block on a rigid foundation.

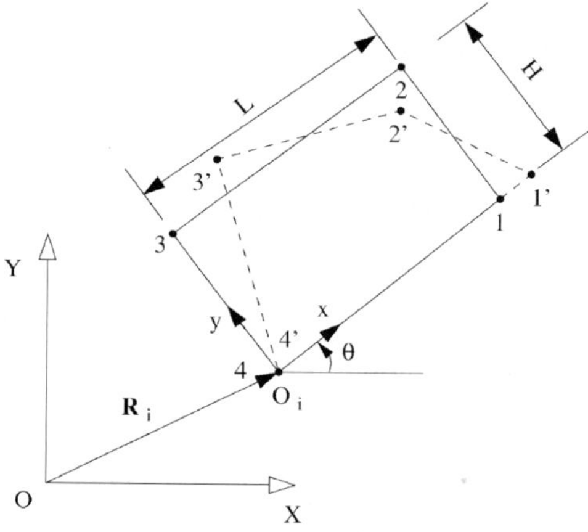

Fig. 4. An elastic block in the undeformed and deformed cases.

coordinate system. Therefore, the position of any arbitrary point on the elastic block can be calculated with respect to the coordinate system O_i through the coordinates of the nodes. The selection of elastic coordinates is $\mathbf{q}_f = (q_{f1_x}, q_{f2_x}, q_{f2_y}, q_{f3_x}, q_{f3_y})$ since they have to be compatible with the rigid coordinates [27]. Rigid and elastic coordinates have to lead to a unique description of position and orientation of the elastic block. This means that node 1 can only move in the x direction with respect to the coordinate system O_i. The vector of generalized coordinates for this example can be considered as $\mathbf{q} = (R_x, R_y, \theta, q_{f1_x}, q_{f2_x}, q_{f2_y}, q_{f3_x}, q_{f3_y})$.

The shape matrix of this elastic block is given by

$$\mathbf{S} = \begin{bmatrix} N_1 & N_2 & 0 & N_3 & 0 \\ 0 & 0 & N_2 & 0 & N_3 \end{bmatrix}, \tag{36}$$

where N_1, N_2 and N_3 are shape functions defined in terms of the length L and the height H of the block

$$N_1 = \frac{x}{L}\left(1 - \frac{y}{H}\right), \qquad N_2 = \frac{x}{L}\frac{y}{H}, \qquad N_3 = \left(1 - \frac{x}{L}\right)\frac{y}{H}. \tag{37}$$

These shape functions have the usual property that they take the value one at their corresponding nodes and the value zero at the other nodes

$$\begin{cases} \text{at node 1:} & N_1 = 1,\ N_2 = 0,\ N_3 = 0, \\ \text{at node 2:} & N_1 = 0,\ N_2 = 1,\ N_3 = 0, \\ \text{at node 3:} & N_1 = 0,\ N_2 = 0,\ N_3 = 1, \\ \text{at node 4:} & N_1 = 0,\ N_2 = 0,\ N_3 = 0. \end{cases} \tag{38}$$

Based on the selected coordinates and the described shape matrix, one can construct the matrix $\mathbf{L}^k$ and start to generate the kinematical relationships of contact for an arbitrary contact point k. After some mathematical manipulations, the matrix $\mathbf{L}^k$ for contact point k can be given as

$$\mathbf{L}^k = \begin{bmatrix} 1 & 0 & -x\sin\theta - y\cos\theta & N_1\cos\theta & N_2\cos\theta & -N_2\sin\theta & N_3\cos\theta & -N_3\sin\theta \\ 0 & 1 & x\cos\theta - y\sin\theta & N_1\sin\theta & N_2\sin\theta & N_2\cos\theta & N_3\sin\theta & N3\cos\theta \end{bmatrix}. \tag{39}$$

In the next step, the derivative of this matrix with respect to time has to be taken and then the matrix $\dot{\mathbf{L}}^k$ can be generated. Afterwards, following Equation (20) all necessary parameters for contact between the elastic block and the rigid foundation are obtained.

After formulating the kinematical relationships of Equation (20), the equations of motion of the elastic block have to be generated. Since there is no explicit constraint in this system, the second row of the system of equations in Equation (23) vanishes and this equation can be rewritten as

$$\mathbf{M} \cdot \ddot{\mathbf{q}} + \mathbf{C} \cdot \dot{\mathbf{q}} + \mathbf{K} \cdot \mathbf{q} = \mathbf{F}_{ext} + \mathbf{F}_v + \mathbf{F}_c. \tag{40}$$

The components of the damping matrix $\mathbf{c}_{ff}$ in matrix $\mathbf{C}$ which correspond to the elastic coordinates can often be chosen as Rayleigh damping in terms of the mass matrix $\mathbf{m}_{ff}$ and the stiffness matrix $\mathbf{k}_{ff}$ associated with the elastic coordinates and using two constant parameters α and β, see [26],

$$\mathbf{c}_{ff} = \alpha\mathbf{m}_{ff} + \beta\mathbf{k}_{ff}. \tag{41}$$

The quantities in Equation (40) are derived following the procedure in [28]. The contact force vector $\mathbf{F}_c$ can be calculated based on the relations in Section 3.1.2 in terms of the Lagrange multipliers $\boldsymbol{\lambda}_N$ and $\boldsymbol{\lambda}_H$. In the last step, the complementarity form of the equations of motion is derived and by solving this complementarity equations, the contact of the elastic block on the rigid foundation can be simulated.

In this work, the linear complementarity equations are solved by the PATH solver which is an algorithm for mixed complementarity problems, see [3, 11]. In order

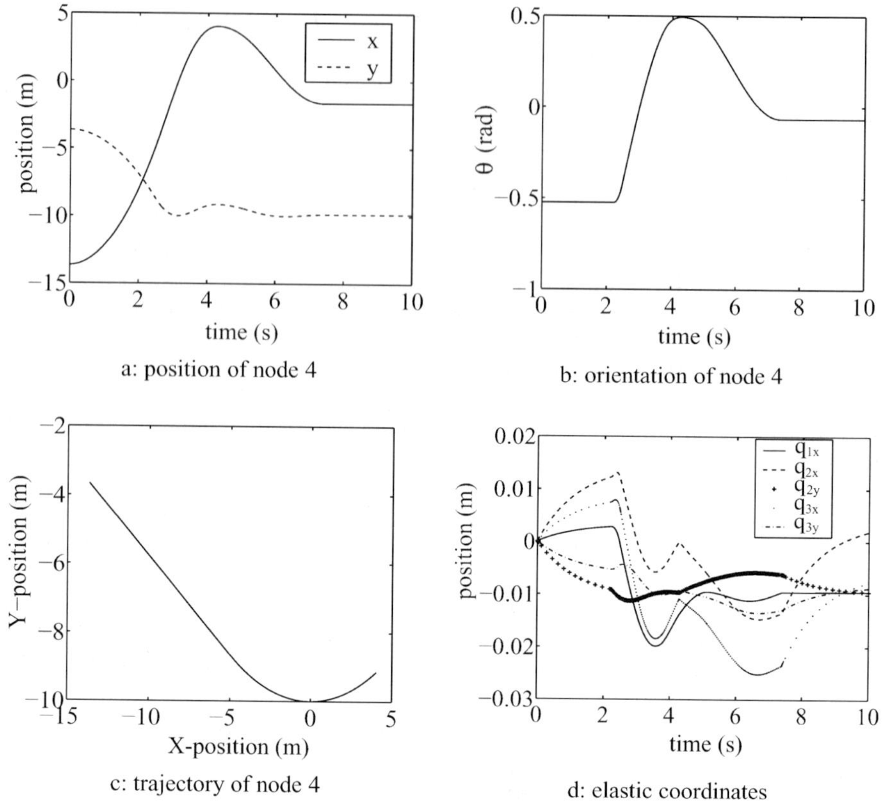

Fig. 5. Contact simulation of the elastic block for $\mu = 0.2$.

to overcome common problems of instability in the integration process of flexible bodies, the equations of motion are integrated using the RADAU5 code [13].

For simulation of this academic test example, the elastic block initially is located according to Figure 3 and the following material and geometrical properties are used:

$$\begin{array}{llll} \text{material:} & E = 1000\ \text{N/m}, & \nu = 0.3, & \rho = 2\ \text{kg/m}^2, \\ \text{geometry:} & H = 1\ \text{m}, & \text{L=2 m (defined in Figure 4)} & . \end{array}$$

According to Figures 5a and 5b it reaches the lowest point of the curved path at $t = 3$ s but it cannot enter the right inclined part since θ does not reach the constant value of 0.523 rad which is the inclination angle of the right inclined part. Instead, it starts to move in the opposite direction at $t = 4$ s. In the end, it comes to rest somewhere on the curved part at $t = 7.6$ s and at this time due to the friction sliding contact switches to sticking one and the block will not have any movement after this time. The trajectory of motion of the elastic block and the elastic coordinates are shown in Figures 5c and 5d. It can be seen that even in the sticking phase there are still some elastic vibrations of the block.

4 Conclusion

In this paper, first some of the most well-known and frequently used formulations in finite element methods and multibody systems dynamics which are applied to incorporate the contact constraints into the governing equations of the system were introduced briefly. Among them, the penalty technique, the Lagrange multiplier method, the augmented Lagrange multiplier method and linear complementarity problem (LCP) formulations were explained. Some special formulations used in multibody system dynamics such as impulse-based approaches, the polygonal contact method and time-stepping methods were also mentioned. Then, a brief reference of contact in granular media and geared systems is given as applications the authors are dealing with.

In Section 3 as the core of this paper a formulation leading to linear complementarity problems for continual contact of planar deformable bodies was presented. The procedure of this formulation started by formulating the kinematics of an arbitrary contact point for an arbitrary pair of contacting bodies. In doing so, relative accelerations of the contact point in normal and tangential direction were used. The effect of deformations was taken into account during the formulation of these accelerations.

In the next step, deformable bodies were modeled by utilizing the well-known moving frame of reference approach assuming small deformations, see [28, 30]. In the last step of this formulation, the complementarity form of the equations of motion was constructed by following the same procedure as described in [24] for rigid body contact.

For validation of the described procedure, the presented formulation was examined through an example. Here, continual contact of a non-guided elastic block on the rigid foundation is investigated and at the end, the results obtained from simulation of this example were given. See also another example in [5] for continual contact of a guided elastic block on a half-circular rigid foundation. At the end, we emphasize once again that this formulation just considers the continual contact case of deformable bodies and for impact calculation this formulation has to be extended [10].

References

1. Anitescu M, Potra FA (2002) A constraint-stabilized time-stepping approach for rigid multibody dynamics with joints, contact and friction. International Journal for Numerical Methods in Engineering 55:753–784

2. Anitescu M, Hart GD (2004) A time-stepping method for stiff multibody dynamics with contact and friction. International Journal for Numerical Methods in Engineering 60:2335–2371
3. Dirske PS, Ferris MC (1995) The PATH solver: A non-monotone stabilization scheme for mixed complementarity problems. Optimization Methods and Software 5:123–156
4. Eberhard P (2000) Kontaktuntersuchungen durch hybride Mehrkörpersystem/ Finite Elemente Simulationen (in German). Shaker Verlag, Aachen
5. Eberhard P, Ebrahimi S (2005) On the use of linear complementarity problems for contact of planar flexible bodies. Proceedings of the ECCOMAS Thematic Conference on Multibody Dynamics, Eds.: J.M. Goicolea, J. Cuadrado, J.C. Garcia Orden, Madrid, Spain
6. Ebrahimi S, Eberhard P (2005) Contact of planar deformable bodies using a linear complementarity formulation. PAMM Proceedings in Applied Mathematics and Mechanics 5:197–198
7. Ebrahimi S, Hippmann G, Eberhard P (2005) Extension of the polygonal contact model for flexible multibody systems. International Journal of Applied Mathematics and Mechanics 1:33–50
8. Ebrahimi S, Eberhard P (2004) Polygonal contact model for elastic bodies. Internal Report IB 140, University of Stuttgart, Institute B of Mechanics
9. Ebrahimi S, Eberhard P (2006) Rigid-elastic modeling of gear wheels in multibody systems, Multibody System Dynamics 16: 55–71
10. Ebrahimi S, Eberhard P (2006) Frictional impact of planar deformable bodies. Submitted to the "Proceedings of IUTAM 2006, Multiscale Problems in Multibody System Contact", Stuttgart, Germany
11. Ferris MC, Munson TS (1999) Interface to PATH 3.0: design, implementation and usage. Computational and Applied Optimization 12:207–227
12. Glocker C (1995) Dynamik von Starrkörpersystemen mit Reibung und Stössen (in German). PhD Thesis, VDI Fortschritt-Berichte, Reihe 18, Nr. 182, VDI-Verlag, Düsseldorf
13. Hairer E, Wanner G (1996) Solving ordinary differential equations II: stiff and differential-algebraic problems. Springer Verlag, Berlin
14. Held M, Klosowski JT, Mitchell JSB (1995) Evaluation of collision detection methods for virtual reality fly-throughs. In Proceedings: 7th Canadian Conference on Computational Geometry, pp. 205–210
15. Hippmann G (2003) An algorithm for compliant contact between complex shaped surfaces in multibody dynamics. Proceedings of the ECCOMAS Thematic Conference on Multibody Dynamics, Ed.: J.A.C. Ambrosio, Lisbon, Portugal
16. Hippmann G (2004) An algorithm for compliant contact between complexly shaped bodies. Multibody System Dynamics 12:345–362
17. Hubbard PM (1993) Interactive collision detection. In IEEE Symposium on Research Frontiers in VR, California, pp. 24–31
18. Johnson KL (1985) Contact mechanics. Cambridge University Press, Cambridge
19. Kahraman A , Singh R (1990) Non-linear dynamics of a spur gear pair. Journal of Sound and Vibration 142:49–75
20. Mauer L (2005) Force element 225 gear wheel. Internal report, Intec GmbH, Wessling, Germany
21. Mirtich B (1996) Impulse-based dynamic simulation of rigid body systems. PhD thesis, University of California, Berkeley

22. Muth B, Müller MK, Eberhard P, Luding S (2004) Contacts between many bodies. Machine Dynamics Problems 28:101–114
23. Nour-Omid B, Wriggers P (1987) A note on the optimum choice of penalty parameters. Communications in Applied Numerical Methods 3:581–585
24. Pfeiffer F, Glocker C (1996) Multibody dynamics with unilateral contacts. John Wiley & Sons, New York
25. Pfeiffer F (2003) The idea of complementarity in multibody dynamics. Archive of Applied Mechanics 72:807–816
26. Schwertassek R, Wallrapp O (1999) Dynamik flexibler Mehrkörpersysteme (in German). Vieweg Verlag, Wiesbaden
27. Schwertassek R, Wallrapp O, Shabana AA (1999) Flexible multibody simulation and choice of shape functions. Nonlinear Dynamics 20:361–380
28. Shabana AA (1998) Dynamics of multibody systems. John Wiley & Sons, New York
29. Shabana AA (2001) Computational dynamics. John Wiley & Sons, New York
30. Shabana AA (1997) Flexible multibody dynamics: review of past and recent developments. Multibody System Dynamics 1:189–227
31. SIMPACK, Commercial MBS Code, www.simpack.de
32. Song P, Pang JS, Kumar V (2004) A semi-implicit time-stepping model for frictional compliant contact problem. International Journal for Numerical Methods in Engineering 60:2231–2261
33. Stewart DE (2005) Convolution complementarity problems with application to impact problems. IMA Journal of Applied Mathematics, doi: 10.1093/imamat/hxh087:1–28
34. Stewart DE, Trinkle JC (1996) An implicit time-stepping scheme for rigid body dynamics with inelastic collisions and coulomb friction. International Journal for Numerical Methods in Engineering 39:2673–2691
35. Wavefront object file specification: www710.univ-lyon1.fr/~jciehl/Public/utiles/maya_obj_spec.pdf
36. Wriggers P (2002) Computational contact mechanics. John Wiley & Sons, Chichester
37. Wriggers P, Simo JC, Taylor RL (1985) Penalty and augmented Lagrangian formulation for contact problems. In: Proceedings of NUMETA Conference, Balkema, Rotterdam
38. Wriggers P (1995) Finite element algorithms for contact problems. Archives of Computational Methods in Engineering 2:1–49
39. Zachmann G (1998) Rapid collision detection by dynamically aligned DOP-tress. Proceedings of IEEE Virtual Reality Annual International Symposium (VRAIS), Atlanta, Georgia

On the Stabilizing Properties of Energy-Momentum Integrators and Coordinate Projections for Constrained Mechanical Systems

Juan C. García Orden[1] and Daniel Dopico Dopico[2]

[1] *Grupo de Mecánica Computacional, Escuela Técnica Superior de Ingenieros de Caminos, Canales y Puertos, Universidad Politécnica de Madrid, c/ Profesor Aranguren s/n, 28040 Madrid, Spain; E-mail: jcgarcia@mecanica.upm.es*

[2] *Laboratorio de Ingeniería Mecánica, Escuela Politécnica Superior, Universidad de La Coruña, c/ Mendizábal s/n, 15403 Ferrol, Spain; E-mail: ddopico@udc.es*

1 Introduction

Several considerations are important if we try to carry out fast and precise simulations in multibody dynamics: the choice of modeling coordinates, the choice of dynamical formulations and the numerical integration scheme along with the numerical implementation. All these matters are very important in order to decide whether a specific method is good or not for a particular purpose.

Some of the most robust methods for real-time dynamics in multibody systems make use of natural or fully Cartesian coordinates in the modeling [11], which are dependent by nature. Different formulations used to solve the equations of motion with dependent coordinates have been developed, such as the widely known method of Lagrange multipliers, the penalty and augmented Lagrangian schemes [4], or velocity transformations [23, 27]. Some of them set a system of differential-algebraic equations (DAE) [8], others set system of ordinary differential equations (ODE).

Generally, it can be said that the dynamic formulation determines the choice of the numerical integrator. In this direction different authors proposed several options to successfully integrate the equations arising from constrained multibody systems, using integrators coming from the field of structural dynamics [9, 11]. Formulations based on penalty and augmented Lagrangian methods have the advantages of being very simple, computationally inexpensive and very robust in the presence of singular configurations or redundant constraints [3].

In [5, 9] the authors proposed the use of augmented Lagrangian techniques with penalty only at position level along with the trapezoidal rule. In order to guarantee the correct satisfaction of constraints, different kinds of velocities and acceleration

J.C. García Orden et al. (eds), Multibody Dynamics. Computational Methods and Applications, 49–67.

projections were proposed. More recently, in [10] the use of augmented Lagrangian techniques with other integrators of the generalized-α family along with projections was proposed, which provides very good behavior for real-time applications. The advantages of the projections are the simplicity and the variety of integrators which can be used with them, since the projections are responsible for maintaining the stability of the formulation.

On the other hand, other authors [12, 17, 18] developed a formulation based on an energy conserving penalty formulation, enforcing constraints at the position level, and applied it to the dynamics of multibody systems parametrized with Cartesian coordinates. In this case, the use of penalty at position level has the advantage of permitting to derive the constraint forces from a potential function: the constraint energy. The formulation includes the employment of an energy-momentum method as integration scheme [19, 25], so that the conservation of the total energy of the system is imposed by construction of the algorithm. Here, the stabilization of the equations of motion arises in a natural manner from the integration scheme.

The outline of this work is as follows. First, and overview of the most common formulations employed for the representation of the dynamics of constrained mechanical systems are presented. Next, the numerical difficulties that pose the different formulations are discussed. These issues prepare the context for the presentation of two proposed methods in the next sections, one of them based on the use of a standard ODE integrator with projections, the other on a conservative scheme. The following section analyzes with more detail their behaviour in terms of the discrete energy balance, and draws some interpretations about their stabilization features. Finally, a representative numerical example is presented, illustrating the most relevant issues introduced in the previous sections.

2 Dynamics of Constrained Mechanical Systems

In this section, we consider the formulation and the numerical solution of the dynamics of a constrained mechanical system; for instance, a set of rigid and deformable bodies linked by joints (represented by a vector of r holonomic constraints $\mathbf{0} = \mathbf{\Phi}(\mathbf{q}, t) \in \mathbb{R}^r$), being $\mathbf{q} \in \mathbb{R}^n$ a set of Cartesian coordinates.

In this work we focus on the different methods to impose constraints, which lead to different formulations for the equations of motion. Several strategies can be used to solve these equations, each of them posing special numerical difficulties that will be addressed with more detail in the following sections.

The three basic formulations considered here are based on Lagrange multipliers, penalty and augmented Lagrangian respectively. Following, a brief review of these three formulations is presented, along with a short description of the methods most commonly used to solve them.

2.1 Lagrange multiplier method

This method leads to an index-3 DAE system, given by:

$$\mathbf{M}\ddot{\mathbf{q}} + \mathbf{\Phi}_{\mathbf{q}}^{\mathrm{T}}\boldsymbol{\lambda} = \mathbf{Q}, \qquad \mathbf{\Phi} = \mathbf{0}, \tag{1}$$

$\mathbf{M}$ being the mass matrix, $\boldsymbol{\lambda} \in \mathbb{R}^r$ the vector of Lagrange multipliers, $\mathbf{Q}(\mathbf{q}, \dot{\mathbf{q}}, t)$ the applied force vector, and denoting by $\mathbf{\Phi}_{\mathbf{q}} \stackrel{\text{def}}{=} \partial\mathbf{\Phi}/\partial\mathbf{q}$.

There are several methods that can be employed to solve the equation system (1):

1. Direct solution with a DAE solver. Backward Differentiation Formula (BDF), Implicit Runge–Kutta (IRK) and collocation methods are examples of numerical integration algorithms that are very efficient on the direct solution of these type of systems [8].
2. Index reduction (index-2). Differentiation of the constraint equation reduces the index by one, resulting in the following index-2 DAE system:

$$\mathbf{M}\ddot{\mathbf{q}} + \mathbf{\Phi}_{\mathbf{q}}^{\mathrm{T}}\boldsymbol{\lambda} = \mathbf{Q}, \qquad \dot{\mathbf{\Phi}} = \mathbf{0}, \tag{2}$$

   which, again, can be solved directly by a suitable DAE solver.
3. Index reduction (index-1). Two differentiations of the constraint equation reduces the index by two, resulting in the following index-1 DAE system:

$$\mathbf{M}\ddot{\mathbf{q}} + \mathbf{\Phi}_{\mathbf{q}}^{\mathrm{T}}\boldsymbol{\lambda} = \mathbf{Q}, \qquad \ddot{\mathbf{\Phi}} = \mathbf{0}. \tag{3}$$

   If desired, a further index reduction may be performed, eliminating the Lagrange multiplier vector $\boldsymbol{\lambda}$ and obtaining a standard ODE system. This can be done taking into account the differential system in (3) and the expression for the second derivative of the constraint $\ddot{\mathbf{\Phi}} = \mathbf{\Phi}_{\mathbf{q}}\ddot{\mathbf{q}} + \dot{\mathbf{\Phi}}_{\mathbf{q}}\dot{\mathbf{q}} + \dot{\mathbf{\Phi}}_t$, with $\mathbf{\Phi}_t \stackrel{\text{def}}{=} \partial\mathbf{\Phi}/\partial t$; after some algebraic manipulations an ODE system results, given by:

$$\mathbf{M}\ddot{\mathbf{q}} = \mathbf{Q} - \mathbf{\Phi}_{\mathbf{q}}^{\mathrm{T}}\left(\mathbf{\Phi}_{\mathbf{q}}\mathbf{M}^{-1}\mathbf{\Phi}_{\mathbf{q}}^{\mathrm{T}}\right)^{-1}\left(\mathbf{\Phi}_{\mathbf{q}}\mathbf{M}^{-1}\mathbf{Q} + \dot{\mathbf{\Phi}}_{\mathbf{q}}\dot{\mathbf{q}} + \dot{\mathbf{\Phi}}_t\right), \tag{4}$$

   which can be solved with any ODE solver.

2.2 Penalty method

This method leads to an ODE system given by:

$$\mathbf{M}\ddot{\mathbf{q}} + \mathbf{\Phi}_{\mathbf{q}}^{\mathrm{T}}(\boldsymbol{\alpha}\mathbf{\Phi}) = \mathbf{Q}, \tag{5}$$

$\boldsymbol{\alpha}$ being the penalty matrix, which is often defined with a single penalty parameter α, such that $\boldsymbol{\alpha} = \alpha\mathbf{1}$, $\mathbf{1}$ being the unit matrix.

This formulation can be interpreted as the perturbed DAE problem given by (1), verifying $\mathbf{\Phi} \to \mathbf{0}$ as $\alpha \to \infty$, and can be solved by a suitable ODE integrator.

Formulation (5) penalizes the constraint at position level only ($\mathbf{\Phi}$), but it may include also the constraint at velocity and acceleration levels ($\dot{\mathbf{\Phi}}$ and $\ddot{\mathbf{\Phi}}$ respectively), taking the more general form:

$$\mathbf{M}\ddot{\mathbf{q}} + \mathbf{\Phi}^{\mathrm{T}}\alpha\left(\ddot{\mathbf{\Phi}} + 2\xi\omega\dot{\mathbf{\Phi}} + \omega^2\mathbf{\Phi}\right) = \mathbf{Q}, \qquad (6)$$

where ω and ξ can be interpreted as the natural frequency and damping ratio of the penalized constraint [11].

2.3 Augmented Lagrangian method

It can be understood as a compromise between the Lagrange multiplier and the penalty method, and leads to and index-3 DAE system given by:

$$\mathbf{M}\ddot{\mathbf{q}} + \mathbf{\Phi}_{\mathbf{q}}^{\mathrm{T}}\boldsymbol{\lambda}^* + \mathbf{\Phi}_{\mathbf{q}}^{\mathrm{T}}(\alpha\mathbf{\Phi}) = \mathbf{Q}, \qquad \mathbf{\Phi} = \mathbf{0}, \qquad (7)$$

where $\boldsymbol{\lambda}^*$ represents the Lagrange multiplier vector. This formulation is commonly set, from an algorithmic point of view, by the Uzawa method, which introduces an iterative scheme for the multipliers. This algorithmic approach, in practice, transforms the DAE system into an ODE system, defining an update for the multipliers given by $\boldsymbol{\lambda}^*_{(k+1)} = \boldsymbol{\lambda}^*_{(k)} + \boldsymbol{\alpha}\mathbf{\Phi}$, verifying $\boldsymbol{\lambda}^* \to \boldsymbol{\lambda}$ as iteration in $\boldsymbol{\lambda}^*$ progresses, being $\boldsymbol{\lambda}$ the exact Lagrange multiplier vector.

Different strategies can be followed to solve the formulation (7). All of them assume the use of Uzawa's method, thus effectively leading to the application of an ODE solver. The basic difference among them is the index of the original DAE system to be solved.

1. Direct solution of the index-3 DAE system (7), applying Uzawa's method and a suitable ODE integrator.
2. Index reduction (index-2), introducing the first derivative of the constraint equation and obtaining:

$$\mathbf{M}\ddot{\mathbf{q}} + \mathbf{\Phi}_{\mathbf{q}}^{\mathrm{T}}\boldsymbol{\lambda}^* + \mathbf{\Phi}_{\mathbf{q}}^{\mathrm{T}}\alpha(2\xi\omega\dot{\mathbf{\Phi}} + \omega^2\mathbf{\Phi}) = \mathbf{Q}\ , \qquad 2\xi\omega\dot{\mathbf{\Phi}} + \omega^2\mathbf{\Phi} = \mathbf{0} \qquad (8)$$

   to be solved applying Uzawa's method and a suitable ODE integrator.
3. Index reduction (index-1), introducing the second derivative of the constraint equation and obtaining:

$$\mathbf{M}\ddot{\mathbf{q}} + \mathbf{\Phi}_{\mathbf{q}}^{\mathrm{T}}\boldsymbol{\lambda}^* + \mathbf{\Phi}_{\mathbf{q}}^{\mathrm{T}}\alpha(\ddot{\mathbf{\Phi}} + 2\xi\omega\dot{\mathbf{\Phi}} + \omega^2\mathbf{\Phi}) = \mathbf{Q}\ , \quad \ddot{\mathbf{\Phi}} + 2\xi\omega\dot{\mathbf{\Phi}} + \omega^2\mathbf{\Phi} = \mathbf{0} \quad (9)$$

   to be solved applying Uzawa's method and a suitable ODE integrator.

All these formulations pose numerical difficulties, which are going to be explored with more detail in the following section, and they will motivate the search for improvements, particularly regarding stability issues.

3 Stability Problems of the Numerical Solution

As previously pointed out, the formulation of the dynamics of a constrained mechanical system pose numerical difficulties. These difficulties are, in general, different for each formulation and solution method, but they are typically related to stability properties of the numerical scheme, and they are going to be briefly exposed in the following paragraphs.

- Direct integration of DAEs of index higher than 1 (formulations (1), (2), (7) and (8)) is usually not recommended for stability reasons. Actually, there are index-2 and index-3 DAEs for which all of the multistep (including BDF) and Runge–Kutta methods are unstable, as pointed out in [8]. In the particular case of a constrained mechanical system described by the index-3 DAE given by (1),[1] or the augmented version (7), direct integration has been reported to show instability problems [9].
- The analytical differentiation of the constraint equations is an unstabilized index reduction (formulations (2), (3), (8) and (9)). Constraints on position, velocity or acceleration levels define invariant manifolds, where the exact solution lies. However, the numerical solution may depart from them, and indeed it usually does due to the referred unstable reduction.
  On the other hand, several numerical experiments by different authors suggest than the solution is more stable on the manifold than off it. This fact justifies the search for methods that enforce the solution to be on the constraint manifold, thus enhancing the stability of the resulting numerical scheme.
- ODE integrators may exhibit severe numerical instabilities for stiff systems, such as those resulting from a penalty formulation (5) or (6), where large penalty parameters are required in order to get a satisfactory constraint enforcement.
  Some integration schemes are better suited for these type of problems, such as implicit Runge–Kutta and BDF methods [20]. However, these methods usually introduce a significant amount of numerical damping, which can be unacceptable for long term simulations. In the context of Hamiltonian systems, energy-momentum methods [19] exhibit very good stability for stiff systems, while exactly conserving the total energy, and are actually a very adequate choice for robust long-term simulations.

These numerical problems motivate the interest in developing algorithms capable of providing stable and accurate solutions for reasonable time steps. Several methods have been proposed in the literature to alleviate these problems for the different formulations (e.g. [1, 2, 7, 8]). To collect and discuss all these different methods is not a simple task, because they are numerous and sometimes application-dependent, and it is out of the scope of this paper.

[1] This is a special DAE form, known in the literature as Hessenberg index-3 type.

Here we restrict ourselves to the analysis of two methods, both of them based on the augmented Lagrangian formulation (7), and discuss their stabilizing properties. The first proposed approach is a coordinate projection method, the second is a conservative formulation, and both will be outlined in the sections that follows.

4 Augmented Lagrangian with Projections

The point of departure of this approach is the index-3 DAE given by expression (7), repeated here for the sake of clarity:

$$\mathbf{M}\ddot{\mathbf{q}} + \mathbf{\Phi}_{\mathbf{q}}^{\mathrm{T}}\boldsymbol{\lambda}^{*} + \mathbf{\Phi}_{\mathbf{q}}^{\mathrm{T}}(\alpha\mathbf{\Phi}) = \mathbf{Q}, \qquad \mathbf{\Phi} = \mathbf{0}.$$

As stated in Section 2, a numerical solution can be obtained combining an ODE integrator with an update formula for the Lagrange multipliers, in a procedure referred to as the Uzawa method, or method of multipliers. Now, for non-linear problems, it is possible to define two main alternatives for the multipliers update scheme in a single time step:

- *Nested iteration*, setting two nested loops; an outer loop for the multiplier $\boldsymbol{\lambda}$, and an interior loop that solve the ODE for a fixed value of the multiplier. This is the most common implementation of this method, originally introduced in the context of constrained optimization [6] and applied in many engineering problems, as contact mechanics [21, 28].
- *Simultaneous iteration* sets only one loop, where the multiplier update is done simultaneously with the iterations required by the ODE solver. This implementation may exhibit stability problems in some applications, caused by the non-differentiability of the update [21], but nevertheless it has been successfully applied to multibody systems [4, 9].

This augmented Lagrangian formulation leads to an exact fulfillment of the original position constraints ($\mathbf{\Phi} = \mathbf{0}$), but usually exhibits an unstable behaviour for moderate time step sizes, even with ODE integrators suited to stiff systems.

As mentioned in Section 3, based on previous results in the literature, the numerical solution of a constrained mechanical system seems to be more stable on the constraint manifold than off it. Based on this fact, the exact enforcement of the constraint not only at position level, but also at velocity and acceleration levels ($\dot{\mathbf{\Phi}} = \mathbf{0}$ and $\ddot{\mathbf{\Phi}} = \mathbf{0}$ respectively), which is not accomplished by the augmented Lagrangian formulation presented here, is foreseen to stabilize the numerical solution.

This enforcement can be accomplished by different methods; one of them is the so-called coordinate projection technique, which is the one selected in this work, and will be outlined in the next paragraphs.

In case of a *velocity projection*, the velocities computed with the ODE integrator ($\dot{\mathbf{q}}^{*}$) are projected onto the velocity constraint manifold to obtain new velocities ($\dot{\mathbf{q}}$), solving a constrained minimization problem given by:

$$\min_{\dot{\mathbf{q}}} \frac{1}{2}(\dot{\mathbf{q}} - \dot{\mathbf{q}}^*)^{\mathrm{T}}\mathbf{A}(\dot{\mathbf{q}} - \dot{\mathbf{q}}^*) \quad \text{such that} \quad \dot{\mathbf{\Phi}} = \mathbf{0}, \tag{10}$$

being $\mathbf{A}$ a definite positive matrix. This constrained minimization problem can be solved with different methods; one of the simplest is penalty, which leads to the solution for $\dot{\mathbf{q}}$ of a linear algebraic system given by:

$$(\mathbf{A} + \alpha\mathbf{\Phi}_{\mathbf{q}}^{\mathrm{T}}\mathbf{\Phi}_{\mathbf{q}})\dot{\mathbf{q}} = \mathbf{A}\dot{\mathbf{q}}^*, \tag{11}$$

being α a penalty parameter.

In case of a *acceleration projection*, the accelerations computed with the ODE integrator ($\ddot{\mathbf{q}}^*$) are projected onto the acceleration constraint manifold to obtain new accelerations ($\ddot{\mathbf{q}}$), solving a constrained minimization problem given by:

$$\min_{\ddot{\mathbf{q}}} \frac{1}{2}(\ddot{\mathbf{q}} - \ddot{\mathbf{q}}^*)^{\mathrm{T}}\mathbf{A}(\ddot{\mathbf{q}} - \ddot{\mathbf{q}}^*) \quad \text{such that} \quad \ddot{\mathbf{\Phi}} = \mathbf{0}, \tag{12}$$

$\mathbf{A}$ being a definite positive matrix.[2] Again, this constrained minimization problem can be solved with penalty, which leads to the solution for $\ddot{\mathbf{q}}$ of a linear algebraic system given by:

$$(\mathbf{A} + \alpha\mathbf{\Phi}_{\mathbf{q}}^{\mathrm{T}}\mathbf{\Phi}_{\mathbf{q}})\ddot{\mathbf{q}} = \mathbf{A}\ddot{\mathbf{q}}^* - \alpha\mathbf{\Phi}_{\mathbf{q}}^{\mathrm{T}}\dot{\mathbf{\Phi}}_{\mathbf{q}}\dot{\mathbf{q}}. \tag{13}$$

5 Conserving Augmented Lagrangian Formulation

The point of departure of this approach is the algorithmic expression of the energy-momentum method [19, 24] applied to a conservative mechanical system given by (5), which enforces a set of holonomic constraints $\mathbf{\Phi}(\mathbf{q})$ with the penalty method [12, 17]:

$$\mathbf{M}\big(\dot{\mathbf{q}}_{n+1} - \dot{\mathbf{q}}_n\big) + \Delta t\,\mathbf{\Phi}_{\mathbf{q}_{n+\beta}}^{\mathrm{T}}\alpha\overline{\mathbf{\Phi}}_{n+\frac{1}{2}} = \mathbf{0}, \tag{14}$$

$$\frac{1}{2}(\dot{\mathbf{q}}_{n+1} + \dot{\mathbf{q}}_n) = \frac{1}{\Delta t}(\mathbf{q}_{n+1} - \mathbf{q}_n),$$

where it has been assumed, with no loss of generality for the following discussion, that will focus on the constraint forces, that the applied forces are null ($\mathbf{Q} = \mathbf{0}$), and denoting $\overline{(\cdot)}_{n+\frac{1}{2}} \stackrel{\text{def}}{=} [(\cdot)_n + (\cdot)_{n+1}]/2$ and $(\cdot)_{n+\beta}$ evaluation at $\mathbf{q}_{n+\beta} \stackrel{\text{def}}{=} \mathbf{q}_n + \beta(\mathbf{q}_{n+1} - \mathbf{q}_n)$.

The parameter $\beta \in [0, 1]$ has to be computed at each time step, imposing that the dot product between the gradient of the constraint and the increment in position verify:

[2] Not necessarily the same employed for the velocity projection.

$$\beta; \qquad \boldsymbol{\Phi}_{\mathbf{q}_{n+\beta}} (\mathbf{q}_{n+1} - \mathbf{q}_n) = \boldsymbol{\Phi}_{n+1} - \boldsymbol{\Phi}_n. \tag{15}$$

Note that when constraints are exactly fulfilled the gradient of the constraint (in other words, the constraint force) is orthogonal to the increment in position. As it will be shown later in Section 6, this condition leads to the exact algorithmic nullity of the work performed by the constraint forces, thus leading to exact conservation of total energy. If the constraint is at most quadratic, it is straightforward to see that $\beta = 1/2$. Besides, if the constraint is generally expressed in terms of an scalar variable (e.g. the distance between points), the constraint force term in (14) can be formulated in a closed form without any additional parameter [12].

In order to obtain a conservative augmented Lagrangian formulation, it is convenient to read (14) as a second order approximation of an integral balance of linear momentum between n and $n + 1$:

$$\int_{t_n}^{t_{n+1}} \mathbf{M}\ddot{\mathbf{q}}\, \mathrm{d}t + \int_{t_n}^{t_{n+1}} \boldsymbol{\Phi}_{\mathbf{q}}^{\mathrm{T}} \boldsymbol{\alpha} \boldsymbol{\Phi}\, \mathrm{d}t = \mathbf{0}. \tag{16}$$

Equations (14) and (16) reveal that the term $\boldsymbol{\alpha\Phi}$ is evaluated as $\boldsymbol{\alpha}\overline{\boldsymbol{\Phi}}_{n+\frac{1}{2}}$ in order to calculate this integral, that leads to the conserving formulation.

On the other hand, it is also possible to understand the augmented Lagrangian method (7) as an extended penalty method, where the penalized constraint $\boldsymbol{\alpha\Phi}$ is corrected at each time step by a set $\boldsymbol{\lambda}^*$ of Lagrange multipliers, which are updated with a scheme given by $\boldsymbol{\lambda}^{*(k+1)} = \boldsymbol{\lambda}^{*(k)} + \boldsymbol{\alpha\Phi}$ with a nested or simultaneous iteration strategy, as discussed in Section 4.

Taking into account these considerations, it is possible to define a conserving algorithm that incorporates the augmented term:

$$\mathbf{M}(\dot{\mathbf{q}}_{n+1} - \dot{\mathbf{q}}_n) + \Delta t\left[\boldsymbol{\Phi}_{\mathbf{q}_{n+\beta}}^{\mathrm{T}} \boldsymbol{\alpha} \overline{\boldsymbol{\Phi}}_{n+\frac{1}{2}} + \boldsymbol{\Phi}_{\mathbf{q}_{n+\beta}}^{\mathrm{T}} \boldsymbol{\lambda}^*\right] = \mathbf{0} \tag{17}$$

$$\frac{1}{2}(\dot{\mathbf{q}}_{n+1} + \dot{\mathbf{q}}_n) = \frac{1}{\Delta t}(\mathbf{q}_{n+1} - \mathbf{q}_n)$$

and accordingly sets an update scheme for the set of multipliers, given by:

$$\boldsymbol{\lambda}_{n+1}^{*(k+1)} = \boldsymbol{\lambda}_{n+1}^{*(k)} + \boldsymbol{\alpha}\overline{\boldsymbol{\Phi}}_{n+\frac{1}{2}}. \tag{18}$$

The proposed algorithm given by (17) and (18) achieve exact conservation of total energy (see [15] for more details) and exact fulfillment of the position constraints, as the augmented Lagrange multipliers set $\boldsymbol{\lambda}_{n+1}^*$ converge to the true Lagrange multipliers set $\boldsymbol{\lambda}_{n+1}$ when its iteration progresses.

Finally, note that the coordinate projection technique described in Section 4 can be applied here too in order to enforce the constraints at the velocity and/or acceleration levels at each time step, but then the energy is no longer conserved. The energy can grow or diminish depending on the way the projection is carried out, which is one of the main topics discussed in the following section.

6 Energetic Considerations

For ODE systems arising from the dynamics of mechanical systems, the stability properties of the numerical methods used to solve them are typically related to the concept of energy. Actually, in the linear case, exact algorithmic energy conservation and unconditional stability are directly related, as it happens, for instance, with the trapezoidal rule, which has both features [16]. However, this direct relationship does not hold for the non-linear case [22], which is the case of the equations resulting from practical multibody systems. Nevertheless, exact conservation of energy (or unconditional energy dissipation) has revealed extremely useful in the design of robust integration schemes, with excellent stability in the non linear case (see [26] and references therein).

With these arguments in mind, it is interesting to analyze how the two proposed methodologies behave in terms of discrete energy balance. As it will be shown next, it comes out that both methods actually controls the energy (thus providing a justification for their stabilization properties) but they do it differently; the projection method provides a means for conserving or dissipating energy, and the conservative approach exactly conserves it.

In order to study both cases from a common point of departure, let us consider a constrained mechanical system, parametrized with a set of coordinates $\mathbf{q} \in \mathbb{R}^n$, subjected to a set of r holonomic constraints $\boldsymbol{\Phi}(\mathbf{q}) \in \mathbb{R}^r$ and with no applied forces. The dynamics of this system is represented by the differential equation:

$$\mathbf{M}\ddot{\mathbf{q}} + \mathbf{Q}_{\Phi}(\mathbf{q}) = \mathbf{0}, \tag{19}$$

$\mathbf{Q}_{\Phi}$ being the constraint force vector, which in the case of the augmented Lagrangian method is given by $\mathbf{Q}_{\Phi} = \boldsymbol{\Phi}_{\mathbf{q}}^{\mathrm{T}}\boldsymbol{\lambda}^* + \boldsymbol{\Phi}_{\mathbf{q}}^{\mathrm{T}}(\alpha\boldsymbol{\Phi})$. Note that the dynamical system represented by (19) is conservative (the total mechanical energy remains constant), since the work performed by holonomic constraints which do not depend explicitly on time is null. This fact does not pose any practical limitation for our purposes and helps the understanding of the developments presented in the rest of the section.

Using an ODE integrator to calculate the solution from t_n to t_{n+1}, combined with the proper Lagrange multiplier update scheme, a solution $\mathbf{q}_{n+1}$ that exactly satisfies the position constraint can be obtained. Consequently, the constraint force at t_{n+1} takes the value $\mathbf{Q}_{\Phi_{n+1}} = \boldsymbol{\Phi}_{\mathbf{q}_{n+1}}^{\mathrm{T}}\boldsymbol{\lambda}_{n+1}$, being $\boldsymbol{\lambda}_{n+1}$ the vector of exact Lagrange multipliers.

A velocity vector $\dot{\mathbf{q}}_{n+1}^*$ is also obtained but, in general, the velocity constraint $\dot{\boldsymbol{\Phi}}_{n+1}$ is not exactly satisfied. In order to move the solution back to the velocity manifold, let us assume that a velocity projection is performed at the end of each time step as explained in Section 4, obtaining a new velocity vector $\dot{\mathbf{q}}_{n+1}$.

The total energy balance ΔE between t_n and t_{n+1} is given by:

$$\Delta E = \frac{1}{2}\dot{\mathbf{q}}_{n+1}^{\mathrm{T}}\mathbf{M}\dot{\mathbf{q}}_{n+1} - \frac{1}{2}\dot{\mathbf{q}}_n^{\mathrm{T}}\mathbf{M}\dot{\mathbf{q}}_n. \tag{20}$$

Note that the total energy balance ΔE given by (20) equals the kinetic energy balance ΔT, which means that $\Delta V = 0$. This is due to the fact that there are not applied forces, the position constraints are exactly satisfied and the position $\mathbf{q}_{n+1}$ does not change under the projection.

Adding and subtracting a term $(1/2)\dot{\mathbf{q}}_{n+1}^{*\mathrm{T}}\mathbf{M}\dot{\mathbf{q}}_{n+1}^{*}$ to equation (20), the following relation is obtained:

$$\Delta E = \underbrace{\frac{1}{2}\dot{\mathbf{q}}_{n+1}^{*\mathrm{T}}\mathbf{M}\dot{\mathbf{q}}_{n+1}^{*} - \frac{1}{2}\dot{\mathbf{q}}_{n}^{\mathrm{T}}\mathbf{M}\dot{\mathbf{q}}_{n}}_{\Delta E_i} + \underbrace{\frac{1}{2}\dot{\mathbf{q}}_{n+1}^{\mathrm{T}}\mathbf{M}\dot{\mathbf{q}}_{n+1} - \frac{1}{2}\dot{\mathbf{q}}_{n+1}^{*\mathrm{T}}\mathbf{M}\dot{\mathbf{q}}_{n+1}^{*}}_{\Delta E_p}, \tag{21}$$

where ΔE_i is the energy variation introduced by the ODE integrator, and ΔE_p the energy variation introduced by the velocity projection.

It is possible to calculate the energy variation introduced by any ODE integrator employing the following preliminary equation:

$$\Delta E_i = \frac{1}{2}(\dot{\mathbf{q}}_{n+1}^{*} + \dot{\mathbf{q}}_n)^{\mathrm{T}}\mathbf{M}(\dot{\mathbf{q}}_{n+1}^{*} - \dot{\mathbf{q}}_n) \tag{22}$$

and using the algorithmic expressions of the method with the original system (19). For instance, for the *trapezoidal rule* the following relations hold:

$$\dot{\mathbf{q}}_{n+1}^{*} + \dot{\mathbf{q}}_n = \frac{2}{\Delta t}(\mathbf{q}_{n+1} - \mathbf{q}_n) \tag{23}$$

$$\dot{\mathbf{q}}_{n+1}^{*} - \dot{\mathbf{q}}_n = -\frac{\Delta t}{2}\mathbf{M}^{-1}\left(\mathbf{Q}_{\Phi_n} + \mathbf{Q}_{\Phi_{n+1}}\right) \tag{24}$$

which introduced in expression (22) gives:

$$\Delta E_i = -(\mathbf{q}_{n+1} - \mathbf{q}_n)^{\mathrm{T}}\,\overline{\mathbf{Q}}_{\Phi_{n+\frac{1}{2}}}, \tag{25}$$

where the notation $\overline{(\cdot)}_{n+\frac{1}{2}} \stackrel{\mathrm{def}}{=} [(\cdot)_n + (\cdot)_{n+1}]/2$ introduced already in Section 5 has been employed again.

Other example is the *implicit midpoint rule*, that introduces an energy variation given by:

$$\Delta E_i = -(\mathbf{q}_{n+1} - \mathbf{q}_n)^{\mathrm{T}}\,\mathbf{Q}_{\Phi_{n+\frac{1}{2}}}, \tag{26}$$

where $(\cdot)_{n+\frac{1}{2}}$ denotes evaluation at the midpoint. Note that, in a general non linear case, $\overline{\mathbf{Q}}_{\Phi_{n+\frac{1}{2}}} \neq \mathbf{Q}_{\Phi_{n+\frac{1}{2}}}$ and $\Delta E_i \neq 0$ can be positive or negative. Note from (25) and (26) that both numerical schemes are the same and exactly conserve energy ($\Delta E_i = 0$) if the constraints are linear.

Finally, using relation (15), it can be shown that the energy variation of the *conserving method* is null, given by:

$$\Delta E_i = -(\mathbf{q}_{n+1} - \mathbf{q}_n)^{\mathrm{T}} \boldsymbol{\Phi}^{\mathrm{T}}_{\mathbf{q}_{n+\beta}} \boldsymbol{\lambda}_{n+1} = 0. \tag{27}$$

Other expressions similar to (25), (26) and (27) can be obtained for other integrators, but this exhaustive description falls out of the scope of the work presented here. It is important to remark that the sign of the energy contribution ΔE_i may not be constant along the simulation, thus increasing or decreasing the total energy, which can in turn affect the numerical stability.

The second contribution to the energy variation is ΔE_p, associated with the velocity projection described in Section 4, and can be obtained solving a minimization problem with a definite positive matrix $\mathbf{A}$ with a penalty method. This leads to the solution for $\dot{\mathbf{q}}_{n+1}$ of the linear algebraic equation system (11), given by:

$$\dot{\mathbf{q}}_{n+1} = \mathbf{P}^{-1}\, \dot{\mathbf{q}}^*_{n+1} \quad \text{with} \quad \mathbf{P} = \left(\mathbf{1} + \alpha \mathbf{A}^{-1} \boldsymbol{\Phi}^{\mathrm{T}}_{\mathbf{q}} \boldsymbol{\Phi}_{\mathbf{q}}\right). \tag{28}$$

Introducing the first expression in (28) in the following relation for ΔE_p:

$$\Delta E_p = \frac{1}{2}(\dot{\mathbf{q}}_{n+1} + \dot{\mathbf{q}}^*_{n+1})^{\mathrm{T}} \mathbf{M} (\dot{\mathbf{q}}_{n+1} - \dot{\mathbf{q}}^*_{n+1}), \tag{29}$$

the following expression is obtained for the energy variation introduced by the velocity projection:

$$\Delta E_p = \dot{\mathbf{q}}^{\mathrm{T}}_{n+1} \mathbf{D} \dot{\mathbf{q}}_{n+1} \quad \text{with} \quad \mathbf{D} = \frac{1}{2}(\mathbf{1} + \mathbf{P})^{\mathrm{T}} \mathbf{M} (\mathbf{1} - \mathbf{P}). \tag{30}$$

Therefore, the effect of projection upon the energy depends of the properties of matrix $\mathbf{D}$, which is the matrix associated to the quadratic form ΔE_p, and governs the damping behaviour of the projection. If this matrix is semidefinite negative, artificial energy growth is avoided in any case, and a significant improvement of the stability of the overall numerical scheme would be expected.

A detailed analysis of the quadratic form (30) can be performed [13], providing a practical assessment about the adequate selection of the projection matrix $\mathbf{A}$, such that artificial energy growth is unconditionally avoided. One preliminary and interesting result of this analysis is that the selection $\mathbf{A} = \mathbf{M}$ introduces unconditional dissipation to any incompatible velocity field (which is a velocity field that falls out of the velocity manifold $\dot{\boldsymbol{\Phi}} = \mathbf{0}$). This property, and the impact that it has over the stability of the resulting numerical algorithm will be observed in the numerical experiment performed in the next section, and perfectly agrees with results previously reported by other authors [5].

7 Numerical Simulation

To better understand the behavior of the formulations presented in Sections 4 and 5, let us present a simple but representative example that poses the essential numerical difficulties typically associated to the constrained dynamics of more complex mechanical systems.

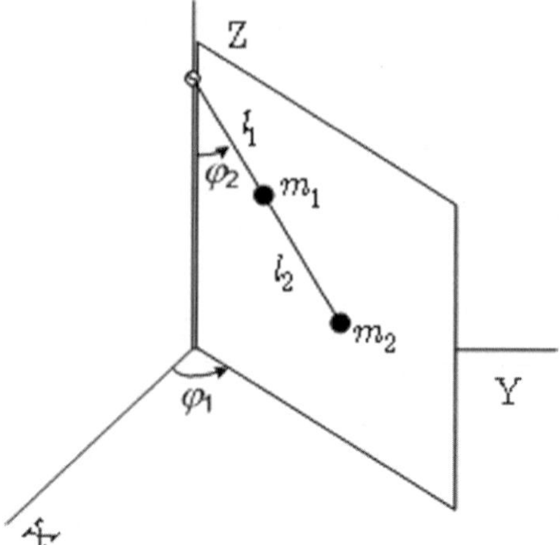

Fig. 1. Numerical simulation: an spherical compound pendulum.

Let us consider a spherical compound pendulum [14], shown in Figure 1 with two particles with masses $m_1 = m_2 = 1$ kg, placed at distances $l_1 = 1$ m and $l_2 = 1$ m on a rigid massless rod of total length $l_1 + l_2$. The system is released from the position $\varphi_1 = 0$, $\varphi_2 = \pi/2$ rad with initial velocities $\dot{\varphi}_1 = 0.5$ and $\dot{\varphi}_2 = 0$ rad/s. The system has two degrees of freedom and it is modeled with six coordinates ($\mathbf{q} \in \mathbb{R}^6$), which are the absolute Cartesian coordinates of both particles. There are five constraint equations; two of them express that distances l_1 and l_2 are constant, and the other three express the alignment of the two segments connecting the particles through a cross product. Note that one of these three equations is redundant, which means that the system has 2 degrees of freedom.

We will use this example to illustrate the main issues discussed in the previous sections; namely:

- the effect of different implementations of the augmented Lagrangian scheme (nested and simultaneous simulation);
- the numerical difficulties associated to the augmented Lagrangian approach when used with a standard ODE integrator, without stabilization;
- the comparison in terms of stability between the conservative integration scheme and the use of a standard ODE integrator with projections;
- the evaluation of a conservative scheme with projections, that will allow to take a deeper look to the energy balance of the projection technique and its effect over stability.

In all the following cases the simulation is carried out for 20 seconds and integrated with 0.025 s of time step. The penalty factor is set to 10^7.

7.1 Augmented Lagrangian schemes: Nested and simultaneous iteration

As explained in Section 4, there are two different possibilities for implementing the augmented Lagrangian schemes in a non linear case:

- *Nested iteration*, with an outer iteration for the Lagrange multipliers and an inner Newton–Raphson iteration.
- *Simultaneous iteration*: with an unique iteration loop for Newton–Raphson, which includes the update of the Lagrange multipliers.

It is observed that the first scheme, in general, leads to a slower convergence and needs more number of iterations. Moreover, small differences in the fulfillment of the constraints are obtained depending on the tolerances imposed to the outer iteration of the first scheme. But this differences have no significance on the response of the solution, or the conservation of energy. If we pay attention to the stability of the methods, it is neither observed a better performance of the nested iteration implementation, since the maximum time steps achieved are similar.

Finally, note that in the nested iteration we have an additional uncertain parameter to take into account: the outer iteration tolerance, which should be supplied by the user and directly determines the accuracy of the constraint fulfillment.

7.2 Stability problems of the index-3 augmented Lagrangian scheme with a standard integrator

To illustrate the problems exhibited by the augmented Lagrangian formulation (7) if no stabilization method is used, the example of the compound pendulum is solved using the trapezoidal rule without projections.

Figure 2 shows the behavior of the energy, and the quadratic norm of the constraints at position $||\mathbf{\Phi}||$, velocity $||\dot{\mathbf{\Phi}}||$, and acceleration $||\ddot{\mathbf{\Phi}}||$ levels along time.

Note that the instability is characterized by the unbounded growth of the violation of the constraint equations at velocity and acceleration levels, together with the unbounded growth of the vibrating energy associated to the constraints. Other integration schemes, such as implicit Runge–Kutta or BDF, better suited for stiff ODE systems than the trapezoidal rule, exhibit qualitatively the same behaviour with slightly larger time steps.

7.3 Augmented Lagrangian stabilized formulations: Coordinate projection vs. conservative formulation

We analyze here two methods, both based on the augmented Lagrange formulation and presented in Sections 4 and 5. These two methods are designed to overcome the stability problems, explained in Section 3 and illustrated in Section 7.2, associated to the numerical solution of the DAE representing the system's dynamics.

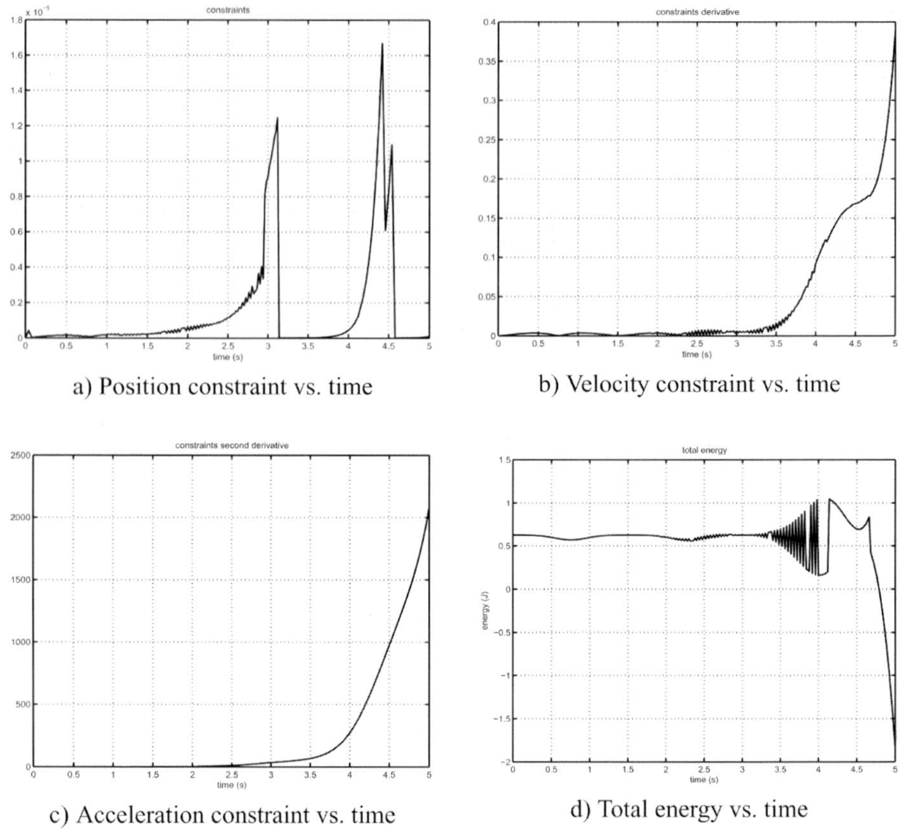

a) Position constraint vs. time

b) Velocity constraint vs. time

c) Acceleration constraint vs. time

d) Total energy vs. time

Fig. 2. Index-3 augmented Lagrangian with trapezoidal rule, without projections.

Figure 3 shows that the scheme with coordinate projections better fulfills the constraint equations at velocity and position levels than expected, while the conservative scheme achieves the exact conservation of the total energy. Nevertheless, the important remark to be made is that both schemes provide an adequate stabilization to the equations, while enforcing accurately the constraints at position level. Note from Figure 3 that a stable simulation is carried out up to 20 s, while Figure 2 shows severe instabilities after the first 3.5 s for the same time step.

If we try to achieve the highest possible time step for a stable simulation during 20 s, we find similar situations for both schemes. In the case of trapezoidal rule with coordinate projection, we can achieve a maximum time step of 0.25 s, while in the case of the conservative formulation we achieve a maximum time step of 0.20 s.

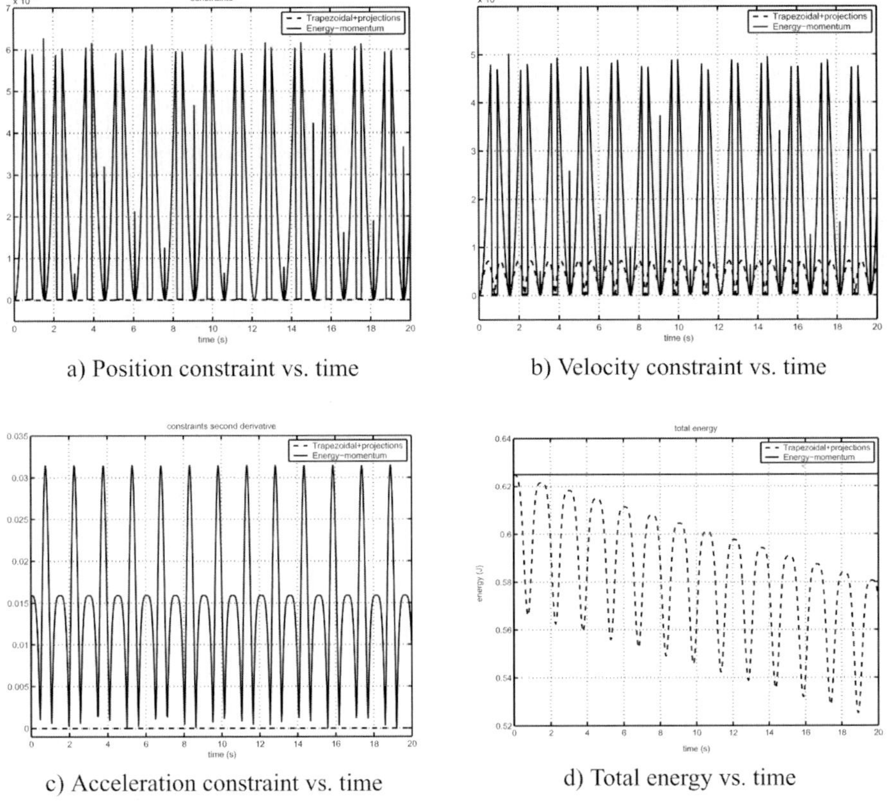

a) Position constraint vs. time

b) Velocity constraint vs. time

c) Acceleration constraint vs. time

d) Total energy vs. time

Fig. 3. Comparison between trapezoidal rule with projections and the energy conserving scheme.

7.4 A conservative scheme with projections

As pointed out at the end of Section 5, the conserving augmented Lagrangian formulation and the projection technique are not incompatible at all. When combined, the resulting scheme has two important features:

- It introduces two stabilization effects; the energy-momentum integrator stabilizes the equations keeping the energy on the system bounded, while the projection stabilize the equations maintaining the solution onto the constraints manifold, at velocity and acceleration levels. As a result, a more stable algorithm is obtained.
- It allows to appreciate clearly the effect of the projections over the energy balance, represented by the contribution ΔE_p in expression (21), since the other contribution ΔE_i is null for the conserving scheme.

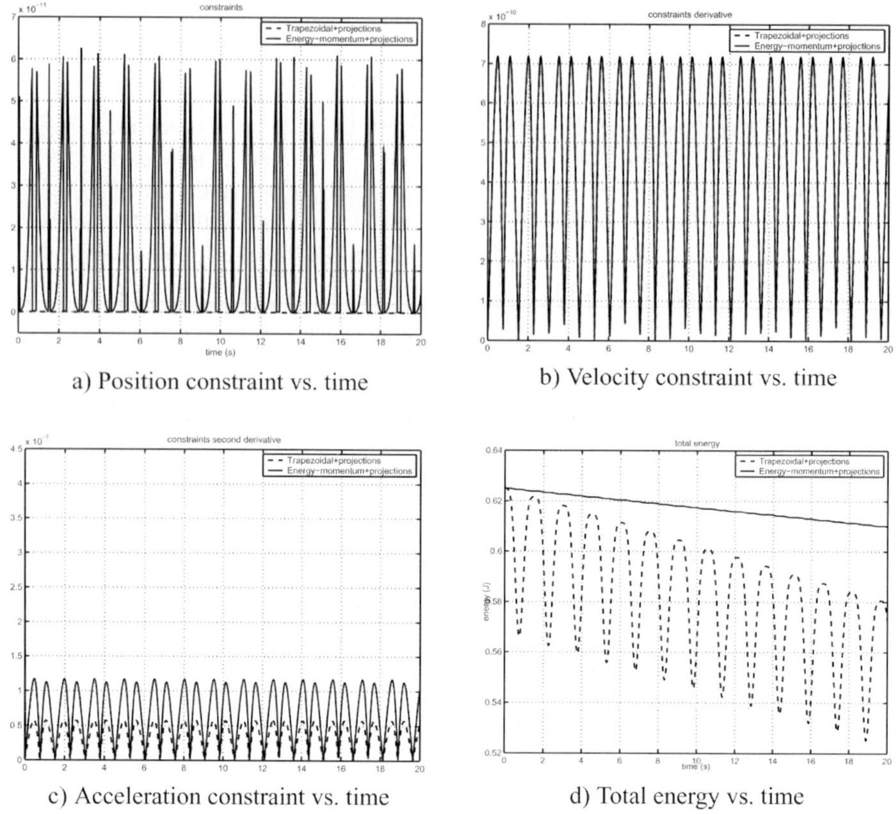

a) Position constraint vs. time

b) Velocity constraint vs. time

c) Acceleration constraint vs. time

d) Total energy vs. time

Fig. 4. Effect of projections over the trapezoidal rule and the conserving scheme.

Figure 4 shows a comparison between the results obtained applying coordinate projections on both the trapezoidal rule and the conserving scheme, using as a projection matrix the system mass matrix $\mathbf{M}$.

Note that energy decreases in both cases, but it does it monotonically only with the conservative method. This result agrees with the theoretical results discussed in Section 6, since the energy dissipation observed in the conserving case entirely comes from the projection phase, which unconditionally dissipates energy when the mass matrix is used to perform the projection. For the trapezoidal rule, the decrease of energy is not monotonic because the contribution of the integration scheme ΔE_i, which takes positive and negative values along the simulation.

8 Conclusions

The main conclusions that can be drawn from the results presented in this work are:

- Several formulations may be used to represent the dynamics of constrained mechanical systems. These formulations differs on the method employed to enforce constraints, commonly based on the Lagrange multiplier, penalty and augmented Lagrangian methods. All these formulations pose numerical difficulties, mainly related to stability; some of these difficulties come from an unstable index reduction of the original index-3 DAE, and some arise from the intrinsic characteristics of the solution method.
- Two methods that alleviate these stability problems have been presented; the use of a standard ODE integrator (such as the trapezoidal rule) with projections, and a conservative integrator. Both exhibit excellent stability characteristics, both comply with the constraints at position level, and are very adequate to carry out robust long term simulations.
- The stability properties of the conserving method can be understood as an effect of the control over the energy. This idea comes as a natural extrapolation of the situation observed in linear conservative systems, where energy conservation implies unconditional stability. This property no longer holds in the nonlinear regime, but nevertheless provides a valid intuitive justification for the observed improvement in the stability performance of conservative schemes.
- The stability properties of the projection (combined with a standard integrator such as the trapezoidal rule or with the conservative scheme) can be understood in two different manners.
  The first is related with its effect over the energy; it has been justified that under some assumptions (mainly based on the selection of the projection matrix) no growth of energy, even unconditional dissipation, can be guaranteed for the projection, justifying its stabilization effect.
  The second interpretation is based on the observation, supported by different studies at the literature, that the numerical solution is more stable on the constraint manifold than off it. Thus, exact fulfillment of the constraints at all levels (position, velocity, acceleration) is expected to improve the overall stability of the algorithm, which is indeed observed at the numerical simulation presented at this work.
- The combination of the conservative scheme with the coordinate projection technique results is a very stable algorithm, very adequate for long term simulations. Besides, provides an optimal framework for a deeper study of the energy control provided by projections, that is naturally related to their stabilization effect, and may lead to obtain practical assessments on the projection matrix selection.
- The two alternative implementations of the Uzawa method for the augmented Lagrangian formulation (nested and simultaneous iteration) lead to similar results

with moderate time steps, with no significant impact on the overall algorithm's stability.

Acknowledgements

Juan C. García Orden would like to acknowledge the support by the Ministerio de Educación y Ciencia from the Spanish Government, within the project DPI-2003-07201 entitled "Modelización robusta de uniones en mecanismos flexibles". Daniel Dopico Dopico would like to acknowledge the support by the Ministerio de Educación y Ciencia from the Spanish Government, within the project DPI-2003-05547-C02-01 entitled "Herramientas de colaboración en dinámica de sistemas multicuerpo" and the co-support by the Galician Government, within the project PGIDT04PXIC16601PN.

References

1. Uri M. Ascher. Stabilization of invariants of discretized differential systems. *Numerical Algorithms*, 14:1–23, 1997.
2. J. Baumgarte. Stabilization of constraints and integrals of motion in dynamical systems. *Computer Methods in Applied Mechanics and Engineering*, 1:1–16, 1972.
3. E. Bayo and A. Avello. Singularity-free augmented lagrangian algortihms for constrained multibody dynamics. *Nonlinear Dynamics*, 9:113–130, 1996.
4. E. Bayo, J. García de Jalón, and M.A. Serna. A modified lagrangian formulation for the dynamic analysis of constrained mechanical systems. *Computer Methods in Applied Mechanics and Engineering*, 71:183–195, 1988.
5. E. Bayo and R. Ledesma. Augmented lagrangian and mass-orthogonal projection methods for constrained multibody dynamics. *Nonlinear Dynamics*, 9:113–130, 1996.
6. Dimitri P. Bertsekas. *Nonlinear Programming*. Athena Scientific, second edition, 2003.
7. Marco Borri, Lorezo Trainelli, and Alessandro Croce. The embedded projection method: a general index reduction procedure for constrained system dynamics. *Computer Methods in Applied Mechanics and Engineering*, 2005.
8. K. E. Brenan, S. L. Campbell, and L. R. Petzold. *Numerical Solution of Initial-Value Problems in Differential-Algebraic Numerical Solution of Initial-Value Problems in Differential-Algebraic*. SIAM, 1996.
9. J. Cuadrado, J. Cardenal, P. Morer, and E. Bayo. Intelligent simulation of multibody dynamics: space-state and descriptor methods in sequential and pararell computing environments. *Multibody System Dynamics*, 4:55–73, 2000.
10. J. Cuadrado, D. Dopico, M.A. Naya, and M. González. Penalty, semi-recursive and hybrid methods for mbs real-time dynamics in the context of structural integrators. *Multibody System Dynamics*, 12:117–132, 2004.
11. J. García de Jalón and E. Bayo. *Kinematic and Dynamic Simulation of Multibody Systems. The Real Time Challenge*. Springer-Verlag, 1994.

12. J. C. García Orden and J. M. Goicolea. Conserving properties in constrained dynamics of flexible multibody systems. *Multibody System Dynamics*, 4:225–244, 2000.
13. J.C. García Orden and D. Dopico Dopico. Energetic considerations of stabilization through coordinate projection in constrained mechanical systems. In preparation, 2006.
14. J.C. García Orden and J.M. Goicolea. An energy-momentum algorithm for flexible multibody systems with finite element techniques. *Computer Assisted Mechanics and Engineering Sciences (Polish Academy of Sciences)*, 8:313–324, 2001.
15. J.C. García Orden and R. Ortega. A conservative augmented lagrangian algorithm for the dynamics of constrained mechanical systems. In C.A. Mota Soares et al. (ed.), *III European Conference on Computational Mechanics. Solids, Structures and Coupled Problems in Engineering*, Lisboa, Portugal, 5–8 June 2006.
16. M. Geradin and D. Rixen. *Mechanical Vibrations*. Wiley, 1997.
17. J. M. Goicolea and J. C. García Orden. Dynamic analysis of rigid and deformable multibody systems with penalty methods and energy-momentum schemes. *Computer Methods in Applied Mechanics and Engineering*, 188(4):789–804, 2000.
18. J. M. Goicolea and J. C. García Orden. Quadratic and higher-order constraints in energy-conserving formulations in flexible multibody systems. *Multibody System Dynamics*, 7:3–29, 2002.
19. O. González and J.C. Simó. On the stability of symplectic and energy-momentum algorithms for non linear hamiltonian systems with symmetry. *Computer Methods in Applied Mechanics and Engineering*, 1996.
20. E. Hairer and G. Wanner. *Solving Ordinary Differential Equations II, Stiff and Differential-Algebraic Problems*. Springer-Verlag, 1991.
21. T.A. Laursen. *Computational Contact and Impact Mechanics. Fundamentals of Modeling Interfacial Phenomena in Nonlinear Finite Element Analysis*. Springer, 2002.
22. M. Ortiz. A note on energy conservation and stability of nonlinear time-stepping algorithms. *Computers and Structures*, 24(1), 1986.
23. M.A. Serna, R. Avilés, and J. García de Jalón. Dynamics analysis of plane mechanisms with lower pairs in basic coordinates. *Mechanism and Machine Theory*, 17:397–403, 1982.
24. J.C. Simó and N. Tarnow. The discrete energy-momentum method. conserving algorithms for nonlinear elastodynamics. *ZAMP*, 1992.
25. J.C. Simó and K.K. Wong. Unconditionally stable algorithms for rigid body dynamics that exactly preserve energy and momentum. *International Journal of Numerical Methods in Engineering*, 31:19–52, 1991.
26. A.M. Stuart and A.R. Humphries. *Dynamical Systems and Numerical Analysis*. Cambridge, 1996.
27. R. Wehage and E.J. Haugh. Generalized coordinate partitioning for dimension reduction in analysis of constrained mechanical systems. *Journal of Mechanical Design*, 104:247–255, 1982.
28. P. Wriggers. *Computational Contact Mechanics*. Wiley, 2002.

Analysis and Improved Methods for the Error Estimation of Numerical Solutions in Solid and Multibody Dynamics

Ignacio Romero[1] and Luis M. Lacoma[2]

[1]*E.T.S. Ingenieros Industriales, Universidad Politécnica de Madrid, José Gutiérrez Abascal, 2, 28006 Madrid, Spain; E-mail: iromero@mecanica.upm.es*
[2]*E.T.S. Ingenieros de Caminos, Universidad Politécnica de Madrid, Ciudad Universitaria s/n, 28040 Madrid, Spain; E-mail: ll@mecanica.upm.es*

1 Introduction

A posteriori error estimators are useful tools in general purpose numerical computations because they provide an automatic, quantitative assessment of the accuracy of the results. Without some sort of error estimation the validity of any numerical results relies solely on the analyst experience and good judgment. While these are also necessary, they fail to be quantitative and are thus prone to mistakes.

In dynamic computations, the ones of interest in this work, error estimators have a second, very practical, application. By estimating the error in every step of the solution, the time step size can be varied in order to be as large as possible while keeping the errors small enough. In large scale simulations of transient multibody systems it is mandatory to use some kind of adaptive time stepping. It is hence crucial to base them on error estimators as accurate and robust as possible.

A posteriori error estimation methods for ordinary differential equations are standard and are treated in many texts on the topic (see, for example, [1] and also the review article [2]). While these techniques are fairly general, we are mostly interested in methods tailored for the second order differential equations that arise in solid, structural, and multibody dynamics. Literature for these specific application is less abundant but can be found, among elsewhere, in the works of [3–10]. Among all the available techniques for error estimation in ODEs, the aforementioned references are all based in asymptotic methods. These seek to compute an approximation to the error made in every step of the calculation by comparing the numerical solution with another, high order accurate approximation to the exact problem. This approach will also be employed in the present article.

J.C. García Orden et al. (eds), Multibody Dynamics. Computational Methods and Applications, 69–89.

Most of the work on error estimators mentioned above focuses on the formulation of accurate *local* error estimates. That is, approximations of the error made in one single time step of the integration. This information suffices for constructing adaptive time stepping strategies but provides little knowledge about the *global* error, the difference between the real and computed solutions. This is the most important error from the point of view of the final results. Much information about it can be obtained, at least for the linear case, as will be described in the present work. While nonlinear problems are of course the most important ones, error estimators which are most effective for linear problems will most likely perform well in nonlinear situations.

In fact, a detailed study of the errors made in the time integration and the relation between local errors and global errors allows to clearly state the necessary and sufficient conditions that any error estimator must satisfy in order to produce accurate global error estimations. As will be explained, this global point of view reveals that considerations based only on local errors do not suffice, and that some of the existing error estimators can provide inaccurate estimations for this reason, even in linear problems.

The analysis we present serves to identify the conditions that guarantee the accuracy of a given error estimator of the asymptotic type. Moreover, it provides the guidelines to formulate new, accurate, and robust estimators. As an example, a new error estimator is proposed for the time integration with Newmark's and HHT methods. The article contains the key ideas that can be employed in order to develop error estimators for other time marching algorithms. Details omitted in the proofs of this article can be found in our previous work [11, 12].

An outline of the rest of the article is as follows. In Section 2, the problem of interest is described and formulated mathematically. The errors made in one time step of the integration, i.e., the local errors, are defined and studied in Section 3. This concept is employed in Section 4 to discuss the global errors made in a complete numerical solution. Next, in Section 5, new error estimators are presented for Newmark's and HHT method, explaining how they are obtained by using the ideas of the previous sections. The performance of these algorithms is illustrated in Section 6 by means of several numerical computations. The article ends in Section 7 with a summary of results and conclusions.

2 Problem Statement

In this work we focus on the numerical analysis and a posteriori error estimation of the equations of flexible multibody systems with no constraints. They can be formulated as a system of ordinary differential equations of dimension N where the independent variable is the time $t \in [0, T]$:

$$\begin{aligned} \mathbf{M}\ddot{\mathbf{U}}(t) + \mathbf{F}_{int}(\mathbf{U}(t), \dot{\mathbf{U}}(t)) &= \mathbf{F}(t)\,, \qquad t \in (0, T), \\ \mathbf{U}(0) &= \mathbf{U}_o, \\ \dot{\mathbf{U}}(0) &= \mathbf{V}_o. \end{aligned} \tag{1}$$

The vector $\mathbf{U}$ contains the unknown displacements; $\mathbf{F}$, $\mathbf{F}_{int}$ are, respectively, the external and internal force vector; $\mathbf{M}$ is the mass matrix, and $\mathbf{U}_o$, $\mathbf{V}_o$ are the initial displacement and velocity vectors.

Among all the numerous time stepping methods that exist to integrate (1) we will concentrate on a class of methods designed specifically for second order systems of stiff equations and originally formulated in the context of structural mechanics. To define this time marching strategy, consider a partition of the time interval $[0, T]$ in N subintervals of the form $\mathcal{I}_n = [t_n, t_{n+1}]$ of length $\Delta t_n = t_{n+1} - t_n$, where $0 \equiv t_o < t_1 < t_2 \cdots < t_N \equiv T$. Given algorithmic approximations of the displacement, velocity, and acceleration at time t_n and denoted respectively $\mathbf{U}_n, \mathbf{V}_n, \mathbf{A}_n$, we solve for the same quantities at the next instant t_{n+1} by solving the following algebraic equations:

$$\begin{aligned} \mathbf{M}\mathbf{A}_{n+1} + \mathbf{F}_{int}(\mathbf{U}_{n+\alpha}, \mathbf{V}_{n+\alpha}) &= \mathbf{F}_{n+\alpha}, \\ \mathbf{U}_{n+1} &= \mathbf{U}_n + \mathbf{V}_n \Delta t_n + \frac{\Delta t_n^2}{2}((1-2\beta)\mathbf{A}_n + 2\beta \mathbf{A}_{n+1}), \\ \mathbf{V}_{n+1} &= \mathbf{V}_n + \Delta t_n((1-\gamma)\mathbf{A}_n + \gamma \mathbf{A}_{n+1}). \end{aligned} \tag{2}$$

In the previous equations, and for the rest of the article, the notation $(\cdot)_{n+\alpha}$ with $0 < \alpha < 1$ is used to denote the convex combination $(1-\alpha)(\cdot)_n + \alpha(\cdot)_{n+1}$. The system of equations (2) must be supplemented with the value of the acceleration at time $t = 0$ which is defined to be:

$$\mathbf{A}_o = \mathbf{M}^{-1}(\mathbf{F}(0) - \mathbf{F}_{int}(\mathbf{U}_o, \mathbf{V}_o)). \tag{3}$$

The three parameters α, β, and γ allow to select among different members of the family, each one with different algorithmic properties. Two specific choices will be of major interest. The two-parameter method resulting by selecting $\alpha = 1$ is known as the Newmark method [13]. The one parameter method that follows from imposing $\beta = (1-\alpha/2)^2, \gamma = 3/2 - \alpha, 0.7 \leq \alpha \leq 1$, is the Hilber–Hughes–Taylor (HHT) algorithm [14].

3 A Posteriori Local Error Estimates

We start our analysis by presenting the type of asymptotic error estimations that can be performed to approximate the error made in one time step of the numerical integration of (1). First, we must establish the norm that will be employed to measure the size of vectors in the analysis.

The most natural way of measuring the size of the quantities that appear in our analysis is to employ the energy norm. This norm combines the strain energy of the displacements and kinetic energy of the velocity to give a non-negative real number that measures their combined size. In linear problems, it can be shown that this norm is a natural norm equivalent to certain Sobolev norm. In order to define the energy norm, we gather displacements and velocities at one instant in one single vector of dimension $2N$. For example, if we define the vectors $\mathbf{z}(t_n)$ y $\tilde{\mathbf{z}}_n$ that hold, respectively, the solutions to problems (1) and (2) at time t_n we can write:

$$\mathbf{z}(t_n) = \begin{Bmatrix} \mathbf{U}(t_n) \\ \dot{\mathbf{U}}(t_n) \end{Bmatrix}, \qquad \tilde{\mathbf{z}}_n = \begin{Bmatrix} \mathbf{U}_n \\ \mathbf{V}_n \end{Bmatrix}, \tag{4}$$

and we can define their energy norms as:

$$\begin{aligned} \|\mathbf{z}(t_n)\|_E^2 &:= \frac{1}{2}\mathbf{U}(t_n) \cdot \mathbf{K}_o \mathbf{U}(t_n) + \frac{1}{2}\dot{\mathbf{U}}(t_n) \cdot \mathbf{M}\dot{\mathbf{U}}(t_n) , \\ \|\tilde{\mathbf{z}}_n\|_E^2 &:= \frac{1}{2}\mathbf{U}_n \cdot \mathbf{K}_o \mathbf{U}_n + \frac{1}{2}\mathbf{V}_n \cdot \mathbf{M}\mathbf{V}_n . \end{aligned} \tag{5}$$

In both of these equations we have defined the initial tangent stiffness matrix

$$\mathbf{K}_o = \left. \frac{\partial \mathbf{F}_{int}(\mathbf{U}, \dot{\mathbf{U}})}{\partial \mathbf{U}} \right|_{(\mathbf{U},\dot{\mathbf{U}})=(\mathbf{U}_o,\mathbf{V}_o)} , \tag{6}$$

to measure the strain energy contribution from the displacements. The tangent stiffness matrix is not constant except for linear problems. However, in order to use the same norm at every instant and, furthermore, save computational time, we will only employ the initial stiffness for the energy norm computation.

To correctly assess the error made by the integration scheme in one single time step – and thus separating it from the error that might have accumulated in previous steps – we define the following initial value problem in the time interval of interest:

$$\begin{aligned} &\mathbf{M}\ddot{\mathbf{u}}^{[n]}(t) + \mathbf{F}_{int}(\mathbf{u}^{[n]}(t), \dot{\mathbf{u}}^{[n]}(t)) = \mathbf{F}(t) , \quad t \in \mathcal{I}_n, \\ &\mathbf{u}^{[n]}(t_n) = \mathbf{U}_n, \\ &\dot{\mathbf{u}}^{[n]}(t_n) = \mathbf{V}_n. \end{aligned} \tag{7}$$

In this problem, the initial values of displacement and velocity are the one calculated by the numerical scheme in the previous time step. Gathering as before the solution of each local problem and its time derivative in a single vector $\mathbf{y}^{[n]}(t) = \langle \mathbf{u}^{[n]}(t), \dot{\mathbf{u}}^{[n]}(t) \rangle$, we define the *local* error as the quantity

$$\mathbf{e}_{n+1} := \mathbf{y}^{[n]}(t_{n+1}) - \tilde{\mathbf{z}}_{n+1}. \tag{8}$$

Local errors, as simple as they might seem, cannot be computed exactly. For that, it would be necessary to exactly know the solution at the end of every single time step. However, by induction, this amounts to solving exactly the original problem (1). Nevertheless, most of the times it suffices to have a good approximation of the error, or its norm, and this can be accomplished by means of an a posteriori error estimation.

A common strategy to obtain useful error estimates consists in replacing $\mathbf{y}^{[n]}(t_{n+1})$ in equation (8) with a numerical solution of the local problem (7) obtained using a numerical method of a high order of accuracy and small computational cost. The high order solution of the local problem at the end of the interval $\mathcal{I}_n$ will be denoted $\tilde{\mathbf{y}}^{[n]}_{n+1}$, and we will later explain how to obtain it for particular situations.

If the time integration algorithm that is employed to solve (1) has order of accuracy k, the local errors must be of order $k+1$, that is:

$$\|\mathbf{e}_{n+1}\|_E = \|\mathbf{y}^{[n]}(t_{n+1}) - \tilde{\mathbf{z}}_{n+1}\|_E = \mathcal{O}(\Delta t_n^{k+1}). \tag{9}$$

The high order solution to the local problem must be significantly more accurate than the time integration algorithm that is being used to solve (1). The approximation $\tilde{\mathbf{y}}^{[n]}_{n+1}$ that we need to provide must be, at least, one order of Δt closer to $\mathbf{y}^{[n]}(t_{n+1})$ than $\tilde{\mathbf{z}}_{n+1}$. If this condition is met the local error in one time step could be approximated (making only an error of size Δt^{k+2}) by

$$\mathbf{e}_{n+1} = \underbrace{\mathbf{y}^{[n]}(t_{n+1}) - \tilde{\mathbf{z}}_{n+1}}_{\mathcal{O}(\Delta t^{k+1})} = \underbrace{\mathbf{y}^{[n]}(t_{n+1}) - \tilde{\mathbf{y}}^{[n]}_{n+1}}_{\mathcal{O}(\Delta t^{k+2})} + \underbrace{\tilde{\mathbf{y}}^{[n]}_{n+1} - \tilde{\mathbf{z}}_{n+1}}_{\mathcal{O}(\Delta t^{k+1})} \approx \tilde{\mathbf{y}}^{[n]}_{n+1} - \tilde{\mathbf{z}}_{n+1}. \tag{10}$$

The error vector

$$\boldsymbol{\theta}_{n+1} := \tilde{\mathbf{y}}^{[n]}_{n+1} - \tilde{\mathbf{z}}_{n+1} \tag{11}$$

in an order Δt^{k+2} approximation of the true local error $\mathbf{e}_{n+1}$ but, in contrast with the later, can be explicitly computed. On the other hand, the quantity

$$\boldsymbol{\eta}_{n+1} := \mathbf{y}^{[n]}(t_{n+1}) - \tilde{\mathbf{y}}^{[n]}_{n+1} \tag{12}$$

has an unknown value. However, its energy norm is known to be of size $\mathcal{O}(\Delta t^{k+2})$ and thus negligible when added to the error estimate $\boldsymbol{\theta}_{n+1}$.

To sum up, in order to obtain an *a posteriori local* error estimate of the type considered, all that is needed is an algorithm that computes a solution $\tilde{\mathbf{y}}^{[n]}_{n+1}$ to every local problem such that:

- it is a high order approximation of the exact solution of the local problem $\mathbf{y}^{[n]}(t_{n+1})$ and,
- it can be computed with a small computational cost.

4 A Posteriori Global Error Estimates

The local error estimates – as presented in Section 3 – are useful quantities, especially for devising adaptive time stepping strategies. However, in most cases, the analyst is even more interested in knowing, or at least having an approximation of, the real error made by the numerical integration scheme at the end of a whole simulation. This error is referred to as the global error. If one employs asymptotic error estimators as the ones previously described, in most situations there is very little that can be done to assess the global error apart from accumulating the local ones. Nevertheless, in linear problems more information can be inferred from local errors and accurate estimations of global errors can be given.

Linear problems are characterized by having an internal force vector which is proportional to displacement and velocity and thus can be written as:

$$\mathbf{F}_{int}(\mathbf{U}, \dot{\mathbf{U}}) = \mathbf{C}\dot{\mathbf{U}} + \mathbf{K}\mathbf{U}. \tag{13}$$

The constant matrices **C** and **K** are, respectively, the damping and stiffness matrices of the system which are symmetric, positive semidefinite. The system of ordinary equations (1) together with the specific form of the internal force (13) has special properties among which is the following stability result:

Theorem 1. *The energy norm of the global error* $\boldsymbol{E}_n := \tilde{\mathbf{z}}_n - \mathbf{z}(t_n)$ *due to the numerical approximation of a linear problem of type* (1) *is bounded by the sum of the norms of the local errors* $\mathbf{e}_n = \mathbf{y}^{[n-1]}(t_n) - \tilde{\mathbf{z}}_n$*, i.e.,*

$$\|\boldsymbol{E}_N\|_E \leq \sum_{n=1}^{N} \|\mathbf{e}_n\|_E. \tag{14}$$

The proof of this result can be found in [11]. This property depends solely on the stability of the continuous equations and is independent of the numerical method employed. From the practical point of view, it is a result that cannot be employed, at least directly, to calculate the global error. The previous bound depends on the local errors which, as explained in Section 3, cannot be computed.

The strategy for deriving useful approximations of the global error consists in employing the local error estimates described in Section 3 to approximate the last error bound. If a procedure is found to compute a posteriori local errors with sufficient accuracy, they can replace the exact local errors in equation (14). While the computed quantity will no longer be a guaranteed upper bound of the global error, it will be a good estimate for it.

To be more specific, let us assume that we are dealing with a numerical method of the type considered which is of order k. This means that, for smooth enough solutions of the differential equations (1), the local errors made by the method are order $k + 1$ and the global error has an order k norm. If the local error of bounds in

(14) are replaced by error estimates $\boldsymbol{\theta}_n$ which are order $k+2$ approximations to the true error then the global error bound will be approximate by a quantity which is no longer a bound, but is an order $k+1$ approximation to it. Since the method itself is only order k the approximated error would be a very good estimate not only of the global error itself, but of a bound for it. Using the notation of Section 3 we have:

$$\|\boldsymbol{E}_N\|_E \leq \sum_{n=1}^{N} \|\mathbf{e}_n\|_E = \sum_{n=1}^{N} \|\boldsymbol{\theta}_n + \boldsymbol{\eta}_n\|_E \leq \sum_{n=1}^{N} \|\boldsymbol{\theta}_n\|_E + \sum_{n=1}^{N} \|\boldsymbol{\eta}_n\|_E \approx \sum_{n=1}^{N} \|\boldsymbol{\theta}_n\|_E. \tag{15}$$

The error estimation obtained by adding up local error estimates is, in contrast to the true global error, computable by numerical means. If the cost of the local error estimate is small compared to the cost of a step in the numerical solution, the cost of the global error estimate will also be small in comparison with the overall cost of the simulation. Moreover, as stressed before, the error estimate computed in this way will be a good approximation to a bound of the true global error. From the point of view of the analyst, this is a desirable property because the reported error assessment will fall in the safe side, never underestimating it.

At the heart of a reliable global error estimator lies an accurate local estimator, one with the properties put forward in Section 3. In the next section we will explain how such an estimator can be computed for the general class of integrators (2).

5 A Procedure for Computing Accurate Error Estimators

In the previous two sections, we have stressed the importance of formulating local error estimates that are both accurate and computationally cheap. As explained, this is necessary both for formulating time adaptive strategies as well as to obtain useful global error estimations.

Within the class of asymptotic error estimates that we are considering in this work, the task of obtaining local error estimators amounts to finding a way to easily compute numerical solutions to the local problems (7) which are at the same time highly accurate. Several alternatives have been proposed in the literature to tackle this problem [3–5, 8]. They differ among each other in their computational cost, their accuracy, and especially their range of applicability. As studied in [11], not all of them are equally valid for general purpose situations. In particular, some of them require that the exact solution of the continuous problem is very smooth, an assumption which is over restrictive and which rules out many interesting mechanical problems.

In this section we present a methodology to obtain error estimators for problem (1) that satisfy the requirements for accuracy, robustness, and cost explained in the previous section. Furthermore, the only smoothness requirement on the exact solution is that it must be two times continuously differentiable. This condition is automatically satisfied by the solution of (1) and thus poses no restriction on the kind of problems that the proposed estimator can be used for.

For the construction of the local error estimator, we restrict ourselves to the numerical integrators of the type (2). The ideas, however, are fairly general and the formulas proposed can be extended to other methods.

Before proceeding, we note that the family of algorithms (2) includes first and second order methods and we need to consider them independently. The main idea is as follows. As explained in Section 3, the only requirement for formulating an accurate error estimator is a method to approximate $\mathbf{y}^{[n]}(t_{n+1})$, the solution of the local problem (7). By the fundamental theorem of calculus this function satisfies:

$$\mathbf{y}^{[n]}(t_{n+1}) = \mathbf{y}^{[n]}(t_n) + \int_{t_n}^{t_{n+1}} \dot{\mathbf{y}}^{[n]}(\tau)\,\mathrm{d}\tau. \tag{16}$$

In the last expression, $\mathbf{y}^{[n]}(t_n)$ is known exactly and its value is the solution obtained by the integrator in the previous time step, i.e., $\tilde{\mathbf{z}}_n$. Thus, in order to approximate $\mathbf{y}^{[n]}(t_{n+1})$, we can use (16) and approximate the integral in that expression. For an order k methods of the ones considered, we propose to find a quadrature Q_n^{n+1}, as simple as possible, that verifies:

$$\left\| \mathsf{Q}_n^{n+1} - \int_{t_n}^{t_{n+1}} \dot{\mathbf{y}}^{[n]}(\tau)\,\mathrm{d}\tau \right\|_E = \mathcal{O}(\Delta t^{k+2}). \tag{17}$$

If we are able to find such rule, an approximation of order Δt^{k+2} to the solution to the local problem could be easily defined as:

$$\tilde{\mathbf{y}}_{n+1}^{[n]} = \tilde{\mathbf{z}}_n + \mathsf{Q}_n^{n+1}. \tag{18}$$

This approximation, if found, is one order more accurate to the solution of the local problem than the solutions provided by any method of the family (2), either first order of second order. Thus, for any of the methods considered, we have

$$\|\boldsymbol{\eta}_{n+1}\|_E = \|\tilde{\mathbf{y}}_{n+1}^{[n]} - \mathbf{y}^{[n]}(t_{n+1})\|_E = \mathcal{O}(\Delta t^{k+2}). \tag{19}$$

If indeed we can compute efficiently the quadrature described then, the vector

$$\boldsymbol{\theta}_{n+1} = \tilde{\mathbf{y}}_{n+1}^{[n]} - \tilde{\mathbf{z}}_{n+1} = \tilde{\mathbf{z}}_n + \mathsf{Q}_n^{n+1} - \tilde{\mathbf{z}}_{n+1}, \tag{20}$$

will be and order Δt^{k+2} estimation of the true local error.

It remains to provide a methodology for the construction of such a quadrature in the context of any of the methods considered. The method should take advantage of the data already available from the time integration so that unnecessary computations are avoided. In order to do this, different approaches must be considered for Newmark and HHT solutions.

5.1 An error estimator for Newmark's method

Newmark's method has, among all the algorithms of type (2), the advantage that it provides, with no additional cost, approximations for the displacement, velocity, and acceleration at every time instant t_n of accuracy $\mathcal{O}(\Delta t^{k+1})$, k being the order of the method. This is in contrast with the HHT method as will be explained later.

Let us consider first the case of $k = 2$. This corresponds to the second order members of the family that can be shown to be the ones with $\gamma = \frac{1}{2}$ (see, for example, [15]). According to equation (17), the quadrature that we must find has to be an order 4 approximation to the integral of expression (16). Making use of the fact that the vectors $\mathbf{U}_{n+1}, \mathbf{V}_{n+1}, \mathbf{A}_{n+1}$ are themselves third order approximations to the displacement, velocity, and acceleration, respectively, at time t_{n+1} we can construct the required quadrature as summarized in Box 1.

Given the solutions of Newmark's method $\mathbf{U}_n, \mathbf{V}_n, \mathbf{A}_n$ and $\mathbf{U}_{n+1}, \mathbf{V}_{n+1}, \mathbf{A}_{n+1}$ at instants t_n and t_{n+1}, respectively:

1. Compute the auxiliary vectors:

$$\begin{aligned} \mathbf{U}^*_{n+1/2} &= \mathbf{U}_n + \frac{\Delta t_n}{2}\mathbf{V}_n + \frac{\Delta t_n^2}{8}\mathbf{A}_n \\ \mathbf{V}^*_{n+1/2} &= \mathbf{V}_n + \frac{3\Delta t_n}{8}\mathbf{A}_n + \frac{\Delta t_n}{8}\mathbf{A}_{n+1} \\ \mathbf{A}^*_{n+1/2} &= \mathbf{M}^{-1}(\mathbf{F}(t_{n+1/2}) - \mathbf{F}_{int}(\mathbf{U}^*_{n+1/2}, \mathbf{V}^*_{n+1/2}) \end{aligned} \tag{21}$$

and the quadrature rule:

$$\mathsf{Q}_n^{n+1} = \frac{\Delta t_n}{6}\begin{Bmatrix}\mathbf{V}_n \\ \mathbf{A}_n\end{Bmatrix} + \frac{4\Delta t_n}{6}\begin{Bmatrix}\mathbf{V}^*_{n+1/2} \\ \mathbf{A}^*_{n+1/2}\end{Bmatrix} + \frac{\Delta t_n}{6}\begin{Bmatrix}\mathbf{V}_{n+1} \\ \mathbf{A}_{n+1}\end{Bmatrix}. \tag{22}$$

2. Compute the high order approximation to the solution of the local problem:

$$\tilde{\mathbf{y}}^{[n]}_{n+1} = \tilde{\mathbf{z}}_n + \mathsf{Q}_n^{n+1}. \tag{23}$$

3. Finally, compute the error estimate

$$\boldsymbol{\theta}_{n+1} = \tilde{\mathbf{y}}^{[n]}_{n+1} - \tilde{\mathbf{z}}_{n+1}. \tag{24}$$

Box 1: Implementation of the local error estimate for Newmark's method.

The auxiliary vectors $\mathbf{U}^*_{n+1/2}$, $\mathbf{V}^*_{n+1/2}$, and $\mathbf{A}^*_{n+1/2}$ are, respectively, third order approximations to the displacement, velocity, and acceleration of the local problem (7) at time $t_{n+1/2}$. In this way, the quadrature (22) is a modified Gauss-Lobatto rule whose order of accuracy is Δt^4, as desired.

For the first order accurate members in the Newmark family, the required order of accuracy for the quadrature Q_n^{n+1} in the error estimator is 3. The same algorithm described in Box 1 can be used. Now the auxiliary vectors would be only second order approximations to the corresponding quantities at time $t_{n+1/2}$ because the displacement, velocity, and acceleration at t_{n+1} are only second order accurate. Hence, the quadrature (22) will now have third order of accuracy, which is enough for estimating the error of a first order method.

We must remark that the specific quadrature rule employed is irrelevant as long as the accuracy requirements for each method are met. For any of the methods of the Newmark family the values of the displacement, velocity, and acceleration at the endpoints of the interval $[t_n, t_{n+1}]$ do not provide enough information to construct an accurate quadrature and hence, additional auxiliary points must be computed at the midpoint $t_{n+1/2}$. The formulas (21) provide accurate enough approximations with a minimum computational cost.

The computation of the error estimator proposed involves a small computational cost. Only the mass matrix $\mathbf{M}$ needs to be factorized in order to compute the acceleration $\mathbf{A}^*_{n+1/2}$. This factorization can be done just once at the beginning of the simulation, even in nonlinear problems. An interesting alternative would be to employ a lumped mass.

5.2 An error estimator for the HHT method

Constructing an error estimate for the HHT method is not as simple as in the case of the Newmark method. The problem with the former is that, as it is well documented in the literature (see, e.g. [16]), the acceleration vector $\mathbf{A}_{n+1}$ provided by the HHT method is not a third order approximation of the acceleration at the end of a local problem (7).

The strategy to construct an error estimator for the HHT method is almost identical to that employed for second order Newmark methods. Formulas in Box 1 would all be valid if we had good approximations for the acceleration at times t_n, t_{n+1} of each integration interval. However, as explained above, this is not the case and we need to construct such approximations from available data. Denoting by $\mathbf{A}_n^{new}$ the third order accurate approximations of the acceleration, we propose a recursive procedure to compute them using only previous values and the acceleration vectors provided by the method itself. The iterative scheme is as follows:

$$\mathbf{A}_o^{new} = \mathbf{A}_o, \qquad \mathbf{A}_{n+1}^{new} = \frac{1}{\alpha}\left(\mathbf{A}_{n+1} - (1-\alpha)\mathbf{A}_n^{new}\right). \tag{25}$$

This formula was proposed in [17] where a proof of its accuracy can be found. Using the modified acceleration vectors $\mathbf{A}_n^{new}$ in Box 1 instead of the algorithmic ones $\mathbf{A}_n$ yields an error estimator of the desired accuracy for the HHT method.

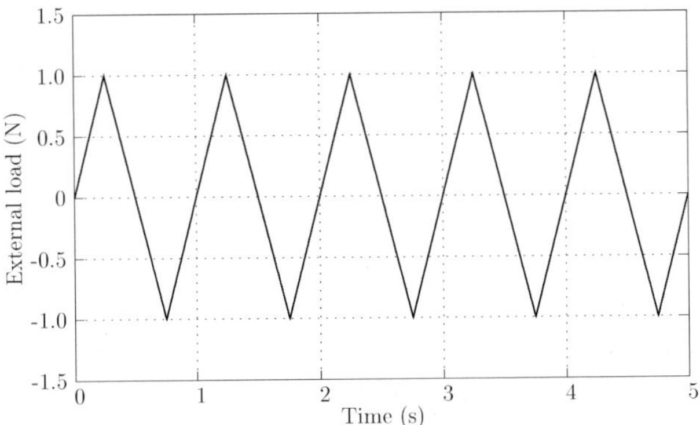

Fig. 1. Triangular loading function.

6 Numerical Simulations

In this section we show a few numerical examples to illustrate the accuracy and robustness of the proposed error estimators. Simulations will be carried out using the trapezoidal rule (Newmark method with $\beta = 0.25$ and $\gamma = 0.5$) and the HHT method summarized in Section 2. In the case of the trapezoidal rule, we shall compare error estimations provided by the algorithm introduced in Section 5 with existing error estimators of the literature such us the ones proposed by Wiberg and Li [5] and Hulbert and Jang [8]. On the other hand, literature about error estimators for the HHT method is scant and the existing ones, e.g. the one described in Geradin and Cardona [9], take only into account the error in displacements, neglecting the contribution of the velocities. Hence, in the case of the HHT method the error estimates will only be compared, when possible, with the exact errors.

6.1 One degree of freedom spring-mass system

The first example is a simple linear model of one degree of freedom, but it is enough to characterize completely the properties of the error estimator in the context of linear problems. The spring-mass system we have chosen has mass $\mathbf{M} = 1$ Kg, stiffness $\mathbf{K} = 6$ N/m, and initial displacement $\mathbf{U}(0) = 1$ m. The initial value problem that describes the behavior of this system is thus:

$$\begin{aligned} &\ddot{\mathbf{U}}(t) + 6\mathbf{U}(t) = \mathbf{F}(t)\ , \quad t \in [0, 5], \\ &\mathbf{U}(0) = 1\ ,\ \dot{\mathbf{U}}(0) = 0. \end{aligned} \tag{26}$$

We consider that the loading function $\mathbf{F}(t)$ has a triangular shape as depicted in Figure 1. The problem is solved with the trapezoidal rule and the HHT method with α equal to 0.8.

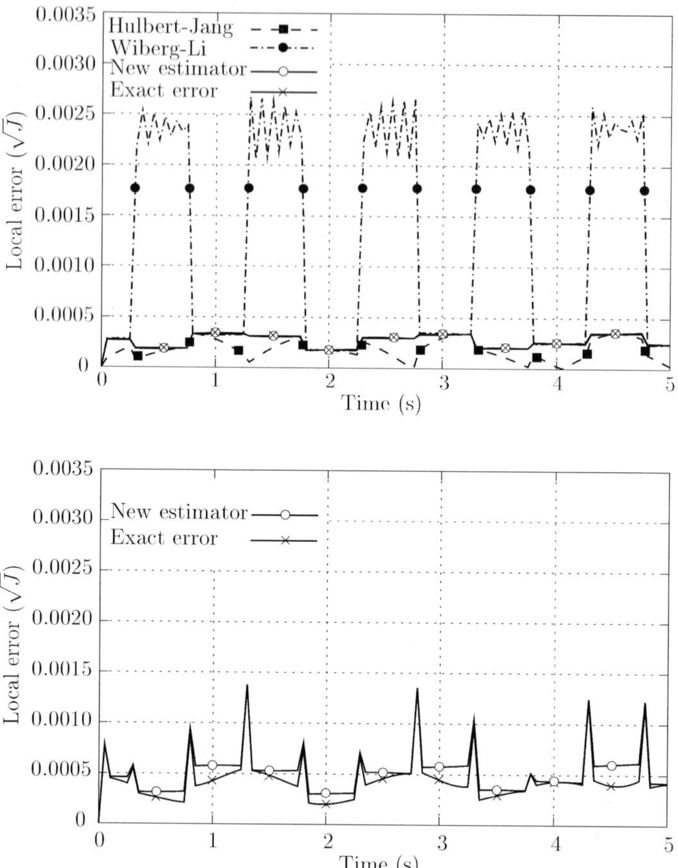

Fig. 2. One dimensional problem. Local error with time step size $\Delta t = 0.05$ s (top: trapezoidal rule; bottom: HHT with $\alpha = 0.8$).

On the top of Figure 2, we show the time evolution of the local errors estimated by the three methods compared and also the true local errors when the trapezoidal rule is employed. The exact local errors can be computed because the problem is very simple but in general it is either impossible or impractical. The error estimator of Wiberg and Li exhibits sharp jumps in the estimated error at instants where the loading function is not differentiable. The error estimator of Hulbert and Jang underestimates the true errors. If the HHT is used (see Figure 2, at bottom), the new estimator computes a local error slightly larger than the true one, keeping a great resemblance.

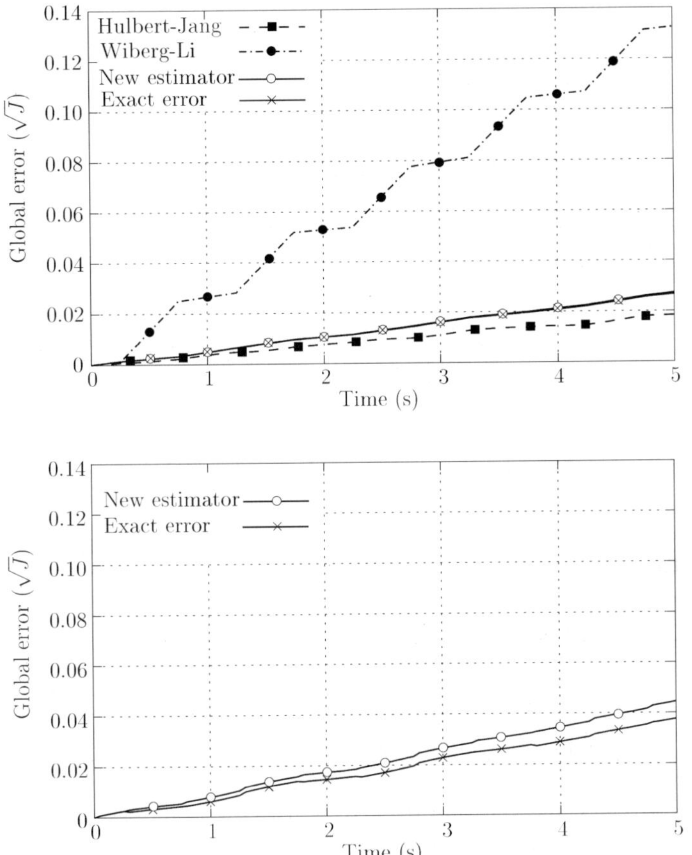

Fig. 3. One dimensional problem. Global error estimates with $\Delta t = 0.05$ s (top: trapezoidal rule; bottom: HHT with $\alpha = 0.8$).

Note that true local errors of the top part of Figure 2 are not equal from the ones in the bottom, since the numerical methods used have been different and, thus, the resulting local problems are different too (the initial conditions change).

In the case of the Newmark method, limitations in the computations of the local errors pointed out above affect the evaluation of the global errors, as can be observed in Figure 3. In contrast, the error estimation method proposed gives an accurate error estimation which, moreover, is an upper bound. If the loading function had been a differentiable function, the error estimator of Wiberg and Li would have given equally good results (see [11] for an explanation of these results).

Figure 4 depicts the convergence of the global errors (in the problem studied) estimated by each of the methods compared. The error estimator of Wiberg and Li

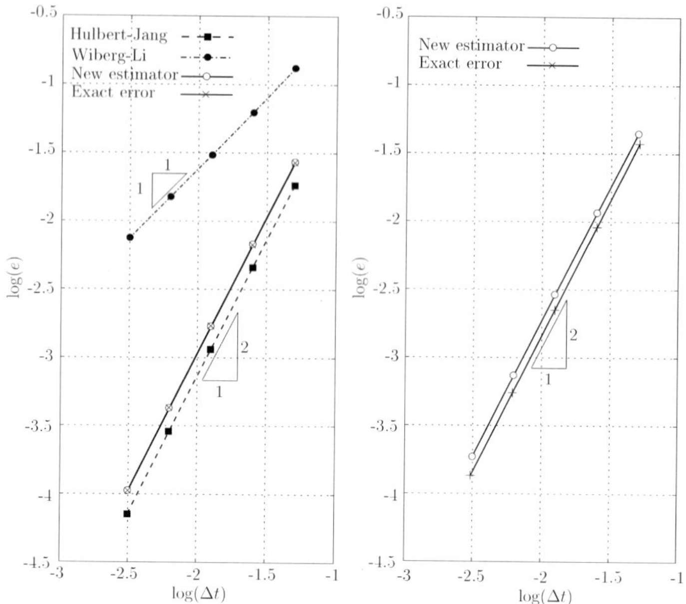

Fig. 4. One dimensional problem. Global errors at instant $t = 5$ s (left: trapezoidal rule; right: HHT with $\alpha = 0.8$).

exhibits an order of accuracy lower than required for the trapezoidal rule. The error estimator of Hulbert and Jang captures the correct order of accuracy, but underestimates the error for all time step sizes. The new error estimate shows the correct order of accuracy and returns upper bounds in all tests and for both time integration methods employed.

6.2 Dynamic behavior of a car suspension

The second example is a more realistic three dimensional multibody system with several degrees of freedom. This mechanism simulates a car suspension and it is composed of five deformable bars, a coil spring and a damper. More concisely, four bars form two parallel isosceles triangles (AEB and CFD) joined by the vertical bar EF. In Figure 5, the exact geometry of the model and position of each point is shown.

The bars undergo large displacements and, thus, the resulting problem is nonlinear. The points A, B, C, D and H are fixed, but they allow rotation around them. Between the points H and F there is a coil spring and a damper, with stiffness $k = 10^4$ N/m and damping $c = 10^3$ Ns/m, respectively. All bars are supposed to be deformable with the following mechanical properties: axial stiffness per unit length $EA = 2.1 \cdot 10^{10}$ N/m, Poisson coefficient $\nu = 0.3$ and density $\rho = 7890$ kg/m^3.

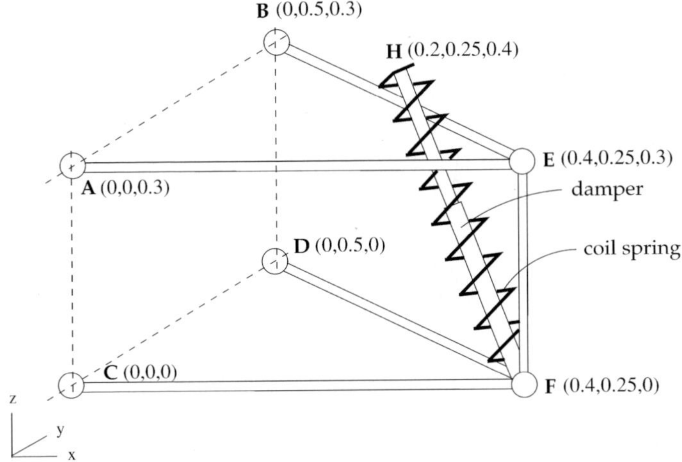

Fig. 5. Geometry of a car suspension (positions of points are expressed in meters).

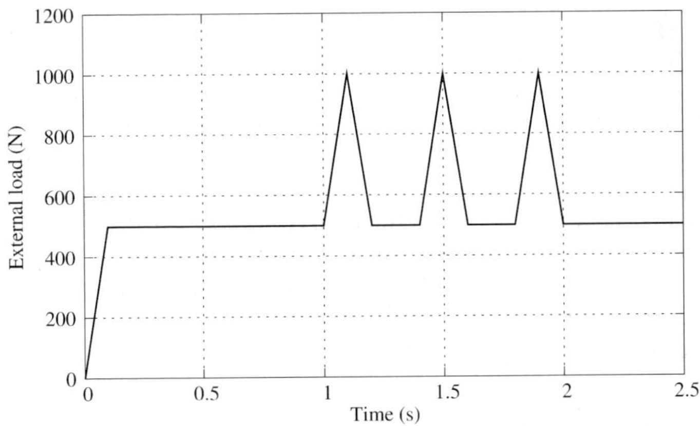

Fig. 6. Evolution in time of external load applied to the car suspension.

In the point F, a vertical external load is applied in the z-direction. Initially, the car suspension is charged up to 500 N. Then, from time $t = 1$ s to $t = 2$ s three consecutive peaks of 1000 N are introduce. Evolution in time of this external load is plotted in Figure 6.

Several transient analysis will be carried out by means of HHT method with three different values for parameter α (0.95, 0.90 and 0.80) and a constant time step $\Delta t = 0.01$ s.

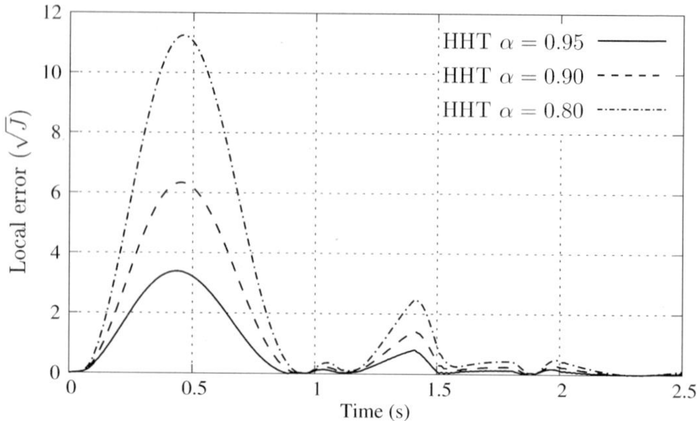

Fig. 7. Car suspension. Local error with time step size $\Delta t = 0.01$ s.

Figure 7 depicts the evolution of local errors computed using the estimator detailed in Section 5.2. It can be observed that smaller the value of parameter α is, larger the local error obtained by the estimator is.

Next, estimates for the global error are compared with "exact" ones obtained solving the problem with a much smaller time step $\Delta t = 0.00001$ s. On the top of Figure 8, the evolution of global error estimates is plotted. Just like in the case of local errors, simulation performed with the HHT method with α equals 0.95 provides more accurate results. Global error estimations for the three different simulations are upper bounds of exact errors, which are presented on the bottom of Figure 8.

In Figure 9 it is shown the order of convergence of global errors. All of them overestimate the exact error being, thus, safe bounds. Also, they present a correct second order of convergence

6.3 Transient simulation of an elastic ball bouncing inside a rigid box

The last simulation describes a more complex nonlinear, elastic system. It describes the motion of a hollow elastic sphere of external radius of 1 m and thickness 0.05 m bouncing inside a square box with rigid walls. The sphere is placed initially in the center of the box, and is released from this position with an initial velocity of 0.2 m/s in a direction that forms a 30 degrees angle with the vertical axis x_3. The box has dimensions $(L_1, L_2, L_3) = (12, 12, 8)$ m. The material of the sphere is nonlinear elastic with Neohookean model of Young's modulus $E = 10^6$ Pa and Poisson ratio $\nu = 0.3$. The finite element model of the sphere has 1106 nodes, 3370 tetrahedral elements for a total of 3318 degrees of freedom. The constraints that keep the sphere within the box boundaries are imposed with a penalty method.

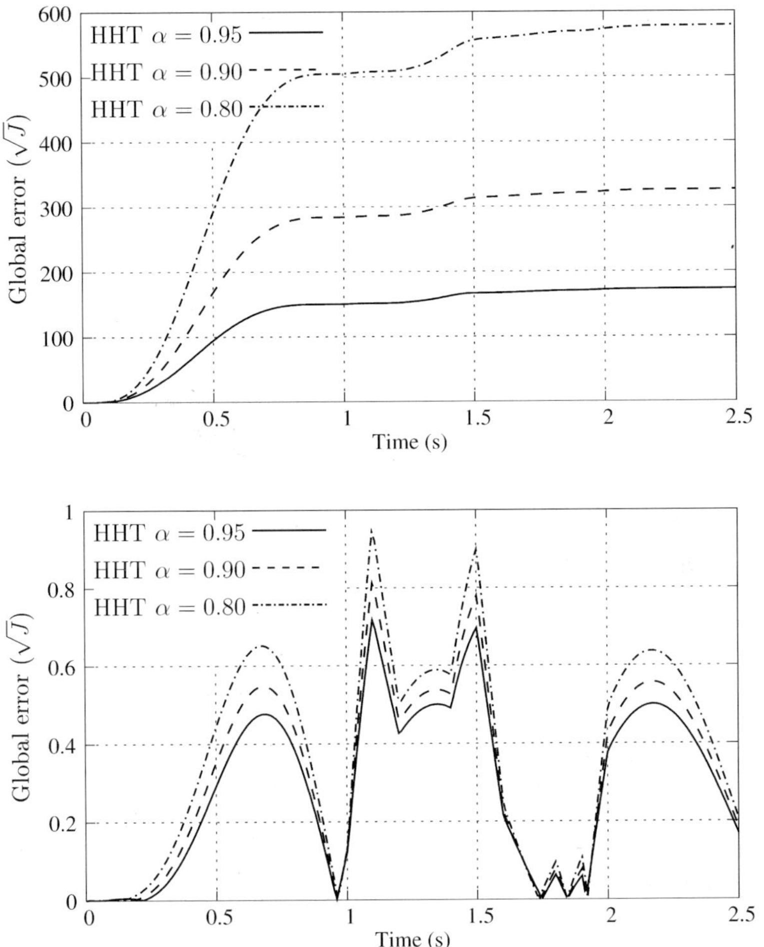

Fig. 8. Car suspension. Global error with time step size $\Delta t = 0.01$ s (top: estimated errors; bottom: exact errors).

The simulated motion consists of the uniform, straight movement of the ball and its impacts against the box walls. Since the ball is elastic, every impact against a wall makes the ball vibrate in addition to modifying its trajectory.

The numerical method chosen to solve this problem is the HHT integrator with parameter $\alpha = 0.7$ and the error estimation is performed with the method described in Section 5.2. The motion is simulated for 100 seconds with several time step sizes. Figure 10 shows the evolution of the estimated local errors when the solution is computed employing a time step of size $\Delta t = 0.5$ s. It can be seen that local errors

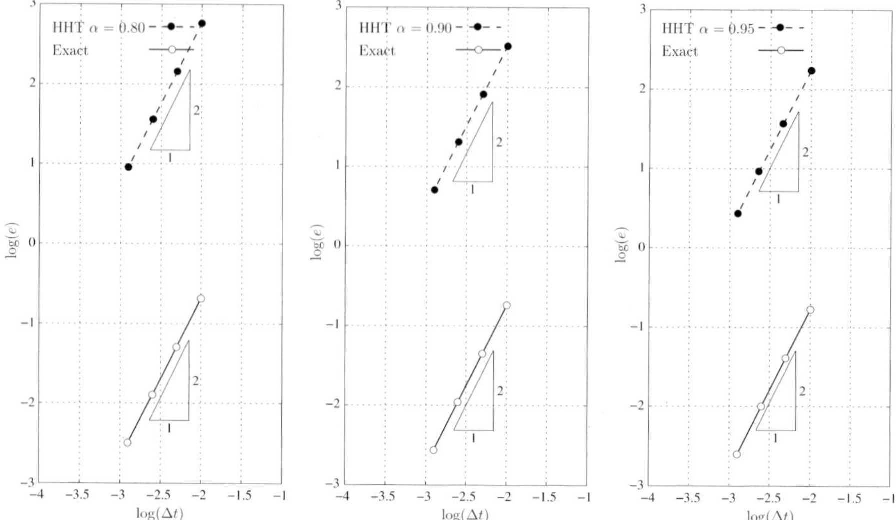

Fig. 9. Car suspension. Global errors at instant $t = 2.5$ s. (Left top: HHT with $\alpha = 0.80$; right top: HHT with $\alpha = 0.90$; bottom: HHT with $\alpha = 0.95$).

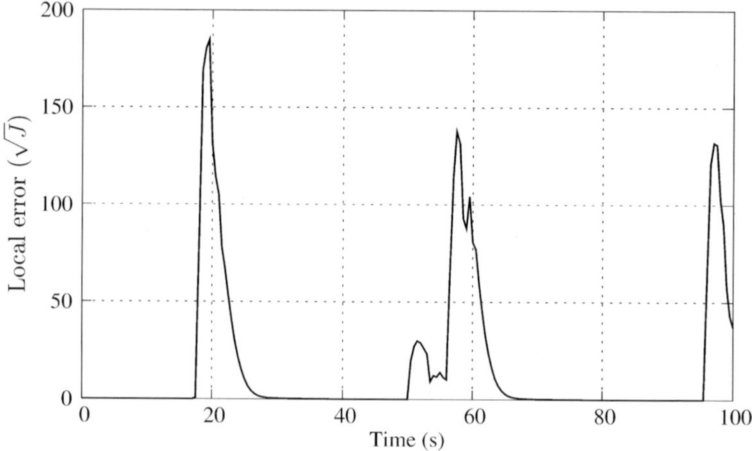

Fig. 10. Elastic ball inside rigid wall. Local errors made by the HHT method with fixed time step size $\Delta t = 0.5$ s, estimated with the proposed error estimator.

are relatively small except for a few time instants that correspond to the ball impact against the walls. Figure 11 depicts the values of the global errors accumulated during this same simulation. As a result of the large local errors in the impacts, the

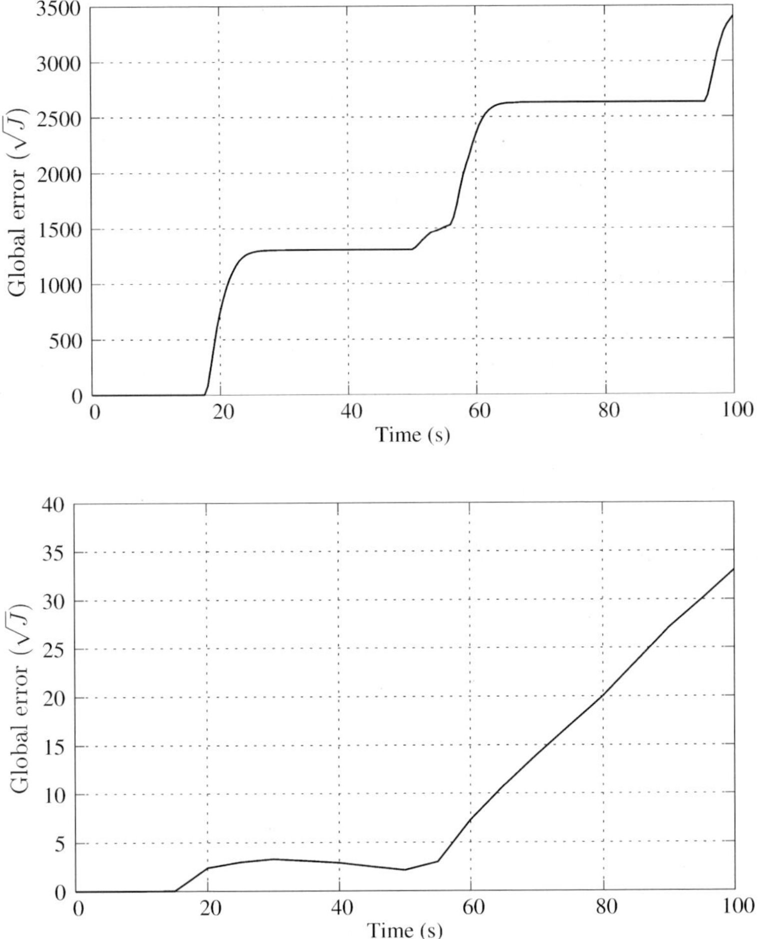

Fig. 11. Elastic ball inside rigid wall. Global error with time step size $\Delta t = 0.5$ s (top: estimated errors; bottom: exact errors).

global error exhibits large growths in these instants, with a negligible increase during the rectilinear motion inside the box.

To evaluate the accuracy of the proposed estimating method we also plot in Figure 11 the evolution of the global error computed by taking as reference value a numerical solution obtained with the HHT but with a much smaller time step size $\Delta t = 0.01$s. The estimated error is quite larger than the "exact" one, but serves as a conservative estimate that can be useful to bound the true error and to guide an adaptive process.

7 Summary and Conclusions

In this work we have presented a class of error estimators suitable for the accuracy assessment of numerical time integration of nonlinear multibody dynamics. The point of departure is a description of the type of errors that appear in the numerical integration of the equations of interest, and their relation. The main conclusion obtain is that, for general purpose, nonlinear problems, the crucial point is the computation of accurate local error estimators which remain valid without demanding unreasonable smoothness of the exact solution.

We have proposed local error estimators for Newmark's and HHT integration schemes. These are two widely employed integration schemes for which there already exist a handful of error estimators. The ones proposed in this work are as accurate as the most advanced ones, with the additional virtue of being valid for solutions without high smoothness. These properties make them very suitable for general purpose multibody simulations.

The analysis presented in this work is not only valid for the two methods studied. In fact, the same ideas could be employed to develop error estimators for other time integration schemes. In the case of linear multistep methods of the type employed in Computational Mechanics, the modifications required to adapt the proposed method to others should be minimal.

Numerical simulations shown demonstrate the good properties of the proposed estimators and their suitability for general purpose multibody simulations.

References

1. E. Hairer, S. P. Nørsett, and G. Wanner. *Solving Ordinary Differential Equations I. Nonstiff Problems*, first edition. Springer Series in Computational Mathematics, Vol. 8 Springer-Verlag, 1987.
2. R.D. Skeel. Thirteen ways to estimate global error. *Numerische Mathematik*, 48:1–20, 1986.
3. O. C. Zienkiewicz and Y. M. Xie. A simple error estimator and adaptive time stepping procedure for dynamic analysis. *Earthquake Engineering and Structural Dynamics*, 20:871–887, 1991.
4. L. F. Zeng, N. E. Wiberg, X. D. Li, and M. Xie. A posteriori local error estimation and adaptive time-stepping for Newmark integration in dynamic analysis. *Earthquake Engineering and Structural Dynamics*, 21:555–571, 1992.
5. N.-E. Wiberg and X. D. Li. A post-processing technique and an a posteriori error estimate for the Newmark method in dynamic analysis. *Earthquake Engineering and Structural Dynamics*, 22:465–489, 1993.
6. X. D. Li and N.-E. Wiberg. Implementation and adaptivity of a space-time finite element method for structural dynamics. *Computer Methods in Applied Mechanics and Engineering*, 156:211–229, 1998.

7. N.-E. Wiberg and X. D. Li. Adaptive finite element procedures for linear and non-linear dynamics. *International Journal for Numerical Methods in Engineering*, 46:1781–1802, 1999.
8. G. M. Hulbert and I. Jang. Automatic time step control algorithms for structural dynamics. *Computer Methods in Applied Mechanics and Engineering*, 126:155–178, 1995.
9. M. Geradin and A. Cardona. *Flexible multibody dynamics: a finite element approach*. John Wiley & Sons, 2001.
10. S. H. Lee and S. S. Hsieh. Expedient implicit integration with adaptive time stepping algorithm for nonlinear transient analysis. *Computer Methods in Applied Mechanics and Engineering*, 81(2):173–182, 1990.
11. I. Romero and L. M. Lacoma. Error estimation for the semidiscrete equations of solid and structural mechanics. *Accepted in Computer Methods in Applied Mechanics and Engineering*, 2004.
12. I. Romero and L. M. Lacoma. A methodology for the formulation of error estimators for time integration in solid and structural dynamics. *International Journal of Numerical Methods in Engineering*, 2005, accepted.
13. N. M. Newmark. A method of computation for structural dynamics. *Journal of the Engineering Mechanics division. ASCE*, 85:67–94, 1956.
14. H. M. Hilber, T. J. R. Hughes, and R. L. Taylor. Improved numerical dissipation for time integration algorihtms in structural dynamics. *Earthquake Engineering and Structural Dynamics*, 5:283–292, 1977.
15. T. J. R. Hughes. Analysis of transient algorithms with particular reference to stability behavior. In T. Belytschko and T. J. R. Hughes (eds), *Computational methods for transient analysis*, pp. 67–155. Elsevier Scientific Publishing, Amsterdam, 1983.
16. S. Erlicher, L. Bonaventura, and O. S. Bursi. The analysis of the generalized-α method for non-linear dynamic problems. *Computational Mechanics*, 28:83–104, 2002.
17. L. M. Lacoma and I. Romero. Error estimation for the hht method in solid dynamics. 2005, in preparation.

A DAE Formulation for the Dynamic Analysis and Control Design of Cranes Executing Prescribed Motions of Payloads

Wojciech Blajer and Krzysztof Kołodziejczyk

Institute of Applied Mechanics, Technical University of Radom, ul. Krasickiego 54, 26-600 Radom, Poland; E-mail: {wblajer,krkolo}@poczta.onet.pl

Abstract. The dynamic behavior and control of cranes executing prescribed motions of payloads are strongly affected by the underactuated nature of the robotic systems, in which the number of control inputs/outputs is smaller than the number of degrees-of-freedom. The outputs are specified in time load coordinates, which, expressed in terms of the system states, lead to servo-constraints on the system. The problem can then viewed from the perspective of constrained motion. It is noticed however that servo-constraints differ from passive constraints in several aspects. Mainly, they are enforced by means of control forces which may have any directions with respect to the servo-constraint manifold, and in the extreme (some of them) may be tangent. A specific methodology must be developed to solve the 'singular' inverse dynamics problem. In this contribution, a theoretical background for the modeling of the partly specified/actuated motion is given. The initial governing equations, arising as index five differential-algebraic equations, are transformed to a more tractable index three form by projecting the dynamic equations into the orthogonal and tangent subspaces with respect to the servo-constraint manifold in the crane velocity space. A simple numerical code for solving the resultant differential-algebraic equations, based on backward Euler method, is then proposed. The feedforward control law obtained this way is enhanced by a closed-loop control strategy with feedback of the actual errors in load position to provide stable tracking of the required reference load trajectory in presence of perturbations. A rotary crane executing a load prescribed motion serves as an illustration. Some results of numerical experiments/simulations are reported.

1 Introduction

Cranes are widely used in transportation and construction. In the industrial practice they are predominantly operated manually, and automatic cranes are comparatively rare. In the crane's conventional mode of operation, the operator actuates different joints, by joysticks and/or buttons, so that to move the load (or hook) from its initial position to its desired final destination in its working space along a trajectory, avoiding obstacles and sway. Even though almost the same paths are often repeated, which allows the operator to 'learn' the maneuver, the cycle time is usually relatively large since the operator has to perform the maneuvers slowly

J.C. García Orden et al. (eds), Multibody Dynamics. Computational Methods and Applications, 91–112.

in order to avoid inertia-induced excitations, and a considerable percentage of the time is spent on maneuvering the load close to the target point. The latter is usually a trial-and-error process, based on feedback provided by the operator's own vision and assessment, and/or hand signals or radio communication from an assistant at the work zone [1]. Automated cranes, after being 'taught' a safe and efficient route between a fixed locations of the source and the target, have a potential to play back that route much faster and more accurately than repeated manual cycles. The high potential of rationalization offered by automatic control systems stimulated an increasing interest and substantial progress in research on modeling and control of cranes [2].

Cranes belong to a class of underactuated/underconstrained systems – the controlled mechanical systems in which the number of control inputs/outputs is smaller than the number of degrees of freedom. Due to the rope flexibility, the undesirable load swing cannot be directly actuated, and advanced feedback control techniques are needed to suppress the swing; see e.g. [2–4]. On the other hand, the control problem of cranes can also be viewed from the perspective of the concept of *flat* systems [5]. The important property of a flat system is that all the state variables and the control inputs can be algebraically expressed in terms of the outputs and their time derivatives up to a certain order. This provides a feedforward control law for sufficiently smooth output functions, and gives guidelines for constructing feedback control schemes [6, 7]. Though such a solution to control problem is theoretically attainable if only the system is controllable, it may be difficult to achieve for more complicated underactuated/underconstrained systems.

In this contribution the problem of dynamics and control of cranes executing prescribed motions of payloads is viewed from the perspective of constrained motion. The control outputs, expressed in terms of the system states, lead to servo-constraints on the system [8, 9]. It is noticed however that servo-constraints differ from passive constraints in several aspects. Mainly, they are enforced by means of control forces, which may have any directions with respect to the manifold of servo-constraints, and in the extreme (some of them) may be tangent. Such a situation arises in the load trajectory tracking control of cranes, and a specific methodology must be developed to solve the 'singular' inverse dynamics problem. A theoretical background for the modeling of the partly specified/actuated motion is given. The initial governing equations, arising as index five differential-algebraic equations, are transformed to a more tractable index three form by projecting the dynamic equations into the orthogonal and tangent subspaces with respect to the servo-constraint manifold in the crane velocity space. A simple numerical code for solving the resultant differential-algebraic equations, based on backward Euler method, is proposed. The feedforward control law obtained this way is then enhanced by a closed-loop control strategy with feedback of the actual errors in load position to provide stable tracking of the required reference load trajectory in presence of perturbations. A rotary crane executing a load prescribed motion serves as an illustration. Some results of numerical simulations are reported.

2 Crane Dynamics and Servo-Constraints

Lumped-mass models of cranes are most often used. The hoisting line is treated as a massless cable, the payload is lumped with the hook and modeled as a point mass, and the cable-hook-payload assembly is modeled as a (spherical) pendulum which is suspended from a point on the support mechanism (trolley-girder, trolley-jib, or a boom, respectively for gantry/overhead, rotary and boom cranes), see e.g. [2,10-14]. The support mechanism moves the suspension point of the line, while the hoisting mechanism changes its length (lifts and lowers the payload). The modeling result is then an n-degree-of-freedom crane model whose position is described by generalized coordinates $\mathbf{q} = [q_1 \ \ldots \ q_n]^T$, and which is enforced, in addition to the applied forces, by m actuator forces/moments $\mathbf{u} = [u_1 \ \ldots \ u_m]^T$, where $m < n$ (the system is underactuated). The crane dynamic equations can be written in the following generic matrix form

$$\mathbf{M}(\mathbf{q})\ddot{\mathbf{q}} + \mathbf{d}(\mathbf{q},\dot{\mathbf{q}}) = \mathbf{f}(\mathbf{q},\dot{\mathbf{q}}) - \mathbf{B}^T\mathbf{u} \tag{1}$$

where $\mathbf{M}$ is the $n \times n$ generalized mass matrix, $\mathbf{d}$ and $\mathbf{f}$ are n-vectors of generalized dynamic and applied forces, and $\mathbf{B}^T$ is the $n \times m$ matrix of influence of control inputs $\mathbf{u}$ on the generalized actuating force vector $\mathbf{f}_u = -\mathbf{B}^T\mathbf{u}$. For the 3D crane models considered in this paper, n and m values are respectively 5 and 3 (for 2D models n and m are 3 and 2).

The desired performance goal of the considered crane model is a desired load trajectory, i.e. the $m = 3$ control outputs are desired (specified in time) load coordinates $\boldsymbol{\gamma}_d(t) = [x_d(t) \ \ y_d(t) \ \ z_d(t)]^T$, where x, y and z are the load coordinates in the inertial coordinate system XYZ (for 2D models $\boldsymbol{\gamma}_d(t) = [x_d(t) \ \ y_d(t)]^T$ and $m = 2$). The outputs, expressed in terms of coordinates $\mathbf{q}$ as

$$\mathbf{c}(\mathbf{q},t) = \boldsymbol{\Phi}(\mathbf{q}) - \boldsymbol{\gamma}_d(t) = \mathbf{0} \tag{2}$$

can be treated as m constraints on the system, called *servo-constraints* [9] (also called *control constraints* [8,15] or *program constraints* [16]) as distinct from *passive* (or *contact*) *constraints* in the classical sense. After twice differentiating the initial constraint equations with respect to time, the constraint conditions at the acceleration level arise as

$$\ddot{\mathbf{c}} = \mathbf{C}(\mathbf{q})\ddot{\mathbf{q}} - \boldsymbol{\xi}(\mathbf{q},\dot{\mathbf{q}},t) = \mathbf{0} \tag{3}$$

where $\mathbf{C} = \partial\boldsymbol{\Phi}/\partial\mathbf{q}$ is the $m \times n$ matrix of servo-constraints, and $\boldsymbol{\xi} = \ddot{\boldsymbol{\gamma}}_d - \dot{\mathbf{C}}\dot{\mathbf{q}}$ is the m-vector of constraint induced accelerations.

Equation (2) is mathematically equivalent to m time-dependent (rheonomic) holonomic constraints on the system, $\mathbf{c}(\mathbf{q},t) = \mathbf{0}$. The resemblance of the load trajectory tracking control problem to the constrained motion case may however be misleading. Assumed Equation (2) represents passive constraints, the generalized actuating force $\mathbf{f}_u = -\mathbf{B}^T\mathbf{u}$ in the Equation (1) would be replaced by the generalized constraint reaction force $\mathbf{f}_c = -\mathbf{C}^T\boldsymbol{\lambda}$, where $\mathbf{C}$ is the constraint matrix defined in Equation (3). While the reactions of (ideal) passive constraints are by assumption orthogonal to the instantaneous manifold of passive constraints $\mathbf{c}(\mathbf{q},t) = \mathbf{0}$ embedded in the n-space related to velocities $\dot{\mathbf{q}}$ [17], the actuating forces may have arbitrary directions with respect to the instantaneous servo-constraint manifold $\mathbf{c}(\mathbf{q},t) = \mathbf{0}$, and in the extreme (some of them) may be tangent [8,16]; see Figure 1 for illustration. In the latter case, qualitatively, not all of the desired outputs can directly be actuated by the control inputs. A measure of the 'control singularity' is the deficiency in rank of the $m \times m$ matrix $\mathbf{P} = \mathbf{C}\mathbf{M}^{-1}\mathbf{B}^T$ which represents the inner product of the constrained and controlled subspaces in the n-space of crane velocities,

$$\text{rank}(\mathbf{C}\mathbf{M}^{-1}\mathbf{B}^T) = p \tag{4}$$

For a more detailed discussion on the problem of realization of servo-constraints and the relevant geometrical interpretations the reader is referred to [8].

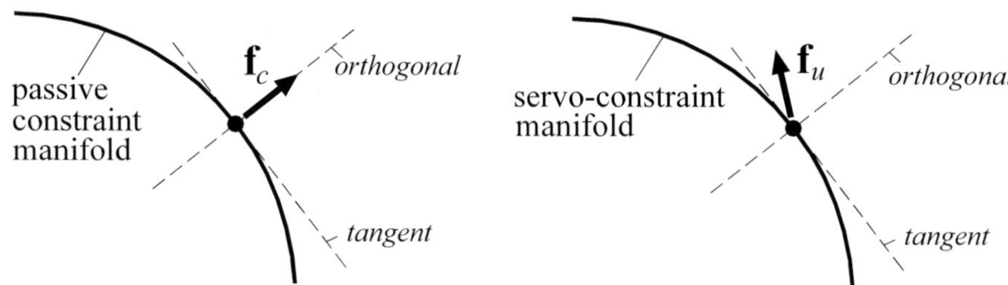

Fig. 1. The reaction of passive constraints and the actuation force with respect to the constraint manifold.

In the case of considered crane models, both 3D and 2D ones, $\text{rank}(\mathbf{P}) = p = 1$, and this means that only one task requirement can be directly (in the orthogonal way) actuated by the available control (the tension force in the rope influenced mainly by the winch torque changing the rope length). The realization of two other task requirements is indirect (tangent), achieved by moving the suspension point of the rope and, in this way, producing the appropriate changes in swing angles. In other words, the direct (orthogonal) control regulates the tension force value in the rope, and the indirect (tangent) control changes its space orientation, and in this (orthogonal-tangent) way the required reactions on the load in all three directions can be produced. The appropriate changes in rope swing angles, due to the tangent realization of servo-constraints, can be viewed as two additional restrictions on the crane configuration, and in this sense the five-degree-of-freedom system is 'fully' specified by three servo-constraints (2), and can explicitly be actuated by three control inputs $\mathbf{u}$. The inference corresponds with the aforementioned concept of flat systems. Nevertheless, the methodology developed in the sequel for solving the dynamics and control of cranes executing a load prescribed motion differs substantially from the formulations that use the flatness concept [5–7]. Prior to the presentation of the new methodology, let us first illustrate the hitherto formulations with an example.

3 Illustration – A Three-Dimensional Rotary Crane

Let us consider the rotary crane model seen in Figure 2. This is a five-degree-of-freedom system, $n=5$, whose position is described by $\mathbf{q}=[\varphi \;\; s \;\; l \;\; \theta_1 \;\; \theta_2]^T$, where φ is the angle of rotation of the girder bridge, s describes the trolley position on the girder, l is the hoisting rope length, and θ_1 and θ_2 are the swing angles as defined in Figure 2. The control inputs are the torque M_b regulating the girder rotation angle φ, the force F actuating the trolley position s, and the winch torque M_w changing the rope length l, $\mathbf{u}=[M_b \;\; F \;\; M_w]^T$. In this meaning, φ, s and l can be regarded as *controlled coordinates*, while θ_1 and θ_2 can be called *uncontrolled coordinates*. The number m of control inputs $\mathbf{u}$ is equal to the number of outputs $\boldsymbol{\gamma}_d(t)=[x_d(t) \;\; y_d(t) \;\; z_d(t)]^T$, $m=3$, and it is smaller than the number of degrees of freedom of the system, $m<n$. We deal thus with an *underactuated system* in a *partly specified motion* [8, 16].

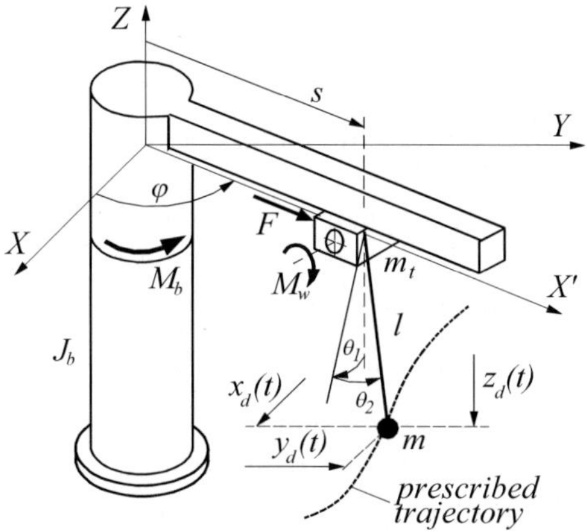

Fig. 2. The rotary crane model.

The derivation of equations of motion for the rotary crane is not a trivial task. An effective method for the derivation is the projection method described in [17]. The starting point are the dynamic equations in $\bar{n}=8$ dependent coordinates $\mathbf{p}=[\varphi \;\; x_t \;\; y_t \;\; z_t \;\; \alpha \;\; x \;\; y \;\; z]^T$, where φ, x, y and z are as defined above, x_t, y_t and z_t are the trolley coordinates in XYZ frame, and α is the winch rotation angle. Treating initially the bridge, trolley, winch (only its rotation) and load unconstrained from each other, their dynamic equations are simply:

$$J_b\ddot{\varphi} = -D_\varphi\dot{\varphi} + M_b \tag{5a}$$

$$m_t\begin{bmatrix}\ddot{x}_t\\ \ddot{y}_t\\ \ddot{z}_t\end{bmatrix} = \begin{bmatrix}-D_s\dot{s}\cos\varphi\\ -D_s\dot{s}\sin\varphi\\ -m_t g\end{bmatrix} + \begin{bmatrix}\cos\varphi\\ \sin\varphi\\ 0\end{bmatrix}F \tag{5b}$$

$$J_w\ddot{\alpha} = -D_\alpha\dot{\alpha} + M_w \tag{5c}$$

$$m\begin{bmatrix}\ddot{x}\\ \ddot{y}\\ \ddot{z}\end{bmatrix} = \begin{bmatrix}0\\ 0\\ -mg\end{bmatrix} \tag{5d}$$

where J_b and J_w are the moments of inertia of the bridge and winch relative their axes of rotation, m_t and m are the trolley and load masses, g is the gravitational acceleration, and D_φ, D_α and D_s are the viscous damping coefficients. The initial dynamic equations in $\mathbf{p}$ can be gathered in the following generic matrix form

$$\overline{\mathbf{M}}\ddot{\mathbf{p}} = \overline{\mathbf{f}} - \overline{\mathbf{B}}^T\mathbf{u} \tag{6}$$

where $\overline{\mathbf{M}}$ is the (diagonal) 8×8 generalized mass matrix, $\overline{\mathbf{f}}$ is the 8-vector of generalized applied forces, both related to $\mathbf{p}$, and $\overline{\mathbf{B}}^T$ is the 8×5 control matrix ($\overline{\mathbf{f}}_c = -\overline{\mathbf{B}}^T\mathbf{u}$ is the 8-vector of generalized control forces related to $\mathbf{p}$). The explicit forms of $\overline{\mathbf{M}}$, $\overline{\mathbf{f}}$ and $\overline{\mathbf{B}}^T$ are easy to deduce from Equations (5).

Three constraints are imposed on the separated subsystems due to the bridge-trolley and winch-load connections. Starting from the above dynamic equations in the dependent coordinates $\mathbf{p}$, the constraint equations given explicitly [17,18] are required for the derivation of the dynamic equations in independent coordinates $\mathbf{q}$. The explicit constraint equations are the relations between the $\overline{n} = 8$ dependent coordinates and the $n = 5$ independent coordinates,

$$\mathbf{p} = \mathbf{h}(\mathbf{q}) \tag{7}$$

where the difference $\overline{n} - n = 8 - 5 = 3$ stands for the number of constraints, more often given implicitly as $\overline{\mathbf{\Phi}}(\mathbf{p}) = \mathbf{0}$ (not required in the sequel), and $\overline{\mathbf{\Phi}}[\mathbf{h}(\mathbf{q})] \equiv \mathbf{0}$ (the implicit constraint equations are satisfied identically when expressed in $\mathbf{q}$). The explicit constraint conditions at the velocity and acceleration levels are then:

$$\dot{\mathbf{p}} = \mathbf{H}(\mathbf{q})\dot{\mathbf{q}} \tag{8}$$

$$\ddot{\mathbf{p}} = \mathbf{H}(\mathbf{q})\ddot{\mathbf{q}} + \boldsymbol{\eta}(\mathbf{q},\dot{\mathbf{q}}) \tag{9}$$

where $\mathbf{H} = \partial\mathbf{h}/\partial\mathbf{q}$ and $\boldsymbol{\eta} = \dot{\mathbf{H}}\dot{\mathbf{q}}$. For the case at hand, the $\overline{n} = 8$ relations of Equation (7) and the $\overline{n}\times n$ (8×5) matrix $\mathbf{H}$ defined in Equation (8) are:

$$\mathbf{p} = \begin{bmatrix} \varphi \\ x_t \\ y_t \\ z_t \\ \alpha \\ x \\ y \\ z \end{bmatrix} = \mathbf{h}(\mathbf{q}) = \begin{bmatrix} \varphi \\ s\cos\varphi \\ s\sin\varphi \\ 0 \\ (l-l_0)/r_w \\ (s+l\sin\theta_2)\cos\varphi + l\cos\theta_2\sin\theta_1\sin\varphi \\ (s+l\sin\theta_2)\sin\varphi - l\cos\theta_2\sin\theta_1\cos\varphi \\ -l\cos\theta_2\cos\theta_1 \end{bmatrix} \tag{10}$$

$$\mathbf{H} = \begin{bmatrix} 1 & 0 & 0 & 0 & 0 \\ -s\sin\varphi & \cos\varphi & 0 & 0 & 0 \\ s\cos\varphi & \sin\varphi & 0 & 0 & 0 \\ 0 & 0 & 0 & 0 & 0 \\ 0 & 0 & 1/r_w & 0 & 0 \\ H_{61} & \cos\varphi & H_{63} & H_{64} & H_{65} \\ H_{71} & \sin\varphi & H_{73} & H_{74} & H_{75} \\ 0 & 0 & H_{83} & H_{84} & H_{85} \end{bmatrix} \tag{11}$$

where: $H_{61} = -(s+l\sin\theta_2)\sin\varphi + l\cos\theta_2\sin\theta_1\cos\varphi$,
$H_{63} = \sin\theta_2\cos\varphi + \cos\theta_2\sin\theta_1\sin\varphi$,
$H_{64} = l\cos\theta_2\cos\theta_1\sin\varphi$,
$H_{65} = l\cos\theta_2\cos\varphi - l\sin\theta_2\sin\theta_1\sin\varphi$,
$H_{71} = (s+l\sin\theta_2)\cos\varphi + l\cos\theta_2\sin\theta_1\sin\varphi$,
$H_{73} = \sin\theta_2\sin\varphi - \cos\theta_2\sin\theta_1\cos\varphi$,
$H_{74} = -l\cos\theta_2\cos\theta_1\cos\varphi$,
$H_{75} = l\cos\theta_2\sin\varphi + l\sin\theta_2\sin\theta_1\cos\varphi$,
$H_{83} = -\cos\theta_2\cos\theta_1$,
$H_{84} = l\cos\theta_2\sin\theta_1$,
$H_{85} = l\sin\theta_2\cos\theta_1$.

The expressions for the 8-vector $\boldsymbol{\eta} = \dot{\mathbf{H}}\dot{\mathbf{q}}$ are rather complex, and will not be reported here for shortness.

Having the explicit constraint equations defined, the $n=5$ dynamic equations of the rotary crane in $\mathbf{q}$ can easily be formulated. Namely, the ingredients of Equation (1) are [17,18]:

$$\mathbf{M} = \mathbf{H}^T\overline{\mathbf{M}}\mathbf{H}\,; \quad \mathbf{d} = \mathbf{H}^T\overline{\mathbf{M}}\boldsymbol{\eta}\,; \quad \mathbf{f} = \mathbf{H}^T\overline{\mathbf{f}}\,; \quad \mathbf{B}^T = \mathbf{H}^T\overline{\mathbf{B}} \tag{12}$$

It is worth noting that only $\overline{\mathbf{M}}$, $\overline{\mathbf{h}}$ and $\overline{\mathbf{B}}^T$ from the initial dynamic equations, and $\mathbf{H}$ and $\boldsymbol{\eta}$ from the explicit constraint equations are required to formulate the crane dynamic equations in $\mathbf{q}$ as defined in Equation (1). In applications, **M**, **d**, **f** and $\mathbf{B}^T$ need not to be obtained in analytical forms, or symbolic computer manipulations can be used for their derivation. Here, we will limit ourselves to the general matrix forms as in Equation (12), and demonstrate only the control distribution matrix $\mathbf{B}^T$ which will be of some use in the sequel,

$$\mathbf{B}^T = \begin{bmatrix} 1 & 0 & 0 \\ 0 & 1 & 0 \\ 0 & 0 & 1/r_w \\ 0 & 0 & 0 \\ 0 & 0 & 0 \end{bmatrix} \quad (13)$$

The servo-constraint equations and the constraint matrix **C** defined in Equations (2) and (3) are:

$$\mathbf{c} = \begin{bmatrix} (s + l\sin\theta_2)\cos\varphi + l\cos\theta_2\sin\theta_1\sin\varphi \\ (s + l\sin\theta_2)\sin\varphi - l\cos\theta_2\sin\theta_1\cos\varphi \\ -l\cos\theta_2\cos\theta_1 \end{bmatrix} - \begin{bmatrix} x_d(t) \\ y_d(t) \\ z_d(t) \end{bmatrix} = \mathbf{0} \quad (14)$$

$$\mathbf{C} = \begin{bmatrix} H_{61} & \cos\varphi & H_{63} & H_{64} & H_{65} \\ H_{71} & \sin\varphi & H_{73} & H_{74} & H_{75} \\ 0 & 0 & H_{83} & H_{84} & H_{85} \end{bmatrix} \quad (15)$$

where $H_{61},\ldots,H_{85}$ are as defined in Equation (11). Again, the analytical expression for $\boldsymbol{\xi} = \ddot{\boldsymbol{\gamma}}_d - \dot{\mathbf{C}}\dot{\mathbf{q}}$ is rather complex and will not be reported here for shortness.

Due to the problem complexity it is impossible to demonstrate plainly that $\text{rank}(\mathbf{CM}^{-1}\mathbf{B}^T) = p = 1$, which can eventually be proved computationally. The physical interpretation of the result, discussed already in Section 2, is the following. The desired load position $\boldsymbol{\gamma}_d(t) = [x_d(t) \;\; y_d(t) \;\; z_d(t)]^T$ in all three directions is regulated by the tension force in the rope, whose value and space orientation must change appropriately. Only the force value can directly be actuated by the available control (mainly by the winch torque M_w, and for $\theta_1 = 0$ and $\theta_2 = 0$ solely by M_w), which is mathematically expressed by the fact that the rank of the 3×3 matrix $\mathbf{P} = \mathbf{CM}^{-1}\mathbf{B}^T$ is reduced to one. Due to the rope flexibility, the required space orientation of the tension force can not be directly actuated by the crane control inputs – it can be actuated only indirectly by adjusting the location of the rope suspension point with respect to the load location $\boldsymbol{\gamma}_d(t)$. In the case of the rotary crane, the desired suspension point location is achieved by appropriate changes in φ and s values, actuated then by M_b and F, respectively. Similar observations relate to the 3D overhead crane model developed in [12], characterized by $n = 5$, $m = 3$ and $p = 1$, and its simplified 2D version studied in [8], which can also be viewed as a rotary crane with φ and θ_1 'frozen', characterized by $n=3$, $m=2$ and $p=1$.

4 Governing Equations and the Solution Code

The initial governing equations of the crane motion in the prescribed motion are formed by n kinematic relations $\dot{\mathbf{q}}=\mathbf{v}$, n dynamic equations, rearranged with the use of the kinematic relation to $\mathbf{M}(\mathbf{q})\dot{\mathbf{v}}+\mathbf{d}(\mathbf{q},\mathbf{v})=\mathbf{f}(\mathbf{q},\mathbf{v})-\mathbf{B}^T\mathbf{u}$, and m servo-constraint equations $\mathbf{c}(\mathbf{q},t)=\mathbf{0}$. All state together $2n+m$ differential-algebraic equations (DAEs) in the same number of $2n$ state variables $\mathbf{q}$ and $\mathbf{v}$ (differential variables), and m control variables $\mathbf{u}$ (algebraic variables). The problem with the DAEs formulated this way is that their *index* is equal to five [19-23], which is a measure of singularity/complexity of a DAE system and determines difficulty in its numerical treatment. In the following a scheme for transforming the initial index-five DAEs to equivalent, numerically more tractable index-three DAEs, is thus developed. An assumption for the further developments is that

$$n-m=m-p=k \tag{16}$$

which can easily be verified for the rotary crane model described in Section 3 ($k=2$), the 3D overhead crane model reported in [12] ($k=2$), and the plane crane model studied in [8] ($k=1$).

Following the projection method [17], the crane dynamic equations can be projected into two complementary subspaces in the crane velocity space, the constrained (specified) and unconstrained (unspecified) ones, defined respectively by the $m\times n$ constraint matrix $\mathbf{C}$ introduced in Equation (3) and its orthogonal complement, an $n\times k$ matrix $\mathbf{D}$ such that

$$\mathbf{C}\mathbf{D}=\mathbf{0} \quad \Leftrightarrow \quad \mathbf{D}^T\mathbf{C}^T=\mathbf{0} \tag{17}$$

For a given $m\times n$ ($m<n$) matrix $\mathbf{C}$, its orthogonal complement, an $n\times k$ ($k=n-m$) matrix $\mathbf{D}$ satisfying Equation (17) can sometimes be guessed (usually for simple cases only) or determined following a numerically oriented code like the scheme followed from the coordinate partitioning method [24]. Namely, assumed $\mathbf{C}$ is of maximal row-rank, $\text{rank}(\mathbf{C})=m$, it can always be factorized to $\mathbf{C}=[\,\mathbf{U} \,\vdots\, \mathbf{W}\,]$ so that $\mathbf{U}$ and $\mathbf{W}$ are the $m\times k$ and $m\times m$ matrices, respectively, and $\det(\mathbf{W})\neq 0$. The orthogonal complement $\mathbf{D}$ to $\mathbf{C}$ can then be found as

$$\mathbf{D}=\begin{bmatrix}\mathbf{I}\\ -\mathbf{W}^{-1}\mathbf{U}\end{bmatrix} \tag{18}$$

where $\mathbf{I}$ is the $k\times k$ identity matrix.

The projection formula of the dynamic equations is [17]

$$\begin{bmatrix} \mathbf{D}^T \\ \mathbf{C}\mathbf{M}^{-1} \end{bmatrix} (\mathbf{M}\dot{\mathbf{v}} + \mathbf{d} - \mathbf{f} + \mathbf{B}^T\mathbf{u}) = \mathbf{0} \tag{19}$$

The projection into the unconstrained subspace, $\mathbf{D}^T\mathbf{M}\dot{\mathbf{v}} + \mathbf{D}^T\mathbf{d} = \mathbf{D}^T\mathbf{f} - \mathbf{D}^T\mathbf{B}^T\mathbf{u}$, represents k differential equations. The projection into the constrained subspace, after using the servo-constraint conditions at the acceleration level defined in Equation (3), leads to $\mathbf{C}\mathbf{M}^{-1}(\mathbf{f} - \mathbf{d}) - \mathbf{C}\mathbf{M}^{-1}\mathbf{B}^T\mathbf{u} - \boldsymbol{\xi} = \mathbf{0}$, which represents m algebraic equations in the system states $\mathbf{q}$ and $\mathbf{v}$, and m control inputs $\mathbf{u}$. Since $\text{rank}(\mathbf{C}\mathbf{M}^{-1}\mathbf{B}^T) = p < m$, these m algebraic equations impose only p independent conditions on $\mathbf{u}$, however, and $k = m - p$ additional restrictions on the crane motion, supplementary to the m original requirements as in Equation (1). They correspond to the discussed before conditions on the value of tension force in the rope and the location of the rope suspension point. In other words, due to the mixed orthogonal-tangent realization of the servo-constraints [8,16], the total number of the original and supplementary motion specifications is $m + k = n$, and thus, in this indirect way, the motion is fully specified. The situation corresponds to flatness [5-7] of the underactuated system in the partly specified motion – all the state variables and control inputs can be expressed in terms of the outputs $\boldsymbol{\gamma}_d(t)$ and their time derivatives. Such an analytical solution, though theoretically attainable, is very difficult to obtain (if possible at all) for the case of a crane.

Using the results of the projection, the governing equations for the crane executing a load prescribed motion can be formed as the following $n + k + m + m$ (for the 3D rotary crane: $5 + 2 + 3 + 3 = 13$) DAEs in $n + n + m$ (for the rotary crane: $5 + 5 + 3 = 13$) variables $\mathbf{q}$, $\mathbf{v}$ and $\mathbf{u}$:

$$\begin{aligned} \dot{\mathbf{q}} &= \mathbf{v} \\ \mathbf{D}^T\mathbf{M}\dot{\mathbf{v}} &= \mathbf{D}^T(\mathbf{f} - \mathbf{d}) - \mathbf{D}^T\mathbf{B}^T\mathbf{u} \\ \mathbf{0} &= \mathbf{C}\mathbf{M}^{-1}(\mathbf{f} - \mathbf{d}) - \mathbf{C}\mathbf{M}^{-1}\mathbf{B}^T\mathbf{u} - \boldsymbol{\xi} \\ \mathbf{0} &= \boldsymbol{\Phi}(\mathbf{q}) - \boldsymbol{\gamma}_d \end{aligned} \quad \Leftrightarrow \quad \begin{aligned} \dot{\mathbf{q}} &= \mathbf{v} \\ \mathbf{H}(\mathbf{q})\dot{\mathbf{v}} &= \mathbf{h}(\mathbf{q}, \dot{\mathbf{q}}, \mathbf{u}, t) \\ \mathbf{0} &= \mathbf{b}(\mathbf{q}, \dot{\mathbf{q}}, \mathbf{u}, t) \\ \mathbf{0} &= \mathbf{c}(\mathbf{q}, t) \end{aligned} \tag{20}$$

The index of the DAEs is equal to three, and their solution are variations in time of state variables of the crane executing the load prescribed motion, $\mathbf{q}_d(t)$ and $\mathbf{v}_d(t)$, and the control $\mathbf{u}_d(t)$ that ensures the realization of motion. The solution encompasses thus both the dynamic analysis and the synthesis of control of the crane executing a load prescribed motion.

A range of DAE solvers has been described in the literature, see e.g. [19-23]. In this paper the simplest possible algorithm, using Euler backward differentiation approximation method, is proposed to solve DAEs (20), in which the derivatives $\dot{\mathbf{q}}$ and $\dot{\mathbf{v}}$ at time $t_{n+1} = t_n + \Delta t$ are approximated by their backward differences, respectively $(\mathbf{q}_{n+1} - \mathbf{q}_n)/\Delta t$ and $(\mathbf{v}_{n+1} - \mathbf{v}_n)/\Delta t$, where Δt is the integration time step. Namely, given $\mathbf{q}_n$ and $\mathbf{v}_n$ at time t_n (note that $\mathbf{u}_n$ is not involved), the values $\mathbf{q}_{n+1}$, $\mathbf{v}_{n+1}$ and $\mathbf{u}_{n+1}$ at time $t_{n+1} = t_n + \Delta t$ can be found as a solution to the following nonlinear algebraic equations:

$$\begin{aligned} \frac{\mathbf{q}_{n+1} - \mathbf{q}_n}{\Delta t} - \mathbf{v}_{n+1} &= \mathbf{0} \\ \mathbf{H}(\mathbf{q}_{n+1}) \frac{\mathbf{v}_{n+1} - \mathbf{v}_n}{\Delta t} - \mathbf{h}(\mathbf{v}_{n+1}, \mathbf{q}_{n+1}, \mathbf{u}_{n+1}, t_{n+1}) &= \mathbf{0} \\ \mathbf{b}(\mathbf{v}_{n+1}, \mathbf{q}_{n+1}, \mathbf{u}_{n+1}, t_{n+1}) &= \mathbf{0} \\ \mathbf{c}(\mathbf{q}_{n+1}, t_{n+1}) &= \mathbf{0} \end{aligned} \tag{21}$$

and in this way the solution can be advanced from time t_n to $t_{n+1} = t_n + \Delta t$.

It is worth noting that, due to the original servo-constraint equations $\mathbf{c}(\mathbf{q},t) = \mathbf{0}$ are involved in Equations (20), the solution obtained according to the scheme of Equation (21) is free from the constraint violation problem, and the truncation errors do not accumulate in time. More strictly, as said before, m algebraic equations $\mathbf{b}(\mathbf{v},\mathbf{q},\mathbf{u},t) = \mathbf{0}$ impose p independent conditions on $\mathbf{u}$, and $k = m - p$ conditions on the system motion, and in particular on its position $\mathbf{q}$. Therefore, the $m + m$ algebraic equations $\mathbf{b}(\mathbf{v},\mathbf{q},\mathbf{u},t) = \mathbf{0}$ and $\mathbf{c}(\mathbf{q},t) = \mathbf{0}$ represent $k + m = n$ explicit equations on the n coordinates $\mathbf{q}$, and thus the solution $\mathbf{q}_d(t)$ is determined with a numerical accuracy of solving the algebraic equations. Then, only $\mathbf{v}_d(t)$ and $\mathbf{u}_d(t)$ are determined with an error followed from the rough backward difference method, which does not accumulate however as the approximation is based on the numerically exact solutions $\mathbf{q}_d(t)$. The proposed simple code leads thus to reasonable and stable solutions.

The inverse simulation control $\mathbf{u}_d(t)$ obtained as a solution to Equations (20) can be used as a feedforward control for the crane executing a load prescribed motion. It should then be enhanced by a feedback control in order to provide stable tracking of the load trajectory in presence of external perturbations and/or modeling inconsistencies. One possibility is to introduce, instead of $\ddot{\mathbf{c}} = \mathbf{0}$, a stabilized form of the servo-constraint conditions, $\ddot{\mathbf{c}} + \alpha\dot{\mathbf{c}} + \beta\mathbf{c} + \gamma \int \mathbf{c}\, dt = \mathbf{0}$, where α, β and γ are the gain values. Then, after replacing the requirement $\mathbf{0} = \mathbf{b}(\mathbf{q},\mathbf{v},\mathbf{u},t)$ in Equations (20) by its stabilized form

$$\mathbf{0} = \mathbf{C}\mathbf{M}^{-1}(\mathbf{f} - \mathbf{d}) - \mathbf{C}\mathbf{M}^{-1}\mathbf{B}^T\mathbf{u} - \xi + \alpha\dot{\mathbf{c}} + \beta\mathbf{c} + \gamma \int \mathbf{c}\, dt = \mathbf{b}_{stab}(\mathbf{v},\mathbf{q},\mathbf{u},t) \tag{22}$$

a hybrid control can be synthesized from such modified DAEs, using the same solution code as in Equation (21). The idea for the crane control with the use of the scheme is shown in Figure 3.

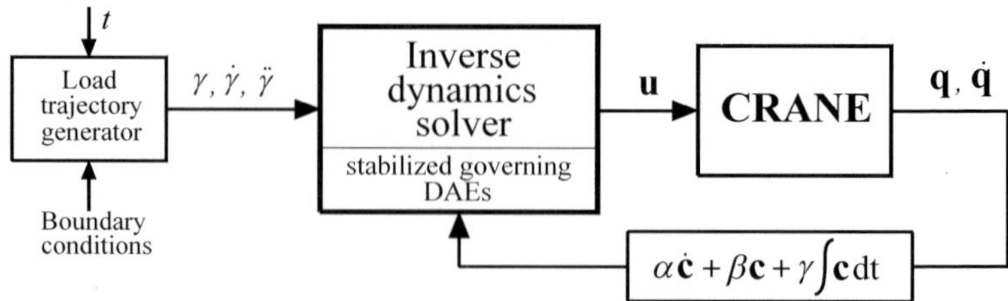

Fig. 3. The idea of the hybrid control scheme.

5 Load Trajectory Modeling

During operation, a duty cycle of a crane consists of moving the load from its initial position to its desired final destination in its working space along a trajectory, avoiding obstacles and sway. This requires motion planning for the load. In the present formulation, the load coordinates in the fixed Cartesian frame *XYZ* need to be specified in time, $\boldsymbol{\gamma}_d(t)=[x_d(t) \quad y_d(t) \quad z_d(t)]^T$, with $x_d(t)$, $y_d(t)$ and $z_d(t)$ being appropriately smooth functions of time. A "rest-to-rest" maneuver is usually required [1,2], i.e. $\dot{\boldsymbol{\gamma}}_d(t_0)=\dot{\boldsymbol{\gamma}}_d(t_f)=0$ and $\ddot{\boldsymbol{\gamma}}_d(t_0)=\ddot{\boldsymbol{\gamma}}_d(t_f)=0$, where t_0 and t_f denote the initial and final times, respectively. One possibility is to synchronize the outputs using a reference function $s(t)$,

$$\boldsymbol{\gamma}_d(t)=\boldsymbol{\gamma}_0+(\boldsymbol{\gamma}_f-\boldsymbol{\gamma}_0)s(t) \tag{23}$$

where $t \in <t_0, t_f>$, and $\boldsymbol{\gamma}_0=[x_0 \;\; y_0 \;\; z_0]^T$ and $\boldsymbol{\gamma}_f=[x_f \;\; y_f \;\; z_f]^T$ are respectively the start and target load positions at time t_0 and t_f. Having $s(t)$ and its time derivatives, the current values $\dot{\boldsymbol{\gamma}}_d(t)$ and $\ddot{\boldsymbol{\gamma}}_d(t)$ used in the mathematical model can be found as $\dot{\boldsymbol{\gamma}}_d(t)=(\boldsymbol{\gamma}_f-\boldsymbol{\gamma}_0)\dot{s}(t)$ and $\ddot{\boldsymbol{\gamma}}_d(t)=(\boldsymbol{\gamma}_f-\boldsymbol{\gamma}_0)\ddot{s}(t)$. For longer traveling distances, the maneuver is often divided into the acceleration (I), steady velocity (II) and deceleration (III) phases. The durations of phases I and III are usually determined by some maximum acceleration/jerk limitations, while the duration of phase II may be consequent to a maximum load velocity limitation. A reasonable proposition for $s(t)$ can be found in [25]. Following the idea contained there, the function $s(t)$ used in this paper was proposed as

$$
\begin{aligned}
s_I(t) &= \frac{1}{\tau-\tau_0}\left(-\frac{5t^8}{2\tau_0^7}+\frac{10t^7}{\tau_0^6}-\frac{14t^6}{\tau_0^5}+\frac{7t^5}{\tau_0^4}\right)\\
s_{II}(t) &= \frac{1}{\tau-\tau_0}\left(t-\frac{\tau_0}{2}\right)\\
s_{III}(t) &= 1+\frac{1}{\tau-\tau_0}\left(\frac{5(\tau-t)^8}{2\tau_0^7}-\frac{10(\tau-t)^7}{\tau_0^6}+\frac{14(\tau-t)^6}{\tau_0^5}-\frac{7(\tau-t)^5}{\tau_0^4}\right)
\end{aligned}
\tag{24}
$$

where $\tau = t_f - t_0$, and τ_0 is the acceleration/deceleration time. Given $\tau = 20\,\text{s}$ and $\tau_0 = 5\,\text{s}$, the reference function defined in Equation (24), and its first and second time derivatives are illustrated in Figure 4.

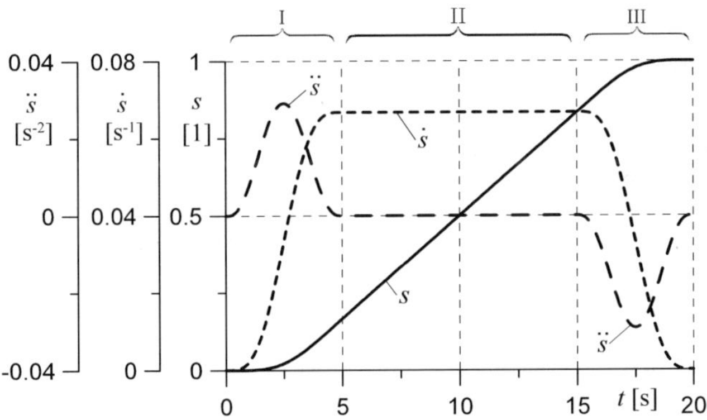

Fig. 4. The reference function and its time derivatives.

The synchronized time functions defined in Equation (23) for the reference load coordinates result in a straight line trajectory from the start to target points. For a rotary crane, the trajectory line cannot cross or pass too close to the tower, however. Therefore, for the tasks in which the start and target load positions are at the opposite sides of the tower, the shape function $s(t)$ should rather specify the load cylindrical coordinates, i.e. $\bar{\gamma}_d(t) = \bar{\gamma}_0 + (\bar{\gamma}_f - \bar{\gamma}_0)s(t)$, $\dot{\bar{\gamma}}_d(t) = (\bar{\gamma}_f - \bar{\gamma}_0)\dot{s}(t)$ and $\ddot{\bar{\gamma}}_d(t) = (\bar{\gamma}_f - \bar{\gamma}_0)\ddot{s}(t)$, where $\bar{\gamma} = [r\ \phi\ z]^T$, and r and ϕ are the polar coordinates. The values $\gamma_d(t)$, $\dot{\gamma}_d(t)$ and $\ddot{\gamma}_d(t)$ required in the model can then be determined following $x_d(t) = r_d(t)\cos\phi_d(t)$ and $y_d(t) = r_d(t)\sin\phi_d(t)$, and the time derivatives of these relations. The two trajectories, the straight line one resulting from Equation (23) and the curvilinear one after imposing $s(t)$ on the polar coordinates, are illustrated in Figure 5.

6 Numerical Simulations

The rotary crane model described in Section 3 served for the simulation purposes and numerical experiments for testing the robustness of developed control strategies. The data used in computations were the following: $m = 100\,\text{kg}$, $m_t = 10\,\text{kg}$, $J_w = 0.1\,\text{kg}\,\text{m}^2$, $r_w = 0.1\,\text{m}$, and $J_b = 480\,\text{kg}\,\text{m}^2$, and the task requirement was to move the load from the start position $\gamma_0 = [5\ \ 0\ \ -5]^T\ \text{m}$ to the target position $\gamma_f = [-2\ \ 2\ \ -2]^T\ \text{m}$, while $\tau = 20\,\text{s}$ and $\tau_0 = 5\,\text{s}$. The time integration step used in simulations was $\Delta t = 0.01\,\text{s}$.

6.1 Inverse simulation study

Two "rest-to-rest" maneuvers were considered: Maneuver 1 along a curvilinear trajectory followed from imposing the reference function from Equation (24) on cylindrical coordinates, and Maneuver 2 along a straight line according to Equation (23). The two reference load trajectories are seen in Figure 5.

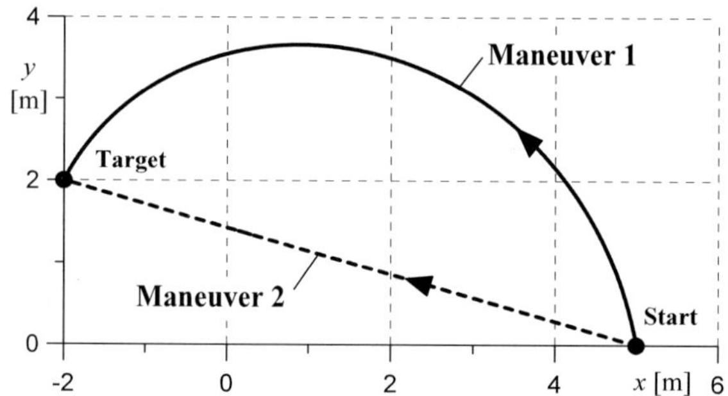

Fig. 5. The straight line and curvilinear trajectories.

Selected results of numerical simulations are seen in Figure 6, obtained for no damping in the system ($D_\varphi = D_\alpha = 0$, $D_s = 0$). Note that only the variations of $l(t)$ and $M_w(t)$ are almost the same for the two maneuvers. All the other motion and control characteristics are qualitatively and quantitatively different. The (nominal) control obtained from the inverse simulation was then used for the direct simulation. The motion pattern of the crane and the load trajectory was repeated with a numerical accuracy.

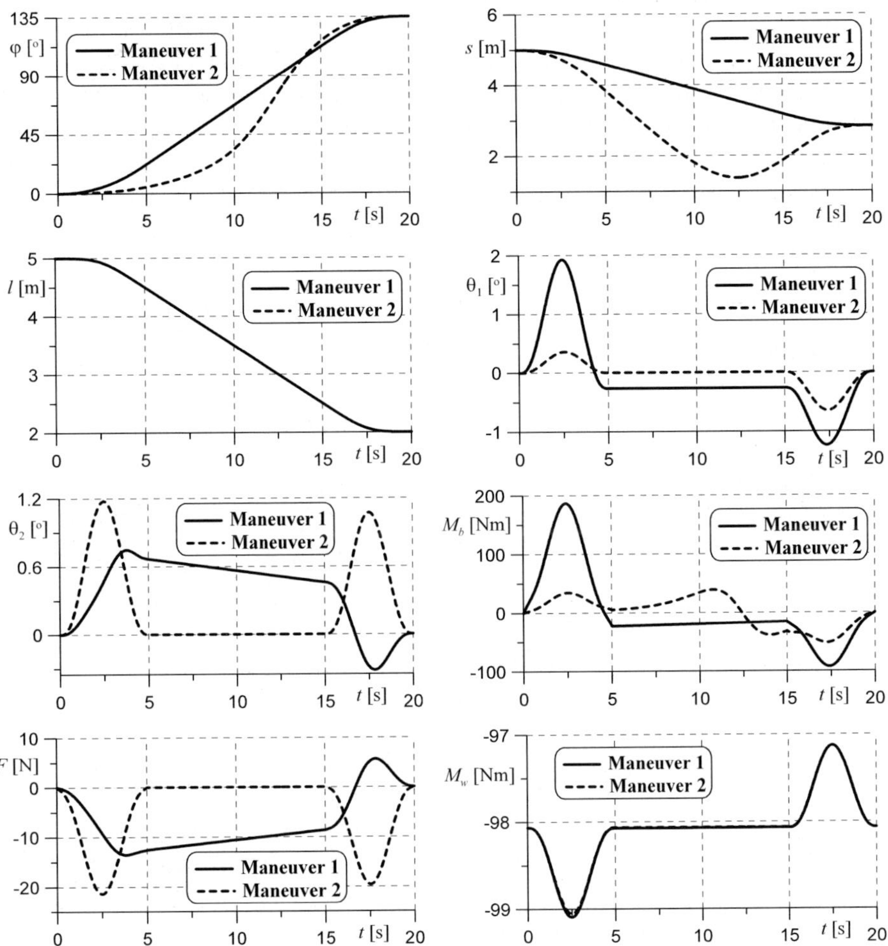

Fig. 6. Crane motion and control of the crane executing Maneuvers 1 and 2.

6.2 Robustness of the hybrid control in perturbed motion

The robustness of the hybrid control proposed in this paper was first tested by applying the inconsistent starting position of the load at t_0, which was placed 0.5 m above its reference position. Moreover, in the mathematical model used for the direct dynamic simulation, the damping related to s and l motions were added, respectively $D_s = 75$ Ns/m and $D_\alpha = 15$ Ns, which were set to zero in the model used for the determination of control. The motion disturbed this way was then stabilized along the reference motion by using the hybrid control. The gain values used assure the critical damping for a PID scheme, i.e. $\alpha^2 = 8\beta$, $32\gamma = \alpha\beta$ [26], and a good choice for β with $\Delta t = 0.01$ s was $\beta = 10$.

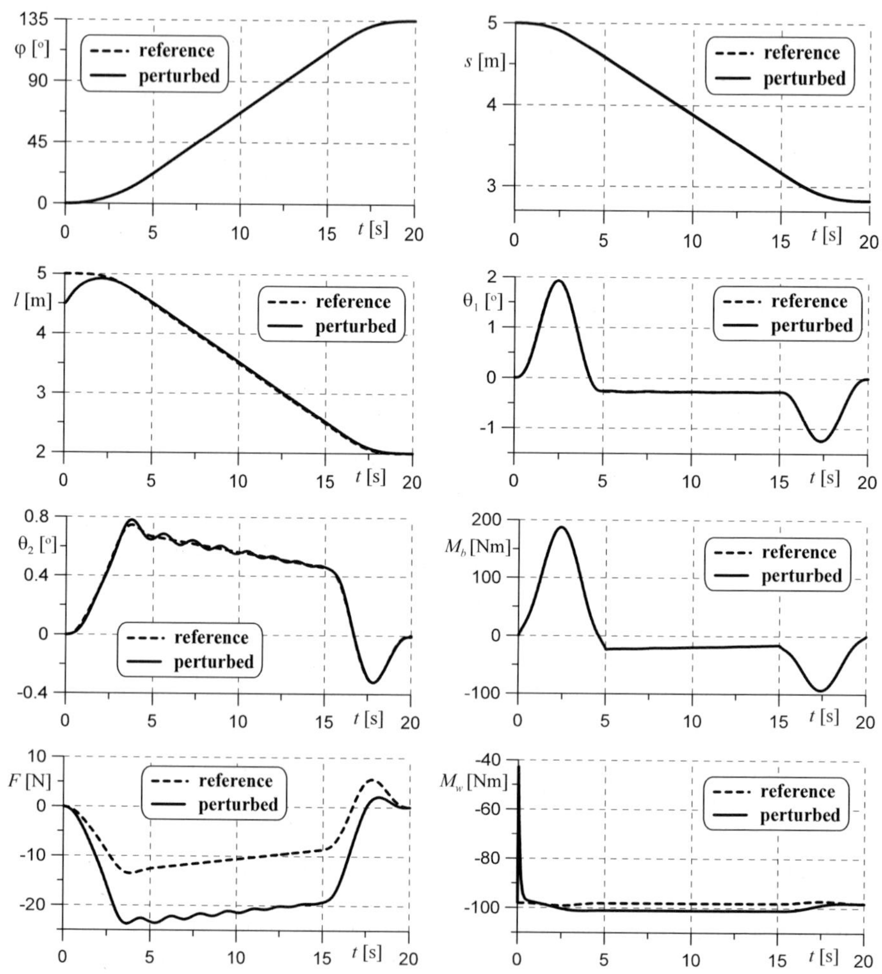

Fig. 7. Simulation of Maneuver 1 in motion perturbed by inconsistent initial load position and the modeling inconsistencies.

Selected results of simulation of Maneuver 1 in the perturbed motion are presented in Fig. 7. It can be seen that there is almost no difference in $\varphi(t)$, $s(t)$, $\theta_1(t)$ and $\theta_2(t)$ between the perturbed and reference motions, and the command $M_b(t)$ is not changed either. The initial difference in the vertical load position is quickly damped to the reference values, and this is achieved by an appropriate abrupt change in M_w value at the beginning of simulation. Then, the modeling inconsistency (additional damping in *s* and *l* motions) is compensated by appropriate differences in $F(t)$ and $M_w(t)$ relative the reference control, and the differences vanish at the target point when $\dot{s} \to 0$ and $\dot{\alpha} = \dot{l}/r_w \to 0$ (the additional damping vanishes).

As seen from Equation (22), the feedback control enhanced in the hybrid control is aimed at minimizing the violations of servo-constraints. The initial inconsistency in the load vertical position is thus effectively damped to zero. By contrast, the compensation of the modeling inconsistencies requires some constraint violation to produce the additional control commands related to $\alpha\dot{\mathbf{c}} + \beta\mathbf{c} + \gamma\int\mathbf{c}\,dt$. Both, the damping features and the range of constraint violations required to overcome modeling inconsistencies are closely related to the values of gain coefficients. In the present experiments the same gain values for all three outputs were used, and as said $\alpha^2 = 8\beta$ and $32\gamma = \alpha\beta$. In Figure 8 the effect of different gain values used is seen. The bigger the gain values, the realization of the perturbed motion is closer to the reference motion. On the other hand, too large gain values lead to instability in simulation, which is a well-known effect in the stabilization of passive constraints as well [26]. The gain values limits and/or their optimal values are dependent mainly on the integration time step Δt, and smaller integration time steps allow for bigger gain values. They are also closely related the system complexity/dimensionality and the type of motion/perturbations simulated. The choice of appropriate gain values is often a trial-and-error process, and disparate gain values for particular outputs are usually involved.

The other numerical experiment relates to external perturbations caused by an additional force applied to the load in the negative sense of *X* direction, not considered in the model used for the determination of control. The force time-profile, which can be regarded as a rough model of a wind blow loads on the payload, is seen in Figure 9. The motion disturbed this way was stabilized along the reference motion of maneuver 2 by using the hybrid control. The gain values used were $\alpha^2 = 8\beta$, $32\gamma = \alpha\beta$, and $\beta = 30$ was chosen.

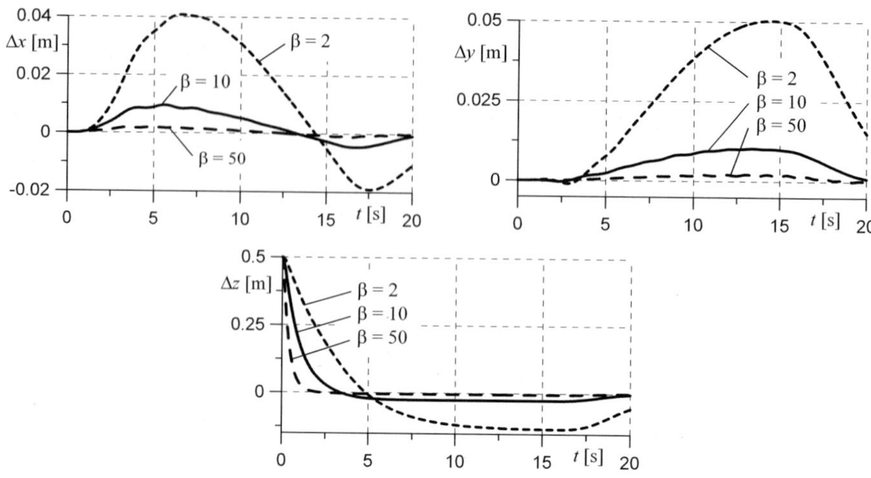

Fig. 8. The differences in the load position in the perturbed motion for different gain values.

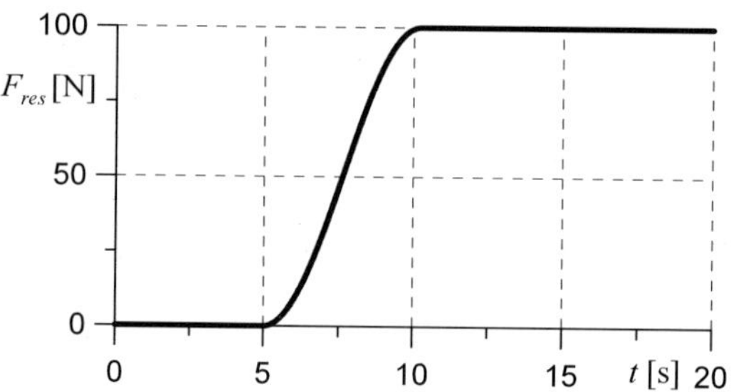

Fig. 9. The perturbation force profile.

The simulation results seen in Figure 10 show that the 'wind load' F_{res} causes the rope bends accordingly, and the trolley position and bridge rotation angle differ slightly from the reference values so that to compensate the rope additional bend and to execute the load prescribed motion. The same happens at the target load position, the rope is out of plumb, and the final trolley position and bridge rotation angle differ from their reference values. The 'wind loads' effects are compensated by appropriate changes in the control commands, which cause that the servo-constraints are realized with a limited accuracy. More strictly, the constraint violations in *Y* and *Z* directions do not exceed $\pm 0.0004\,\mathrm{m}$ during the whole motion, while the violation in *X* direction, required to produce the required feedback control commands, is a up to $3.3\,\mathrm{cm}$. This inaccuracy in the load position can be enlarged/diminished by applying smaller/bigger gain values, see Figure 11.

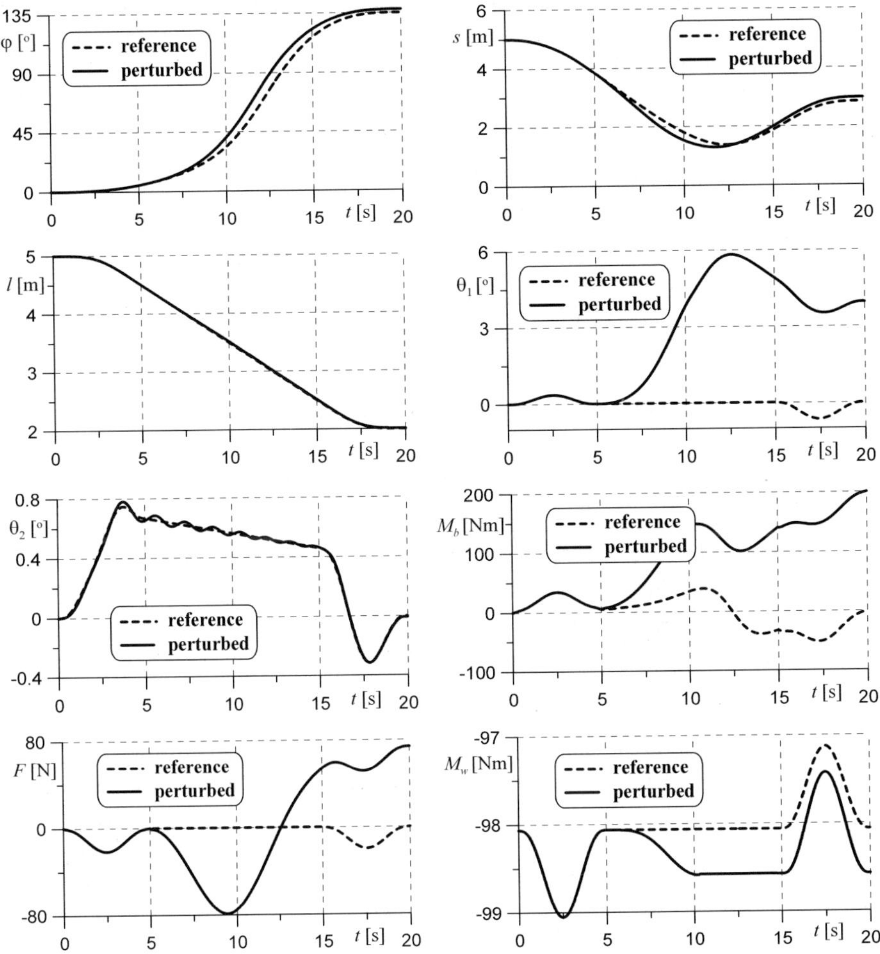

Fig. 10. Simulation of Maneuver 2 in motion perturbed by the "wind load".

7 Conclusions

This work presents a mathematical model for the dynamic analysis and control synthesis of cranes executing prescribed motions of payloads. The developed DAE formulation holds good for a variety of cranes (overhead cranes, rotary cranes, boom cranes) in which the payload is modeled as a point mass suspended by a massless and inextensible cable. The cranes belong to underactuated mechanical systems, in which the number of control inputs is smaller than the number of degrees of freedom.

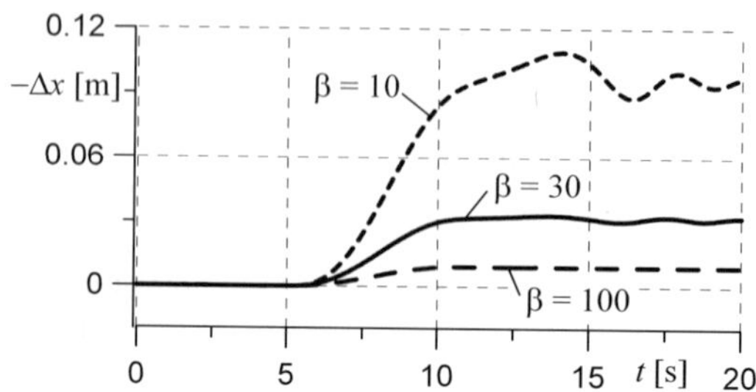

Fig. 11. The difference in the load x-position in the wind-affected motion.

The problem of realization of the load prescribed motion is viewed from the perspective of constrained motion. The m control outputs (specified in time load coordinates), expressed in terms of the n system coordinates, are treated as servo-constraints on the system. Mixed orthogonal-tangent realization of the servo-constraints by the available crane control is discovered, and the tangent realization leads to $n-m$ additional conditions on the crane motion. In this sense, the motion of the cranes can be explicitly prescribed by the m outputs, and explicitly controlled by the available m control inputs.

The governing equations for the dynamics and control of the crane executing a load prescribed motion are formulated as index-three DAEs in the crane state variables and control variables. An simple, effective and stable method for solving the governing DAEs were proposed and tested through numerical experiments.

The solution to the governing DAEs are motion characteristics of the crane executing the prescribed load trajectory and the control commands ensuring the motion realization. The obtained feedforward control law was then enhanced by a closed-loop control strategy with feedback of the actual errors in the load position. The hybrid control law is determined using the governing DAEs modified slightly to the stabilized form involving a PID scheme for the load position errors.

The imposed load trajectory are specified in time load coordinates, modeled as a rest-to-rest maneuver with the use of time dependent reference function $s(t)$. A possible extension of the approach is to sketch the trajectory with a set of successive points in space and then to interpolate/approximate it by spline functions, using $s(t)$ as a specified time-function of the arc length parameter.

The developed mathematical model was tested through numerical simulations. It was shown that the proposed codes enable one for effective inverse simulation studies related to a wide range of load trajectories. Robustness of the hybrid control law in the motion perturbed by an inconsistent initial load position, some modeling inconsistencies, and some external perturbations, was proved.

References

1. Rosenfeld Y, Shapira A (1998) Automation of Existing Tower Cranes: Economic and Technological Feasibility. Automation in Construction 7(4): 285-298
2. Abdel-Rahman EM, Nayfeh AH, Masoud ZN (2003) Dynamics and Control of Cranes: A Review. Journal of Vibration and Control 9(7): 863-908
3. Corriga G, Giua A, Usai G (1998) An Implicit Gain-Scheduling Controller for Cranes. IEEE Transactions on Control Systems Technology 6(1): 15-20
4. Cho S-K, Lee H-H (2002) A Fuzzy-Logic Antiswing Controller for Three-Dimensional Overhead Cranes. ISA Transactions 41(2): 235-243
5. Fliess M, Lévine J, Martins P, Rouchon P (1995) Flatness and Defect of Non-Linear Systems: Introductory Theory and Examples. International Journal of Control 61(6): 1327-1361
6. Bourdache-Siguerdidjane H (1995) Optimal Control of a Container Crane by Fliess Linearization. Journal of Computer and Systems Sciences International 33(5): 82-88
7. Maier T, Woernle C (1999) Flatness-Based Control of Underactuated Cable Suspension Manipulators. In Proceedings of DETC'99, ASME Design Engineering Technical Conference, September 12-15, Las Vegas, Nevada, USA (DETC99/VIB-8223)
8. Blajer W, Kolodziejczyk K (2004) A Geometric Approach to Solving Problems of Control Constraints: Theory and a DAE Framework. Multibody System Dynamics 11(4): 343-364
9. Bajodah AH, Hodges DH, Chen Y-H (2005) Inverse Dynamics of Servo-Constraints Based on the Generalized Inverse. Nonlinear Dynamics 39(1/2): 179-196
10. Ghigliazza RM, Holmes P (2002) On the Dynamics of Cranes, or Spherical Pendula with Moving Supports. International Journal of Non-Linear Mechanics 37(7):1211-1221
11. Sakawa Y, Nakazumi A (1985) Modeling and Control of a Rotary Crane. Journal of Dynamic Systems, Measurement, and Control 107(2): 200-206
12. Lee H-H (1998) Modeling and Control of a Three-Dimensional Overhead Crane. Journal of Dynamic Systems, Measurement, and Control 120(4): 471-476
13. Chin C, Nayfeh AH, Abdel-Rahman EM (2001) Nonlinear Dynamics of a Boom Crane. Journal of Vibration and Control 7(2): 199-220
14. Sawodny O, Aschemann H, Lahres S (2000) An Automated Gantry Crane as a Large Workspace Robot. Control Engineering Practice 10(10): 1323-1338
15. Rosen A (1999) Applying the Lagrange Method to Solve Problems of Control Constraints. Journal of Applied Mechanics 66(4): 1013-1015
16 Blajer W (1997) Dynamics and Control of Mechanical Systems in Partly Specified Motion. Journal of the Franklin Institute 344B(3): 407-426
17. Blajer W (2001) A Geometrical Interpretation and Uniform Matrix Formulation of Multibody System Dynamics. ZAMM 81(4): 247-259
18. Schiehlen W (1997) Multibody System Dynamics: Roots and Perspectives. Multibody System Dynamics 1(2): 149–188
19. Gear CW, Petzold LR (1984) ODE Methods for the Solution of Differential/Algebraic Equations. SIAM Journal of Numerical Analysis 21(4): 716-728

20. Brenan KE, Campbell SL, Petzold LR (1989) Numerical Solution of Initial-Value Problems in Differential-Algebraic Equations. Elsevier, New York.
21. Gear CW (1990) An Introduction to Numerical Methods of ODEs and DAEs. In Haug EJ and Deyo RC (eds.) Real-Time Integration Methods for Mechanical System Simulations, NATO ASI Series, Vol. F69, Springer-Verlag, Berlin: 115-126.
22. Petzold LR (1990) Methods and Software for Differential-Algebraic Systems. In Haug EJ and Deyo RC (eds.) Real-Time Integration Methods for Mechanical System Simulations, NATO ASI Series, Vol. F69, Springer-Verlag, Berlin: 127-140.
23. Ascher UM, Petzold LR (1998) Computer Methods for Ordinary Differential Equations and Differential-Algebraic Equations. SIAM, Philadelphia.
24. Wehage RA, Haug EJ (1982) Generalized Coordinate Partitioning for Dimension Reduction in Analysis of Constrained Dynamic Systems. Journal of Mechanical Design 116(4): 1058-1064.
25. Aschemann H (2002) Optimale Trajektrienplanung sowie modelgestützte Steuerung für einen Brückenkran. Fortschritt-Berichte VDI, Reihe 8: Meß-, Steuerungs- und Regelungstechnik, Nr. 929, Düsseldorf.
26. Ostermayer G-P. On Baugarte Stabilization for Differential Algebraic Equations. In Haug EJ and Deyo RC (eds.) Real-Time Integration Methods for Mechanical System Simulations, NATO ASI Series, Vol. F69, Springer-Verlag, Berlin: 193-207.

Neural-Augmented Planning and Tracking Pilots for Maneuvering Multibody Dynamics

Carlo L. Bottasso, Alessandro Croce and Domenico Leonello

Dipartimento di Ingegneria Aerospaziale, Politecnico di Milano, Via La Masa 34, 20156 Milano, Italy; E-mail: {carlo.bottasso,alessandro.croce,domenico.leonello}@polimi.it

Abstract We propose a methodology for extending the applicability of multibody-based comprehensive analysis codes to the maneuvering regime, with specific application to the flight of rotorcraft vehicles.

Maneuvers are here mathematically described in a concise yet completely general form as optimal control problems, each maneuver being defined by a specific form of the cost function and by suitable constraints on the vehicle states and controls. In principle, by solving the maneuver optimal control problem, one could determine the trajectory and the control time histories that steer the vehicle model, while minimizing the cost and satisfying the constraints. Unfortunately, optimal control problems are prohibitively expensive to solve for detailed comprehensive models of rotorcraft vehicles denoted by a large number of structural degrees of freedom and possibly sophisticated aerodynamics.

In order to make the problem computationally tractable, our formulation makes use of two models of the same vehicle. A coarse level flight mechanics model is used for solving the trajectory optimal control problem. Being based on a reduced model of the vehicle with only a few degrees of freedom, the resulting non-linear multi-point boundary value problem is computationally feasible. Next, the fine scale comprehensive model is steered in closed loop, tracking the trajectory computed at the flight mechanics level using a receding horizon model predictive controller. This amounts to a standard time marching problem for the comprehensive model, which is therefore also computationally feasible. The flight mechanics model is iteratively updated for ensuring close matching of the trajectories flown by the two models, by resorting to a neural adaptive element. This two-level procedure enables the simulation using comprehensive models of arbitrary complexity of maneuvers of possibly long duration, with general constraints on the vehicle inputs and outputs.

The new procedures are demonstrated with the help of numerical applications.

1 Background and Motivation

Modern comprehensive rotorcraft modeling tools are geared towards the evaluation of performance, vibrations, loads, stability and response of rotorcraft systems. These

J.C. García Orden et al. (eds), Multibody Dynamics. Computational Methods and Applications, 113–135.

solution procedures provide computer implementations of high fidelity aeroelastic mathematical models of the vehicle, which are useful for solving simulation problems relevant to all phases of the design and testing processes [1, 17, 20, 25]. Great progress has been made in recent years towards the comprehensive simulation of rotorcraft, mainly due to the continuous advancement in the structural dynamics and aerodynamics computational kernels. State of the art procedures are now based on non-linear dynamics formulations, such as multibody finite element based methods, which provide the ability to model in detail the most complex part of the vehicle, the rotor system. Furthermore, modern codes implement a variety of specific rotorcraft aerodynamic models, including wake and stall models, unsteady aerodynamics and multiple lifting surface interactions. Computational fluid dynamics (CFD) tools are also attracting increasing interest.

These high fidelity aeroelastic mathematical models of rotorcraft systems are currently primarily focused on the analysis of the hover and forward flight regimes. For example, detailed aeroelastic models can be used for evaluating the flutter boundaries and the vibratory levels in steady trimmed flight. Specialized procedures are available to determine the constant-in-time control inputs that trim the aircraft model, typically either in wind-tunnel or free-flight modes [13]. On the other hand, the study and simulation of maneuvering flight is typically performed only with flight mechanics models [10, 16]. These models have far fewer degrees of freedom than the aeroelastic models. In fact, the vehicle is often modeled as a rigid body and the rotor is typically described using blade element theory with wake corrections. The aeroelastic and flight mechanics models are two mathematical idealizations of the same physical system, that however differ on the scale resolution. While the high fidelity aeroelastic models are able to render fine scale details of the solution, as for example the time response of each single blade, the flight mechanics models are blind to these small scales. However, they still capture the coarser scales of the physical processes involved in the gross motion of the vehicle, and in this sense are able to synthesize its flight mechanics characteristics.

Helicopters and tilt-rotors perform complex, highly dynamic and often three-dimensional maneuvers, both in normal operating conditions and during emergencies. Maximum loads and other limiting factors ranging from vibrations to noise are often encountered when operating in the unsteady regime or near the boundaries of the flight envelope, as shown for example in [14] with respect to the noise emission characteristics. More often than not, the critical quantities of interest in maneuvering flight are captured only on fine scale (aeroelastic) models, rather than coarse (flight mechanics) ones. *However, current comprehensive codes have very limited capabilities in their ability to fly specific maneuvers.* Therefore, there is a need to extend the applicability of comprehensive codes to the unsteady flight regime.

Two main areas need to be substantially improved for enabling accurate simulation of maneuvering rotorcraft:

1. Comprehensive codes need to be coupled with time-accurate airload models that can capture, among other effects, the distortion of the wake geometry caused by the maneuver, the interaction of the shed vortex filaments with the fuselage and the tail rotor, dynamic stall effects on the retreating blade, etc. The recent work of Ribera and Celi [24] goes exactly in this direction, proposing the coupling of the time-accurate free-wake model of Bhagwat and Leishman [5] with a flexible blade comprehensive code [26]. A free-wake model applicable to maneuvering flight is also described in [27], and has been included in CHARM [17]. First principle CFD approaches would also be appropriate for capturing the relevant aerodynamic effects during maneuvers, although the computing cost of fully resolved time-accurate simulations for the hundreds or thousands of rotor revolutions that take place during a typical maneuver are still prohibitive and unrealistically large to be profitably used in the design process.
2. Comprehensive multibody-based codes need to be augmented with procedures for computing the controls that pilot the virtual model along a given maneuver (Maneuvering Multibody Dynamics, MMBD [9]). In fact, presently these codes are primarily intended for the sole computation of the system response resulting from *known* time histories of the control inputs. For example, time marching simulations of an arrested descent and of a roll reversal were described in [24], in both cases using prescribed values of the stick inputs. Similarly, Brentner et al. [14] simulated a pull-up using CAMRAD [20], again with given control inputs. Simulations based on pre-assigned inputs can produce invaluable information on the transient behavior of the vehicle, and perfectly answered the specific goals of the cited references, but it is clear that this approach is somewhat limited to fairly simple maneuvers. In fact, *time dependent* control inputs that fly high fidelity virtual models of rotorcraft along complex, aggressive and three-dimensional maneuvers are, in general, very difficult if not impossible to determine based on simple, trial and error procedures. The use of known, experimentally measured controls might alleviate but will not solve this problem. In fact, due to inevitable inaccuracies present in even the most sophisticated simulations, the rotorcraft model will not be capable of following the trajectory flown by the actual rotorcraft during the flight test. This fact is by now well known for the case of steady forward flight: when comparing predictions with experimental data, better correlation is obtained when the model is trimmed, i.e. when the model produces the same forces and moments as the actual rotor, than when identical control inputs are used.

Since comprehensive codes are primarily design and testing tools, the lack of specific maneuver modeling capabilities represents a significant limitation of the current state of the art of rotorcraft computer assisted simulation. In the present work, we try to address this problem, exclusively with respect to the latter of the two central issues mentioned above, by proposing a general procedure for the determination of the time dependent control inputs that will fly an aeroelastic rotorcraft model along a maneuver. The proposed methodology is general and is applicable to any aeroelastic rotorcraft simulation tool, although it will be here demonstrated for the multibody finite element approach of Bauchau et al. [1]. Furthermore, the same framework can be applied to maneuvering multibody dynamics problems involving vehicles other than rotorcraft, such as for example fixed wing aircrafts, automobiles, motorcycles or sailboats.

2 Solution Procedures for Maneuvering Flight

In order to design computational procedures for the problem of maneuvering flight, it is first necessary to provide a way of mathematically defining maneuvers. To address this problem, we define here a maneuver as a time history of control inputs and the resulting associated time history of vehicle states that take the vehicle model from an initial state to a final one, the latter possibly known only in part, according to some criterion and while satisfying the equations of dynamic equilibrium and all the necessary input and output constraints (limits on actuator authority, flight envelope boundaries, etc.).

The controls that will fly a maneuver are a-priori unknown and need to be determined. A constructive way of accomplishing this goal is through the solution of an optimal control problem [15]. The optimal control problem is defined in terms of a cost function, which is typically a vehicle performance index, that specifies the criterion used for flying the maneuver (minimum power, minimum time, etc.). The equations of motion of the vehicle are regarded as constraints of the problem, which is in general also subjected to various additional input and output constraints that complete the definition of the maneuver, and, for example, translate the flight envelope limitations of the aircraft and all the necessary safety and operational requirements. The solution of this optimal control problem yields the control time histories that fly the vehicle according to the prescribed criteria, together with the complete flight path. A possible example would be the determination of a 30 degree turn in minimum time (which represents the cost function, in this case), without exceeding the operational limits of the aircraft and without exceeding a given loss of altitude (which, together with the equations of motion of the vehicle, represent the problem constraints). Another cost function that is often used in flight mechanics for the purpose of defining maneuvers is some norm of the control deflections from a trim state; this situation would correspond to a maneuver with minimum "control effort".

This approach provides a general procedure for defining arbitrary maneuvers and, at the same time, computing time histories of vehicle states which are compatible with the associated control time histories. The solution of optimal control problems is however potentially expensive, especially if detailed models of high dimensionality are used. Therefore this approach, although perfectly suited for studying maneuvers in a purely flight mechanics setting as done, for example, in [16, 10], is not directly applicable per se to high fidelity aeroelastic models of rotorcraft systems. Furthermore, controllability issues might lead to ill-posed problems, for example when the system is elastic.

To address these issues, we propose here the Multi-Model Steering Algorithm (MMSA). This approach blends the technology of comprehensive multibody-based codes with flight mechanics models. In fact, two models of the same vehicle are used: the coarse scales are represented by a "reduced" flight mechanics model, while the fine scales are captured by an aeroelastic comprehensive model of the same aircraft. The maneuver optimal control problem is solved at the coarse flight mechanics level, and it is therefore inexpensive. Since steering in open-loop is prone to instabilities, we use a receding horizon formulation of MMSA, that can be interpreted as an application of model-based predictive control [18]. The idea is to track the trajectory generated at the flight mechanics level with the aeroelastic model. To this effect, an open-loop optimal control problem is solved over a finite horizon using the coarse model. The resulting control policy is implemented in the aeroelastic model for a short period of time, until a new optimization problem is solved again on a time-shifted horizon. The repeated application of open-loop optimal control brings feedback into the system, allowing the aeroelastic model to track the computed trajectory. Since at the fine level the control time histories are known from the solution of the coarse problem over the prediction window, the fine level solution becomes a classical forward dynamics integration, and it is therefore also of acceptable computational cost. The coarse reduced model is progressively adapted in order to guarantee small tracking errors. Adaption is based on the on-line training of a neural element that is used to augment the reduced model.

In the receding horizon formulation of MMSA, the reduced flight mechanics model plays a double role: it is used at the motion planning level for producing the reference trajectory, and it is also used at the trajectory tracking level for implementing the model predictive controller. Notice that arbitrary input and output constraints can be handled in a straightforward manner at both of these stages of the solution. This decomposition of the problem in two layers, path planning and path tracking, mimics the architecture of modern control systems [19], and enables the simulation of complex maneuvers of long duration in the proximity of the flight envelope.

This work is organized as follows. At first we describe the vehicle models in Section 3; the reduced flight mechanics model is presented in Section 3.1, the fine scale aeroelastic model in Section 3.2, and the relationship between the two in Section 3.3. Next, Section 4 discusses the solution of maneuver optimal control problems using flight mechanics models by means of a direct transcription approach. Section 5 presents the MMSA, while in Section 6 we discuss the on-line adaption of the reduced model of Section 3.1 to improve its predictive capabilities. Finally, a numerical example in Section 7 and conclusions and outlook in Section 8 complete the paper.

3 Maneuvering Rotorcraft Models

3.1 Flight mechanics models of rotorcraft vehicles

In this work we consider the two-dimensional longitudinal flight mechanics model described in detail in [11]. The model is valid for both helicopters and tilt-rotors and, being purely longitudinal, it is clearly limited to null or low turn rates. The dynamic equilibrium conditions expressed in a fixed inertial system write

$$\dot{V}_X = \frac{1}{m}(N_r F_{X_{MR}} + F_{X_{TR}} + F_{X_A}), \tag{1a}$$

$$\dot{V}_Z = \frac{1}{m}(N_r F_{Z_{MR}} + F_{Z_{TR}} + F_{Z_A} + mg), \tag{1b}$$

$$\dot{q} = \frac{1}{I}(N_r M_{Y_{MR}} + M_{Y_{TR}} + M_{Y_A}), \tag{1c}$$

$$\dot{X} = V_X, \tag{1d}$$

$$\dot{Z} = V_Z, \tag{1e}$$

$$\dot{\Theta} = q, \tag{1f}$$

where X, Z (positive downward) are the components of the position vector of the vehicle center of gravity, V_X and V_Z their time rates, Θ (positive nose up) is the pitch angle and q the pitch rate, while m and I are the aircraft mass and pitch moment of inertia, respectively, and g is the acceleration of gravity. $F_{X_{MR}}$, $F_{Z_{MR}}$, $M_{Y_{MR}}$ are the components of the forces and moments generated by each of the N_r main rotors, i.e. rotors that generate thrust in the longitudinal plane of the aircraft. Similarly, $F_{X_{TR}}$, $F_{Z_{TR}}$, $M_{Y_{TR}}$ are the force and moment components of the tail rotor. For a helicopter $N_r = 1$, while for a tilt-rotor $N_r = 2$ and $F_{X_{TR}} = F_{Z_{TR}} = M_{Y_{TR}} = 0$. Finally, F_{X_A}, F_{Z_A} and M_{Y_A} are the forces and moment generated by all other aerodynamic surfaces of the vehicle.

For helicopters it is necessary to include the effects of the tail rotor thrust to balance the main rotor torque, in order to accurately evaluate the total power required for flight. To this end, the helicopter model is enriched by three (approximate) algebraic equations expressed in the body attached frame (x, y, z) that enforce roll, yaw and lateral equilibrium of the vehicle:

$$F_{y_{MR}} + F_{y_{TR}} + mg \sin \Phi = 0, \tag{2a}$$

$$M_{z_{MR}} + M_{z_{TR}} + M_{z_V} = 0, \tag{2b}$$

$$M_{x_{MR}} + M_{x_{TR}} + M_{x_V} = 0, \tag{2c}$$

where Φ is the aircraft bank angle.

The vehicle equations of equilibrium (1a–1f, 2a–2c) are augmented by a power balance equation, that writes

$$\dot{\omega} = \frac{\eta_{MR}}{J_p} \left(\frac{1}{N_r} \frac{P}{\omega} - \frac{Q_{MR}}{\eta_{MR}} - r_t \frac{Q_{TR}}{\eta_{TR}} \right), \tag{3}$$

where ω is the main rotor angular velocity and J_p its polar moment of inertia, $P(t)$ is the power produced by the engine(s) at time t. Q_{MR}, Q_{TR} are the torques due to the generic main and tail rotors, respectively, and η_{MR}, η_{TR} the mechanical efficiency of their transmissions; finally, r_t is the ratio between tail and main rotor rotational speeds. Clearly, for a tilt-rotor, we set $Q_{TR} = 0$.

The rotor forces and torques are expressed in terms of the piloting controls using classical blade element theory [21]. The detailed expressions of these quantities are here omitted for brevity, but can be found in [11]. These aerodynamic constitutive equations complement the equations of equilibrium and the power balance equations, and allow one to write the reference model of the vehicle in compact form as

$$\boldsymbol{f}_{\text{ref}}(\dot{\boldsymbol{y}}, \boldsymbol{y}, \boldsymbol{u}) = 0, \tag{4}$$

where $\boldsymbol{y} \in \mathbb{R}^{n_{s,FM}}$ is the set of flight mechanics state variables, $\boldsymbol{u} \in \mathbb{R}^{n_{u,FM}}$ are the flight mechanics controls, and $\boldsymbol{f}_{\text{ref}} : \mathbb{R}^{n_{s,FM}} \times \mathbb{R}^{n_{s,FM}} \times \mathbb{R}^{n_{u,FM}} \rightarrow \mathbb{R}^{n_{s,FM}}$. The vehicle state vector is defined as $\boldsymbol{y} = (X, Z, \Theta, V_X, V_Z, q, \omega)$, $n_{s,FM} = 7$, while the controls are represented for a helicopter by $\boldsymbol{u} = (\theta_{0_{MR}}, \theta_{0_{TR}}, A_1, B_1, P)$, $n_{u,FM} = 5$. For a tilt-rotor we have the additional controls δ_H for the horizontal stabilizer and i_m for the nacelle tilt, but no tail rotor collective $\theta_{0_{TR}}$, so that $n_{u,FM} = 6$ in this case.

The analytical model expressed by equation (4) is here termed a *reference* model, in the sense that it can be used for computing a first approximation to the flight mechanics behavior of the fine scale aeroelastic model described in the next section. The reference model will be later on augmented in Section 6 with an adaptive element, with the goal of improving its prediction capabilities.

3.2 Multibody modeling of rotorcraft vehicles

In this work, the fine scale aeroelastic modeling of rotorcraft is based on the comprehensive multibody dynamics analysis code described in [1]. The formulation is cast within the framework of non-linear finite element based multibody dynamics methods, and the element library includes rigid and deformable bodies, joint elements, including unilateral contact conditions, active element models, including engine and actuator models, and sensors and controls [2, 3, 7, 8].

The finite element formulation of flexible structural elements includes both non-linear models and modal-based approaches. The formulations of beams and shells are composite-ready, i.e. they support the modeling of complex cross sections made of laminated composite materials, and are geometrically exact, i.e. they account for arbitrarily large displacements and finite rotations, but are limited to small strains [2]. Structural modes can be imported from external general finite element codes and can be connected with the rest of the multibody model.

The equations of equilibrium are written in a Cartesian inertial frame. Constraints are modeled using the Lagrange multiplier technique, which leads to systems of equations that are highly sparse, although not of minimal size. The crucial advantage of this approach is that it can treat arbitrarily complex aircraft configurations. After spatial discretization of the flexible components using the finite element method, the equations of dynamic equilibrium can be written as

$$\frac{\mathrm{d}(\boldsymbol{M}\boldsymbol{w})}{\mathrm{d}t} - \boldsymbol{f}_i - \boldsymbol{c}_{,\boldsymbol{q}}\boldsymbol{\lambda} - \boldsymbol{f}_e = 0, \tag{5a}$$

$$\boldsymbol{N}\dot{\boldsymbol{q}} - \boldsymbol{w} = 0, \tag{5b}$$

$$\boldsymbol{c} = 0, \tag{5c}$$

where $\boldsymbol{q} \in \mathbb{R}^{n_q}$ are the generalized coordinates, $\boldsymbol{w} \in \mathbb{R}^{n_q}$ the velocities, $\boldsymbol{p} = \boldsymbol{M}(\boldsymbol{q})\boldsymbol{w} : \mathbb{R}^{n_q} \times \mathbb{R}^{n_q} \to \mathbb{R}^{n_q}$ the system momenta, $\boldsymbol{f}_i(\boldsymbol{q}, \boldsymbol{w}) : \mathbb{R}^{n_q} \times \mathbb{R}^{n_q} \to \mathbb{R}^{n_q}$ the discretized internal and inertial forces, $\boldsymbol{f}_e(\boldsymbol{q}, \boldsymbol{w}, \boldsymbol{u}, t) : \mathbb{R}^{n_q} \times \mathbb{R}^{n_q} \times \mathbb{R}^{n_q} \times \mathbb{R}_+ \to \mathbb{R}^{n_q}$ the external forces, that also include the effects of the system controls $\boldsymbol{u} \in \mathbb{R}^{n_u}$, $\boldsymbol{c}(\boldsymbol{q}, t) : \mathbb{R}^{n_q} \times \mathbb{R}_+ \to \mathbb{R}^{n_c}$ are the holonomic constraints that model the mechanical joints of the system, and finally $\boldsymbol{\lambda} \in \mathbb{R}^{n_c}$ the associated Lagrange multipliers. The generalized coordinates are composed of linear displacements $\boldsymbol{d} \in \mathbb{R}^{n_l}$ and rotation parameters $\boldsymbol{r} \in \mathbb{R}^{n_r}$, $\boldsymbol{q} = (\boldsymbol{d}^T, \boldsymbol{r}^T)^T$, $n_l + n_r = n_q$. Similarly, the generalized velocities are $\boldsymbol{w} = (\boldsymbol{v}^T, \boldsymbol{\omega}^T)^T$, where $\boldsymbol{v} \in \mathbb{R}^{n_l}$ are the linear velocities and $\boldsymbol{\omega} \in \mathbb{R}^{n_r}$ the angular velocities. Then matrix $\boldsymbol{N}$ is defined as

$$\boldsymbol{N} = \begin{bmatrix} \boldsymbol{I} & 0 \\ 0 & \boldsymbol{S} \end{bmatrix}, \tag{6}$$

where $\boldsymbol{S}$ is a parameterization dependent matrix that for each node in the system relates the time rates of the rotation parameters $\boldsymbol{r}$ to the angular velocities $\boldsymbol{\omega}$. The case of non-holonomic constraints is not covered here for the sake of brevity, but can be easily addressed.

The code includes several solution procedures, including static analyses under steady external, aerodynamic and inertial loads, dynamic response from given initial conditions and prescribed loads and control inputs $\boldsymbol{u}(t)$, stability and flutter, and trim analyses. The numerical integration in time of the equations of motion (5a, 5c) is based on an energy decaying scheme that ensures unconditional numerical stability in the non-linear regime, a numerical property that gives superior robustness to the procedures [8].

The software implements models for computing aerodynamic contributions to $\boldsymbol{f}_e$, both through built-in features and through interfaces to external codes. Airloads can be computed by simple lifting line models based on two-dimensional wing theory and table-look-up procedures, and include classical corrections for sweep, unsteady motion and stall. Wake effects are based on the dynamic inflow model of Peters [22]. Prescribed airloads, as obtained from experimental measurements, can also be used. Furthermore, interfaces to external airloads computation modules are provided.

The finite element based multibody dynamics formulation implemented in the code provides a general and flexible paradigm for the modeling of maneuvering helicopters. In particular, the modular nature of the code allows for the development of hierarchies of models providing increasing levels of detail, as required by the analysis, and has the flexibility to describe novel configurations of arbitrary topology. This formulation can accommodate variable rotor speeds, non-periodic responses and large elastic motions. Furthermore, it allows for the modeling of fuselage/rotor/tail rotor interactions in conventional rotorcraft configurations, and rotor/wing/fuselage dynamics in tilt-rotor configurations, through the coupling of fully non-linear rotor models with linearized, modal-based fuselage models.

3.3 Notation and relationships between the models

To ease the notation, it is convenient to rewrite the multibody equations (5a–5c) in a more compact fashion as

$$\frac{\mathrm{d}}{\mathrm{d}t}(\boldsymbol{B}\widetilde{\boldsymbol{x}}) - \boldsymbol{b}(\widetilde{\boldsymbol{x}}, \boldsymbol{\lambda}, \widetilde{\boldsymbol{u}}) = 0, \tag{7a}$$

$$\boldsymbol{c}(\widetilde{\boldsymbol{x}}) = 0, \tag{7b}$$

where

$$\boldsymbol{B} = \begin{bmatrix} \boldsymbol{M} & \boldsymbol{0} \\ \boldsymbol{0} & \boldsymbol{N} \end{bmatrix}, \quad \boldsymbol{b} = ((\boldsymbol{f}_i + \boldsymbol{c}_{,\boldsymbol{q}}\boldsymbol{\lambda} + \boldsymbol{f}_e)^T, \boldsymbol{w}^T)^T. \tag{8}$$

The multibody state vector is now defined as $\widetilde{\boldsymbol{x}} = (\boldsymbol{w}^T, \boldsymbol{q}^T)^T \in \mathbb{R}^{n_{s,MB}} = \mathbb{R}^{2n_q}$, and the multibody controls are $\widetilde{\boldsymbol{u}} \in \mathbb{R}^{n_{u,MB}}$. Here and in the following we will use the symbol $\widetilde{(\cdot)}$ to distinguish quantities of the multibody model from quantities of the flight mechanics model.

For future reference, it is important to note the following facts related to states and controls of the two models.

First of all, it is clear that there are in general many more multibody than flight mechanics states, i.e. $\mathbb{R}^{n_{s,MB}} \gg \mathbb{R}^{n_{s,FM}}$. From the multibody states $\widetilde{\boldsymbol{x}}$ it is always possible to compute some quantities (here termed multibody outputs) $\widetilde{\boldsymbol{y}} \in \mathbb{R}^{n_{s,FM}}$, that have the same physical meaning of the flight mechanics states $\boldsymbol{y}$ (recall, representing the rigid body generalized positions, generalized velocities and the rotor angular velocity). The actual form of this mapping from the multibody to the flight mechanics models will depend on the specific details of the former, but here it will suffice to formally indicate this operation as

$$\widetilde{\boldsymbol{y}} = \mathcal{S}(\widetilde{\boldsymbol{x}}). \tag{9}$$

For example, if the fuselage is modeled as a rigid body in the multibody model, then the operator $\mathcal{S}(\cdot)$ will be a simple boolean identification of the fuselage rigid body degrees of freedom within the state vector $\widetilde{x}$. On the other hand, if a more refined flexible fuselage model is used, then the same operator will compute some form of average position, orientation and velocity of the vehicle. Later on we will use the outputs $\widetilde{y}$ to verify whether or not the multibody and flight mechanics models fly the same maneuver, i.e. to check whether $\widetilde{y} \approx y$.

Regarding the controls, it should be noted that their number in the two models will typically be the same, i.e. $\mathbb{R}^{n_{u,MB}} = \mathbb{R}^{n_{u,FM}}$. Nonetheless, the controls in the two models might have a different physical meaning. For example, the main rotor collective $\theta_{0_{MR}}$ that appears among the flight mechanics controls $\boldsymbol{u}$ might correspond in the multibody control vector $\widetilde{\boldsymbol{u}}$ to the linear translation of an actuator connected to the swash-plate. Here again, it is not possible to specify this mapping further without knowing the specific details of the models. It will suffice, however, to simply formally write this mapping between the controls as

$$\widetilde{\boldsymbol{u}} = \mathcal{C}(\boldsymbol{u}). \tag{10}$$

4 Computation of Rotorcraft Trajectories Using Flight Mechanics Models

In this work, we base the definition of a rotorcraft maneuver on the formulation of an optimal control problem, as discussed in the introduction. Numerical solution procedures for computing rotorcraft trajectory optimization problems are described in detail in [10], and are more concisely reviewed in the present section.

4.1 Formulation of the optimal control problem for rotorcraft trajectories

The flight mechanics equations of dynamic equilibrium and accompanying kinematic equations for a rotorcraft can be written as

$$\boldsymbol{f}(\dot{\boldsymbol{y}}, \boldsymbol{y}, \boldsymbol{u}, \boldsymbol{p}^*) = 0. \tag{11}$$

Note that, compared to the equations of the reference flight mechanics model that we gave in (4), we have added here the dependence of the equations on some parameters $\boldsymbol{p}^* \in \mathbb{R}^{n_p}$. These parameters are obtained by adaptation of the model, as discussed later on in Section 6, in order to guarantee close matching between the trajectories flown by the flight mechanics and aeroelastic models.

The rotorcraft maneuver is defined over the temporal domain $\Omega = (T_0, T) \subset \mathbb{R}$ with boundary $\Gamma = \{T_0, T\}$, where the final time T is often unknown. The problem of optimal control is to determine the controls $\boldsymbol{u}$, the states $\boldsymbol{y}$ and possibly the final time T that minimize a cost function

$$J = \phi(\boldsymbol{y}, \boldsymbol{u}, t)\big|_T + \int_{T_0}^{T} L(\boldsymbol{y}, \boldsymbol{u}, t)\,\mathrm{d}t, \tag{12}$$

subject to the state equations (11) and to various possible additional constraints, as required by the problem at hand. The exact nature of the constraint conditions depends on the maneuver. For example, they might specify initial and/or final conditions, or might provide operational and flight enveloped limits. Collectively, a specific form of the cost function (12) and of the accompanying constraints effectively defines a certain maneuver. In general, the constraints can be classified as boundary conditions

$$\boldsymbol{\psi}(\boldsymbol{y}(T_0)) \in [\boldsymbol{\psi}_{0_{\min}}, \boldsymbol{\psi}_{0_{\max}}], \tag{13a}$$

$$\boldsymbol{\psi}(\boldsymbol{y}(T)) \in [\boldsymbol{\psi}_{T_{\min}}, \boldsymbol{\psi}_{T_{\max}}], \tag{13b}$$

constraints on states and controls

$$\boldsymbol{g}(\boldsymbol{y}, \boldsymbol{u}, t) \in [\boldsymbol{g}_{\min}, \boldsymbol{g}_{\max}], \tag{14}$$

integral conditions on states and controls

$$\int_T \boldsymbol{h}(\boldsymbol{y}, \boldsymbol{u}, t)\, \mathrm{d}t \in [\boldsymbol{h}_{\min}, \boldsymbol{h}_{\max}], \tag{15}$$

and upper and lower bounds

$$\boldsymbol{y} \in [\boldsymbol{y}_{\min}, \boldsymbol{y}_{\max}], \tag{16a}$$

$$\boldsymbol{u} \in [\boldsymbol{u}_{\min}, \boldsymbol{u}_{\max}]. \tag{16b}$$

For generality, all these conditions are expressed as inequality constraints in the form $x \in [x_{\min}, x_{\max}]$, i.e. $x_{\min} \leq x \leq x_{\max}$. Equality constraints are enforced by simply selecting $x_{\min} = x_{\max}$.

In certain cases, the total range Ω can be broken into p sequential phases (sub-domains), $T_0 < T_1 < \ldots < T_{p-1} < T_p \equiv T$, $p \geq 1$. The generic phase i is defined on the interval $\Omega^i = [T_i, T_{i+1}]$, $i = 0, \ldots, p-1$, while T_j, $j = 1, \ldots, p-1$, is the generic internal event. In each phase, different system governing equations and/or constraints might apply. Furthermore, the final time and the event locations might be unknown. An example involving a multi-phase problem will be discussed in the section on numerical applications. For clarity, we consider here a single phase problem, and drop the phase-dependent notation. However, all the following discussion is easily extended to the more general multi-phase case.

4.2 Transformation of the optimal control problem into a parameter optimization problem via the finite element method

The governing equations of optimal control could be obtained by first augmenting the cost function J with the equations of motion (11) and the constraints (13a–15) through the use of Lagrange multipliers, and then imposing the stationarity of the augmented cost [15]. The resulting two-point boundary value problem could be solved with a suitable discretization method.

This is however nor necessary nor desirable. In fact, practical methods for optimal control are currently based on a discretization process that renders the problem finite-dimensional, followed by a parameter optimization problem. Therefore, instead of deriving first the equations of optimal control and then discretizing them with a suitable numerical method, the process is in effect reversed: first one discretizes the equations of equilibrium, the constraints and the cost function, so that the problem from infinite dimensional becomes finite dimensional; next, one optimizes the discrete problem. This approach, called the direct method, is simpler than the one based on the derivation of the optimal control equations, since these never need to be computed. Furthermore, this approach also presents a number of numerical advantages which generally make it more robust than other strategies [4].

The temporal discretization of the problem can be obtained with initial value (shooting and multiple shooting) or boundary value techniques. In this work we use the latter approach, since it can be used even if the system to be controlled is inherently unstable. This is important in the present context, since rotorcraft vehicles are in fact usually unstable. Using a boundary value approach, elements (time steps) are assembled in order to cover the whole temporal domain. For this purpose, here we use the Discontinuous Petrov–Galerkin (DPG) finite element method [12]. This formulation enjoys desirable numerical properties, being of maximal order, algebraically stable and symplectic for Hamiltonian problems [6]. The lowest order member of this family of methods is the familiar second-order (implicit symmetric) mid-point rule.

To define the discrete equations, we let $\mathcal{T}_h$ be a grid of $\overline{\Omega}$, K denoting a generic element. More precisely, we consider a partition $T_0 \equiv t_0 < t_1 < \ldots < t_{n-1} < t_n \equiv T$ composed of $n \geq 1$ intervals $T^i = [t_i, t_{i+1}]$ of size h^i, $i = 0, \ldots, n-1$. Since T itself can be unknown in general, it is convenient to introduce a mapping of time onto a fixed domain parameter s, i.e. $s : (T_0, T) \mapsto (0, 1)$; for example, we can choose $s = t/(T - T_0)$, $0 \leq s \leq 1$, so that the generic step length is now $h^i = (T - T_0)(s_{i+1} - s_i)$, $i = 0, \ldots, n-1$.

Using the finite element method, the infinite dimensional solution fields $\boldsymbol{y}(t)$ and $\boldsymbol{u}(t)$ are approximated with functions $\boldsymbol{y}_h$ and $\boldsymbol{u}_h$ chosen within suitable finite dimensional spaces. The same functions restricted to the generic element K are noted $\boldsymbol{y}_h|_K$ and $\boldsymbol{u}_h|_K$. The functions $\boldsymbol{y}_h$ and $\boldsymbol{u}_h$ can be expressed in terms of finite element shape functions and of discrete (nodal) finite element degrees of freedoms $\boldsymbol{y}_d$ and $\boldsymbol{u}_d$. In the following, having chosen a member of the finite element family and hence having chosen the shape functions, we will regard $\boldsymbol{y}_h$ and $\boldsymbol{u}_h$ as sole functions of the nodal values, so that the functional dependencies $\boldsymbol{y}_h = \boldsymbol{y}_h(\boldsymbol{y}_d)$, $\boldsymbol{u}_h = \boldsymbol{u}_h(\boldsymbol{u}_d)$ are understood.

By applying the DPG finite element method, the system governing equations (11) are transformed into a set of residual equations over each grid element, i.e.

$$\boldsymbol{\xi}_h(\boldsymbol{y}_h|_K, \boldsymbol{u}_h|_K, \boldsymbol{p}^*, T) = 0, \quad \forall K \in \mathcal{T}_h, \tag{17}$$

equations that can be collectively written as

$$\boldsymbol{\xi}_h(\boldsymbol{y}_h, \boldsymbol{u}_h, \boldsymbol{p}^*, T) = 0 \text{ on } \mathcal{T}_h. \tag{18}$$

The functional dependence of these equations on T reflects the fact that the time step size h^K is unknown for an unknown final time problem, as previously recalled.

Through this process, the system dynamic equations (11) are transcribed into constraints of the parameter optimization problem. All other problem constraints and bounds, equations (13a–15), (16a, 16b), are expressed in terms of the same finite dimensional functions $\boldsymbol{y}_h$ and $\boldsymbol{u}_h$ and are appended to equation (17) as additional linear, non-linear or bound constraints, as appropriate. All these constraint conditions can be collectively written as

$$\boldsymbol{\varphi}_h(\boldsymbol{y}_h, \boldsymbol{u}_h, \boldsymbol{p}^*, T) \in [\boldsymbol{\varphi}_{\min}, \boldsymbol{\varphi}_{\max}] \text{ on } \mathcal{T}_h. \tag{19}$$

Finally, the cost function J defined in (12) is expressed in terms of the discrete problem unknowns, yielding the discrete objective $J_h = J_h(\boldsymbol{y}_h, \boldsymbol{u}_h, T)$. This procedure defines a finite-dimensional non-linear programming (NLP) problem:

$$\begin{aligned} &\min_{\boldsymbol{y}_d, \boldsymbol{u}_d, T} J_h, \quad J_h = \boldsymbol{\phi}(\boldsymbol{y}_h, \boldsymbol{u}_h, t)\big|_T + \int_{T_0}^{T} L(\boldsymbol{y}_h, \boldsymbol{u}_h, t)\, \mathrm{d}t, \\ &\text{s.t.: } \boldsymbol{\varphi}_h(\boldsymbol{y}_h, \boldsymbol{u}_h, \boldsymbol{p}^*, T) \in [\boldsymbol{\varphi}_{\min}, \boldsymbol{\varphi}_{\max}]. \end{aligned} \tag{20}$$

Using this approach, the optimality conditions of the discrete NLP problem converge to the optimality conditions of the optimal control problem as the grid is refined ($h \to 0$) and the number of discrete optimization variables goes to infinity ($n \to \infty$) [4].

4.3 Additional implementation issues

A refinement procedure is used for "boot-strapping" the solution, alleviating the need for accurate initial guesses. At first, a rough initial guess is associated to a crude grid, and the corresponding NLP problem is solved. The computed solution is then projected onto a finer grid, and used as initial guess for the subsequent NLP problem. The procedure is continued until sufficient grid refinement has been achieved to yield converged results. This procedure is used because on coarse grids fine scale details of the solution are not captured; this will usually imply a faster convergence of the NLP problem, especially if the initial guess is poor, i.e. the tentative solution is far from the converged one. If a fine grid is used starting from a poor initial guess, the fine details captured by the grid will tend to slow down or even prevent convergence.

An additional robustness issue is related to the scaling of unknowns in the numerical optimization procedures. These are in fact notoriously sensitive to badly scaled problems. To address this issue, we rewrite the governing equations (11) as

$$\bar{f}(\bar{y}', \bar{y}, \bar{u}, p^*) = 0, \tag{21}$$

where $(\cdot)' = \mathrm{d}\cdot/\mathrm{d}\bar{t}$ indicates a derivative with respect to a non-dimensional time $\bar{t} = S^t t$, while $\bar{y}$ and $\bar{u}$ are scaled states and controls, respectively, that are defined as

$$\bar{y} = S^y y, \tag{22a}$$

$$\bar{u} = S^u u, \tag{22b}$$

and where

$$\bar{f}(\cdot,\cdot,\cdot,\cdot) = S^{y^{-1}} f(S^t S^{y^{-1}}\cdot, S^{y^{-1}}\cdot, S^{u^{-1}}\cdot, \cdot). \tag{23}$$

In the previous relations, $S^y = \mathrm{diag}(S_i^y)$, $i = 1, \ldots, n_{s,FM}$, is a (diagonal) matrix of weights that scale the state variables with respect to one another. Similarly, $S^u = \mathrm{diag}(S_i^u)$, $i = 1, \ldots, n_{u,FM}$ is the analogous scaling matrix for the controls. The scaling coefficients are chosen so as to obtain states and controls that are all approximatively of order $\mathcal{O}(1)$.

Without proper corrective actions, the control time histories computed through the optimal control problem often tend to show a somewhat rough (e.g., bang-bang) behavior, jumping from one saturation bound to the other. This implies infinite or very high, and therefore unrealistic, actuation speed and actuation power. This is due to the fact that the flight mechanics models are quasi-steady in the controls, i.e. the controls u are purely algebraic variables that lack proper dynamics. This lack of modeling detail is desirable in many flight mechanics applications, and is justifiable on the grounds of a time scale separation argument. At the same time, the procedures are now blind to the intrinsic limitations of real actuators, such as for example limited control velocities, limited actuation power, etc. In order to ensure smooth computed control time histories, techniques for incorporating approximate knowledge on the actuator dynamics can be used, as proposed in [11]. The most straightforward way of obtaining smooth controls is by using control velocities as part of the optimization constraints and objective function. This derived field can be obtained through a Galerkin projection [11].

5 The Multi-Model Steering Algorithm

The multibody and flight mechanics models of the vehicle described in the previous section are now combined into a single algorithm, whose final goal is to compute the controls that steer the multibody model according to a user-specified criterion.

At first, the discrete maneuver optimal control problem (20) is solved using the flight mechanics model of the aircraft. The solution is not expensive to compute, since a coarse model with relatively few degrees of freedom is used. This problem yields a to-be-tracked time history of flight mechanics states y_h^*.

Next, this trajectory is tracked with the aeroelastic model. The tracking problem is formally identical to the maneuver defining optimal control problem described in

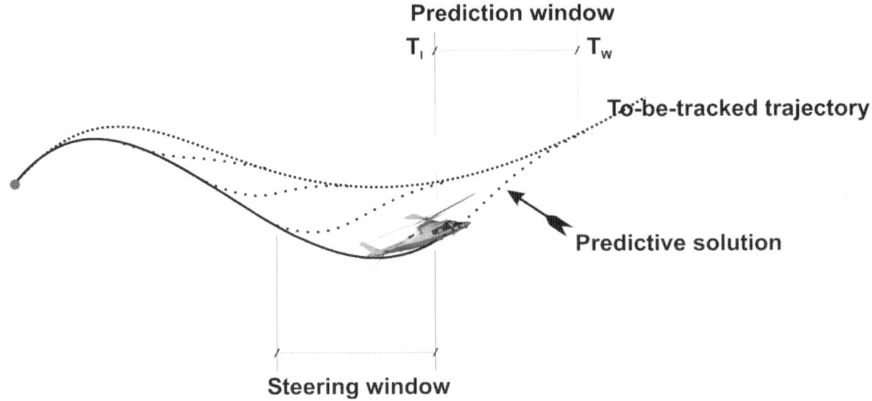

Figure 1. Receding horizon formulation of the multi-model steering algorithm.

the previous section, except for two facts: first, it is defined over a prediction window rather than the full maneuver, and second the optimization cost is a function of the tracking error, i.e. of the distance between to-be-tracked and predicted states. As the simulation proceeds, the window shifts forward in time, as depicted in Figure 1. This approach is known in the controls literature as model predictive or receding horizon control [18]. As a matter of fact, we can interpret here the aeroelastic model as the "real" system to be controlled (the plant), the reduced flight mechanics model as the plant non-linear model, while the trajectory generated by solving the global optimal control problem is the tracking output for the receding horizon controller.

The idea of using a receding horizon approach is particularly attractive here, since it allows for the feed-back stabilization of the system without the need to develop additional computational tools and without the need to derive linearized models of the vehicle. In fact, the same code which solves the optimal control planning problem is used for solving the optimal control tracking one. Furthermore, a model predictive controller can handle in a straightforward manner actuator limitations and other input and output constraints, which are of critical importance here and that would be difficult to account for with other approaches.

As the aeroelastic model tracks the planned trajectory, its reduced model is adapted on-line, as described below. Hence, as the predictions based on the reduced model improve thanks to its on-line adaption, the tracking error of the aeroelastic model decreases. Once the planned trajectory has been fully tracked by the aeroelastic model, the global tracking error is evaluated:

$$\varepsilon = \frac{\displaystyle\int_{T_0}^{T} \parallel \boldsymbol{y}_h^* - \widetilde{\boldsymbol{y}}_h \parallel \, \mathrm{d}t}{\displaystyle\int_{T_0}^{T} \parallel \boldsymbol{y}_h^* \parallel \, \mathrm{d}t}, \tag{24}$$

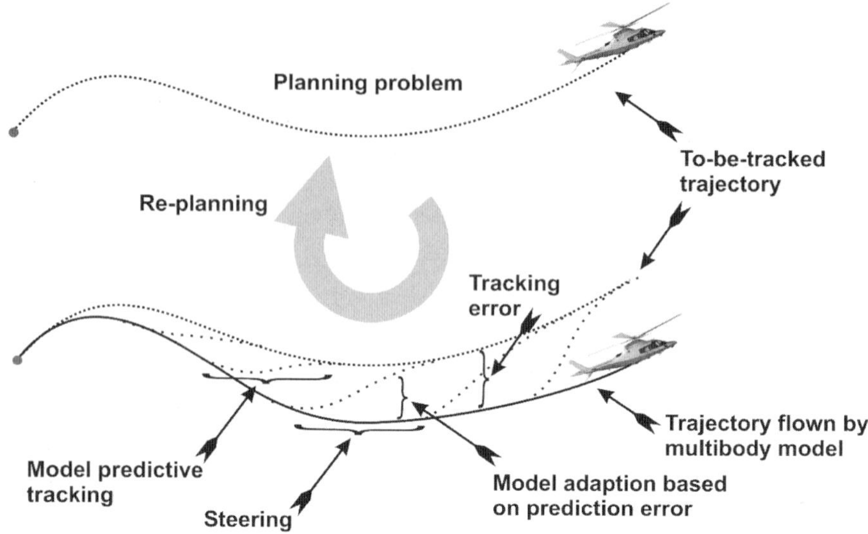

Figure 2. Planning/tracking-steering/adaption iterations.

where $\|\cdot\|$ is some dimensionally consistent norm, or simply the 2-norm if one uses the non-dimensionalization expressed by equation (21). If the error ε is greater than a given tolerance, a new iteration of the MMSA is initiated. This time, using the updated value of the parametric reduced model, the trajectory optimization problem is now solved a second time and re-tracked. The whole procedure is continued until convergence of the tracking error below a user-prescribed tolerance, as schematically depicted in Figure 2.

6 Adaptive Reduced Model

In order to fly a rotorcraft virtual prototype with a predictive controller along a given track, as previously discussed, one needs a reduced model that is capable of representing the system behavior as accurately as possible. Following [9] and [13], we augment the reference flight mechanics model (4) with an adaptive neural element, whose role is to model the error between reference model and plant.

Using this approach, the analytical reference model (4) is augmented as follows

$$\boldsymbol{f}_{\text{ref}}(\dot{\boldsymbol{y}}, \boldsymbol{y}, \boldsymbol{u}) - \boldsymbol{d}(\boldsymbol{y}^{(n)}, \dots, \boldsymbol{y}, \boldsymbol{u}) = 0, \tag{25}$$

where $\boldsymbol{y}^{(n)}$ indicates the derivative of order n of the reduced model states with respect to time (here limited to $n = 1$ at present). The unknown function $\boldsymbol{d}$ represents the defect of the reference model when $\boldsymbol{u} = \widetilde{\boldsymbol{u}}$ and $\boldsymbol{y} = \widetilde{\boldsymbol{y}}$.

Figure 3. Helicopter obstacle avoidance problem.

Function $\boldsymbol{d}$ belongs to an infinite-dimensional class of non-linear functions, which is here approximated with a single-hidden-layer feedforward neural network:

$$\boldsymbol{d}(\dot{\boldsymbol{y}}, \boldsymbol{y}, \boldsymbol{u}) \approx \boldsymbol{d}_p(\dot{\boldsymbol{y}}, \boldsymbol{y}, \boldsymbol{u}, \boldsymbol{p}) = \boldsymbol{W}^T \boldsymbol{\sigma}(\boldsymbol{V}^T \boldsymbol{x} + \boldsymbol{a}) + \boldsymbol{b}, \tag{26}$$

where $\boldsymbol{x} = (\dot{\boldsymbol{y}}^T, \boldsymbol{y}^T, \boldsymbol{u}^T)^T$ are the network inputs, $\boldsymbol{\sigma}$ a vector of sigmoid activation functions, and the reduced model parameters are defined as the synaptic weights and biases of the network, $\boldsymbol{p} = (\dots, W_{ij}, \dots, V_{ij}, \dots, a_i, \dots, b_i, \dots)^T$. Using (26) and (25), the parametric reduced model (11) is found to be

$$\boldsymbol{f}(\dot{\boldsymbol{y}}, \boldsymbol{y}, \boldsymbol{u}, \boldsymbol{p}) = \boldsymbol{f}_{\text{ref}}(\dot{\boldsymbol{y}}, \boldsymbol{y}, \boldsymbol{u}) - \boldsymbol{d}_p(\dot{\boldsymbol{y}}, \boldsymbol{y}, \boldsymbol{u}, \boldsymbol{p}) = 0. \tag{27}$$

The network is trained on-line using the local information provided at each steering of the plant by the error between predicted and effectively realized values of the outputs. An updated value of the parameters is obtained by using the steepest-descent search direction.

7 Numerical Application to an Obstacle Avoidance Maneuver

We consider an All Engines Operative (AEO) maneuver for a helicopter flying in proximity of the ground in a hostile environment. The vehicle is in straight level flight at 30 m/s and must avoid an obstacle of 30 m of height, going back to its original low altitude flight condition in minimum time, in order to minimize its exposure to, for example, enemy fire. The overall maneuver is therefore composed of a violent pull-up followed by a similarly violent pull-down, as shown in Figure 3.

The problem can be formulated as a multi-phase optimal control problem on the domain $\Omega = (T_0, T)$, with unknown final time T. The unknown internal event T_1, $T_1 \in [T_0, T]$, corresponds to the instant where the vehicle passes over the obstacle. The cost function for this problem can be written as

$$J = T + w \frac{1}{T - T_0} \int_{T_0}^{T} \left(\dot{B}_1^2 + \dot{\theta}_{0_{MR}}^2 \right) \mathrm{d}t. \tag{28}$$

The first term enforces the minimum time condition, while the second term penalizes high cyclic and collective rates. The weight w is the tunable factor, which for this problem is assumed to be $w = 100$. The cost function of the tracking problem was defined as

$$J^t = \int_{T_I}^{T_W} \left((X - X^*)^2 + (Z - Z^*)^2 + (\Theta - \Theta^*)^2\right) \mathrm{d}t$$
$$+ \int_{T_I}^{T_W} \left((V_X - V_X^*)^2 + (V_Z - V_Z^*)^2 + (q - q^*)^2\right) \mathrm{d}t$$
$$+ w^t \int_{T_I}^{T_W} \left(\dot{B}_1^2 + \dot{\theta}_{0_{MR}}^2\right) \mathrm{d}t, \tag{29}$$

where the starred quantities denote the to-be-tracked states, while the last integral term penalizes large control rates. Bounds on states, controls and control rates were also enforced during the tracking problem, which is one of the key features of this predictive control scheme.

The initial conditions at time T_0 for the optimization problem are given by the trim conditions at 30 m/s. Throughout the maneuver we impose bounds on the angular velocity $\omega \in [207, 207]$ rpm, on the controls $\theta_{0_{MR}} \in [-5°, 20°]$, $\theta_{0_{TR}} \in [-10°, 30°]$, $A_1 \in [-12°, 12°]$, $B_1 \in [-20°, 20°]$, and on the control rates $\dot{\theta}_{0_{MR}}$, $\dot{\theta}_{0_{TR}}$, $\dot{B}_1$, $\dot{A}_1 \in [-16°\ \mathrm{s}^{-1}, 16°\ \mathrm{s}^{-1}]$. Furthermore, the maximum available power is limited as $P_{\max} = 2500$ hp, and the power rate is limited according to $\dot{P}(t) \leq 500\ \mathrm{hp\ s}^{-1}$.

The solution to the planning problem was computed on a grid of 80 time elements, generated by successive uniform refinement of an initial 20 element grid. Tracking and steering windows were selected of length 2 and 0.2 s, respectively. The activation frequency of the controller is small enough to capture the short period mode of the vehicle, which is approximately equal to 1 s.

Figure 4 illustrates the effect of the local adaption on the final compatibility between the planned trajectory and the effectively realized one. The procedure is ended after 7 planning/tracking-steering/adaption iterations, since no further model improvement is realized at this point. Figure 5 show the fuselage pitch vs. time before and after the iterations. Lines marked with the $\triangle$ symbol correspond to the solution of the planning problem, while the solid lines correspond to the one realized by the multibody model.

Both figures clearly indicate the convergence of planned and tracked solutions. Notice that the final solution is quite different, for example even in terms of duration, from the first one, indicating that a substantial error in planning is possible with the initial (reference) reduced model. Excellent tracking performance is obtained at convergence, illustrating the success of the overall procedure.

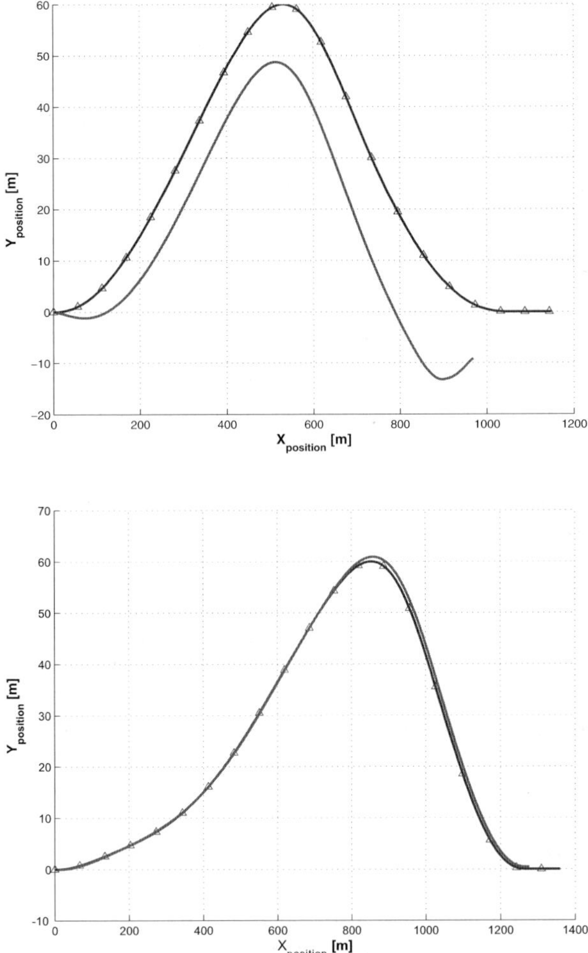

Figure 4. Helicopter obstacle avoidance maneuver. Trajectory flown by the reduced (△ symbols) and multibody (solid line) models, before (top), after (bottom) 7 planning/tracking-steering/adaption iterations.

8 Conclusions

It is well known that steady flight analysis of rotorcraft requires good trim procedures in order to produce accurate results. In this work, we have argued that similar time dependent trim processes are required for maneuvering flight simulation. We have here proposed a methodology that meets this need by blending aeroelasticity, flight

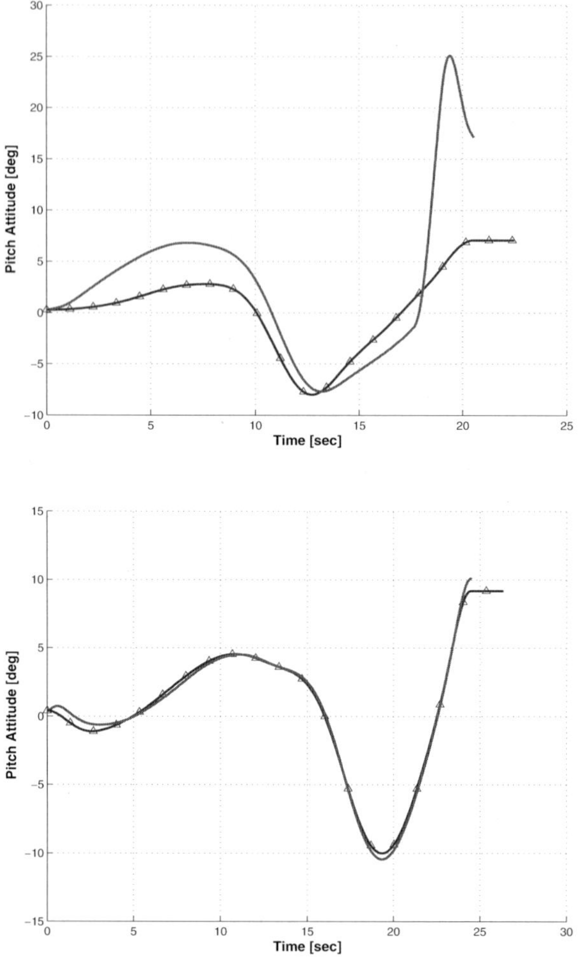

Figure 5. Helicopter obstacle avoidance maneuver. Fuselage pitch for the reduced (△ symbol) and multibody (solid line) models, before (top), after (bottom) 7 planning/tracking-steering/adaption iterations.

mechanics, trajectory optimization and optimal control, and that provides a general and flexible paradigm. The algorithms described herein expand the applicability of comprehensive aeroelastic codes towards the unsteady flight regimes, moving beyond the sole steady flight case. The ability to study maneuvering flight could be of importance in certain applications, especially for high-performance aggressive vehicles, since maximum loads, vibratory levels, noise and other critical design parameters are often encountered in this flight regime. The methodology was demonstrated on specific flight mechanics and aeroelastic models, but it is general and could be applied to models other than the ones here described.

The work conducted so far seems to be promising. The preliminary examples that were presented indicate that the multi-model procedures are indeed capable of computing control time histories that fly an aeroelastic model along unsteady maneuvers. In particular, the use of a coarse flight mechanics model allows one to solve a trajectory optimization problem without incurring in overwhelming computational costs. Furthermore, the receding horizon formulation of the algorithm provides a relatively straightforward way of tracking the planned path, while accounting for output and input constraints.

Improvements to this methodology could and should be pursued in several areas. From the modeling point of view, more sophisticated unsteady aerodynamic models should be used in maneuvering flight for accounting for the distortion processes undergone by the wake. Here again a hierarchical approach of increasing modeling complexity could be used, for example employing the reduced order models of dynamic wake distortion of Prasad et al. [23] at the flight mechanics level, while the time-accurate free-wake models of Bhagwat and Leishman [5] could be used at the aeroelastic level.

Acknowledgement

The authors acknowledge the help of Yalcin Faik Sümer in the preparation of the numerical examples.

References

1. Bauchau OA, Bottasso CL, Nikishkov YG (2001) Modeling Rotorcraft Dynamics with Finite Element Multibody Procedures. Mathematics and Computer Modeling, 33:1113–1137
2. Bauchau OA, Bottasso CL, Trainelli L (2003) Robust Integration Schemes for Flexible Multibody Systems. Computer Methods in Applied Mechanics and Engineering, 192:395–420
3. Bauchau OA, Rodriguez J, Bottasso CL (2001) Modeling of Unilateral Contact Conditions with Application to Aerospace Systems Involving Backlash, Freeplay and Friction. Mechanics Research Communications, 28:571–599
4. Betts JT (2001) Practical Methods for Optimal Control Using Non-Linear Programming. Philadelphia: SIAM
5. Bhagwat MJ, Leishman JG (2001) Stability, Consistency and Convergence of Time Marching Free-Vortex Rotor Wake Algorithms. Journal of the American Helicopter Society, 46:59–71

6. Bottasso CL (1997) A New Look at Finite Elements in Time: A Variational Interpretation of Runge–Kutta Methods. Applied Numerical Mathematics, 25:355–368
7. Bottasso CL, Bauchau OA (2001) Multibody Modeling of Engage and Disengage Operations of Helicopter Rotors. Journal of the American Helicopter Society, 46:290–300
8. Bottasso CL, Bauchau OA (1999) On the Design of Energy Preserving and Decaying Schemes for Flexible, Nonlinear Multibody Systems. Computer Methods in Applied Mechanics and Engineering, 169:61–79
9. Bottasso CL, Chang C-S, Croce A, Leonello D, Riviello L (2004) Adaptive Planning and Tracking of Trajectories for the Simulation of Maneuvers with Multibody Models. Computer Methods in Applied Mechanics and Engineering, 195:7052–7072
10. Bottasso CL, Croce A, Leonello D, Riviello L (2005) Optimization of Critical Trajectories for Rotorcraft Vehicles. Journal of the American Helicopter Society, 50:165-177
11. Bottasso CL, Croce A, Leonello D, Riviello L (2005) Rotorcraft Trajectory Optimization with Realizability Considerations. Journal of Aerospace Engineering, 18:146–155
12. Bottasso CL, Micheletti S, Sacco R (2002) The Discontinuous Petrov-Galerkin Method for Elliptic Problems. Computer Methods in Applied Mechanics and Engineering, 191:3391–3409
13. Bottasso CL, Riviello L (2005) Trimming of Multibody Rotorcraft Models by a Neural Model-Predictive Auto-Pilot. In: Proc. of the ECCOMAS Thematic Conference on Multibody Dynamics, Madrid, Spain
14. Brentner KS, Perez G, Bres GA, Jones HE (2002) Toward a Better Understanding of Maneuvering Rotorcraft Noise. In: Proc. of the 58th American Helicopter Society Annual Forum, Montreal, Canada
15. Bryson AE, Ho YC (1975) Applied Optimal Control. New York: Wiley
16. Carlson EB, Zhao YJ (2001) Optimal Short Takeoff of Tiltrotor Aircraft in One Engine Failure. Journal of Aircraft, 39:280–289
17. CHARM (Comprehensive Hierarchical Aeromechanics Rotorcraft Model). `http://www.continuum-dynamics.com/products/charm/index.html`. NJ: Continuum Dynamics, Inc.
18. Findeisen R, Imland L, Allgöwer F, Foss BA (2003) State and Output Feedback Nonlinear Model Predictive Control: An Overview. European Journal of Control, 9:190–207
19. Frazzoli E, Dahleh MA, Feron E (2000) A Hybrid Control Architecture for Aggressive Maneuvering of Autonomous Aerial Vehicles. In: System Theory: Modeling, Analysis and Control. Djaferis TE, Schick IC (Eds.). SECS 518. Dordrecht: Kluwer Academic Publishers
20. Johnson W (1992-1997) CAMRAD II, Comprehensive Analytical Model of Rotorcraft Aerodynamics and Dynamics. Palo Alto (CA): Johnson Aeronautics
21. Johnson W (1994) Helicopter Theory. New York: Dover Publications
22. Peters DA, He CJ (1995) Finite State Induced Flow Models. Part II: Three-Dimensional Rotor Disk. Journal of Aircraft, 32:323–333
23. Prasad JVR, Zhao J, Peters DA (2002) Helicopter Rotor Wake Distorsion Models for Maneuvering Flight. In: Proc. of the 28th European Rotorcraft Forum, Bristol, UK
24. Ribera M, Celi R (2004) Simulation Modeling of Unsteady Maneuvers Using a Time Accurate Free Wake. In: Proc. of the 60th Annual Forum of the American Helicopter Society, Baltimore, MD
25. Rutkowski M, Ruzicka GC, Ormiston RA, Saberi H, Jung Y (1995) Comprehensive Aeromechanics Analysis of Complex Rotorcraft Using 2GCHAS. Journal of the American Helicopter Society, 40:3–17

26. Theodore CR, Celi R (2002) Helicopter Flight Dynamic Simulation with Refined Aerodynamic and Flexible Blade Modeling. Journal of Aircraft, 39:577–586
27. Wachspress DA, Quackenbush TR, Boschitsch AH (2003) First-Principles Free-Vortex Wake Analysis for Helicopters and Tiltrotors. In: Proc. of the 59th Annual Forum of the American Helicopter Society, Phoenix, AZ

Computational and Design Aspects in Multibody Software Development

Paolo Mantegazza, Pierangelo Masarati, Marco Morandini, and Giuseppe Quaranta

Dipartimento di Ingegneria Aerospaziale, Politecnico di Milano, via La Masa 34, 20156 Milano, Italy; E-mail: {mantegazza,masarati,morandini,quaranta}@aero.polimi.it

Abstract. It is often perceived that one key issue in the development of modern, high-performance numerical software is the need to find a good trade-off between modularity, extensibility and performance requirements. This paper discusses how the need to add real-time simulation capabilities to an existing general-purpose multibody analysis software, and the resulting need for performance improvements, pushed an overall performance improvement and capability extension within an existing modular generic programming environment.

1 Introduction

This chapter discusses significant computational aspects of general-purpose multibody simulation, in the light of modern programming techniques and of the experience gained with the free multibody analysis software MBDyn [14]. General-purpose multibody analysis software, as opposed to specialized software, means that a single software tool is able to perform rather different types of analysis and to address rather different problems. Examples are: real-time hardware-in-the-loop simulation for robotics [2], large multidisciplinary interactional problems involving CFD aerodynamics [20], trajectory optimization [3], and more.

As a consequence, it is worthless to address efficiency by specializing the software only for a given application; on the contrary, given the importance of software reuse and commonality of tools, efficiency must be pursued despite the inevitable trade-offs dictated by flexibility of use.

As a matter of fact, it is often perceived that one key issue in the development of modern, high-performance numerical software is the need to find a good trade-off between modularity, extensibility and performance requirements. Fortran code is believed to be very fast, possibly at the expense of extensibility. On the other side, C++ and Object-Oriented (OO) codes can be designed to be easily extensible, but this can impact performances. However, recent, and even not-so recent achievements in compiler technology [8], together with a better understanding of the potentiality

J.C. García Orden et al. (eds), Multibody Dynamics. Computational Methods and Applications, 137–158.

offered by some C++ programming paradigms, as the metatemplate programming [25], have shown that flexible, high-performance numerical code can be developed in C++. Some key point in OO programming are

- code reuse: the use of templates and/or inheritance allows to easily use the same codebase for different tasks;
- template metaprogramming [26]: allows to overcome some of the limitations of the OO languages, and to reduce the number of temporaries;
- expressiveness: it is possible to code complex tensorial or matrix formulæ without the need to explicitly call Basic Linear Algebra Subroutines (BLAS)-like functions, to allocate memory and/or to write explicit loops;
- classes, inheritance and virtual methods: a class allows to define an abstract concept; this, together with virtual inheritance, eases the task of building libraries of different objects, all derived from the same base class and offering a set of standard, well-defined services; an example can be a base class that defines the basic interface of a linear solver, and of many derived classes that wraps existing solvers; if the interface is well-designed, the code can use any of the available solvers, without the need to rebuild the application, to change the code, and without any impact on run-time performance;
- exception handling: it is possible to deal with unusual conditions without cluttering the code;

Successful examples of scientific, object-oriented codes are the Deal II finite element library [4], the Overture code framework [10], and the pioneering Diffpack library [1, 6]. The multibody code MBDyn is another successful example of an object-oriented scientific code.

Key aspects recently addressed in the enhancement of the MBDyn software are described in this paper. Among these, the implementation of different

a. linear solution strategies
b. assembly strategies
c. nonlinear solution strategies
d. parallelization strategies

under a common object-oriented framework, allows to use the optimal combination of the above based on the properties of the problem under analysis, resulting in very efficient solution of highly demanding problems, like:

- real-time hardware-in-the-loop simulation, with scheduling and interprocess communication delegated to the Real-Time Application Interface (RTAI, `http://www.rtai.org/`) patch for the Linux OS; key issue is to minimize the worst-case solution for the single time step;
- path-planning and optimization in general; key issue is to minimize the overall solution time, and the need to interact with external modules for control (e.g. Scicos, Simulink and so on);

- "rapid prototyping", i.e. the capability to quickly implement and efficiently analyze incomplete problems, where configuration dependent forces are only modeled as residuals, thus losing the second-order convergence properties of direct solvers; the presented success story shows the effectiveness of matrix-free nonlinear solution algorithms when solving the tire-ground interaction problem of a landing aircraft [9].

The discussed features have been implemented with minimal impact on the structure of the code, resulting in outstanding solution efficiency improvements for all the different target problems.

The problems of interest can be split in three categories: dense (1 to 50 equations; typically best dealt with by dense matrices and solvers), very small (50 to 2000; fill-in $\leq 10\%$) and small (2000 to 20000; fill-in $\leq 5\%$). Much larger problems greatly benefit from classical sparse problem handling, but are definitely out of the scope of typical multibody analysis.

2 The General-Purpose Real-Time Challenge

Real-time simulation poses very stringent requirements on participating software. As a consequence, it may not be trivial to provide real-time capabilities to existing software.

A preliminary requirement is related to the capabilities of the operating system (OS), which must be able to provide deterministically bounded worst case scheduling latencies. This is not the case for many, if not all, general-purpose OSes, despite occasional claims of somewhat limited real-time capabilities. One remarkable example is the Linux OS, at least up to version 2.4 of the kernel; in any case, even later versions do not provide very stringent scheduling latencies, of the order of a few tens of microseconds, as required by very specialized applications, like high-speed machinery control.

Simulation software may not have such stringent requirements, unless very specialized applications are addressed. However, hardware-in-the-loop simulation requires the simulation software that emulates the experiment to match the same scheduling requirements of the rest of the system (e.g. A/D, D/A adapters, control systems and so).

Recently, the RTAI extension to the Linux OS, also developed at the "Dipartimento di Ingegneria Aerospaziale" of the University "Politecnico di Milano" and distributed as free software under the GPL license, has gained sufficient maturity to find a clear position in the implementation of industrial scale real-time control systems. It provides a framework consisting in native, as well as POSIX system calls within a reliable worst-case latency well below a hundred microseconds (typically $20 \div 25$).

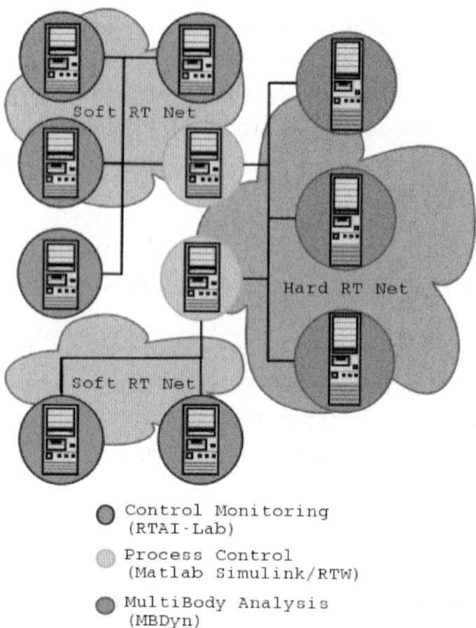

Fig. 1. Distributed real-time network layout.

Usually, the real-time simulation is addressed by means of dedicated software because of its intrinsic needs of speed performances and constant worst-case time step computational costs. This approach may suffer from lack of generality, because dedicated formulations may not allow a wide variety of model libraries and require specific implementation of features typically available in general purpose software.

In the present work the opposite approach is illustrated, consisting in real-time enabling a general-purpose software [2, 19]. The multibody analysis software MBDyn [14, 15], also developed at the "Dipartimento di Ingegneria Aerospaziale" of the University "Politecnico di Milano", has been used.

MBDyn is implemented in C++, and may link a considerable number of standard mathematical libraries, especially linear solvers. Since RTAI delegates some kernel-related operations to the Linux kernel upon request by the task under execution, to allow MBDyn to be run without incurring in non real-time scheduling all direct system calls had to be disabled or, if essential, they had to be wrapped by the corresponding RTAI interfaces.

In detail, the results writes on disk have been disabled, and memory allocation has been anticipated so that it is completed and locked before entering the real-time scheduling loop, or, whenever convenient, it has been moved to pre-allocated memory pools.

Finally, essential I/O with other real-time tasks, usually dealt with by regular system sockets in non real-time simulations, has been delegated to RTAI native mailboxes. Despite the naming analogy with UNIX native mailbox infrastructure, the RTAI mailboxes implicitly take care of remote process communication, much like UNIX sockets do. As a consequence, an inter-process communication enabled task can transparently interact with local as well as remote processes, and thus participate in distributed real-time simulations, as illustrated in Figure 1. Local scheduling and communication is entirely handled by RTAI; remote real-time communication occurs via ethernet (most 100Mbit interfaces suffice) and may require an additional layer provided by RTNet (`http://www.rts.uni-hannover.de/rtnet/`).

The remaining changes related to real-time enabling consisted in adding few selected calls to enter hard real-time mode and to synchronize with the rest of the tasks participating in the simulation, either by direct kernel scheduling or by means of semaphores triggered by other processes.

The challenge, to make the real-time solution useful with realistic models, consists in obtaining as much performance as possible in the worst-case scenario, without impacting the versatility of the general purpose software. A significant spin-off is that performance improvements are appreciated also for non real-time runs; this side-advantage is obtained if the performances improvements can be exploited also by the batch runs. It is anticipated that most of the directions explored within this work were beneficial, in terms of performances, for non real-time runs; only few of them, however, gave appreciable advantages to real-time simulations.

3 Computational Issues and Selected Improvements

The software MBDyn has been designed from the beginning in a modular manner, because one of its purposes is to allow independent researchers to investigate new solution approaches by designing and interchanging software components: the problem, by designing new elements; linear and nonlinear solvers, integration schemes, and more. The time integration of nonlinear problems is delegated to a sequence of layers consisting in:

- a "nonlinear problem data manager" (DM); actually, this is split into a nonlinear problem, i.e. an object that defines problems of the form

  $$\mathbf{F}(\mathbf{x}) = \mathbf{0},$$

  and a data manager, i.e. the object that contains and handles the model, because the latter has been specialized to different parallelization strategies described in Section 3.4;
- a "nonlinear solver" (NLS), initially based on the Newton–Raphson iterative procedure, but later extended to include iterative, matrix-free solution approaches;
- a "step solver" (SS), which takes care of advancing the solution step by step, predicting the new guess and controlling the convergence of the nonlinear problem solution;

```
NLS::Solve() {
    // Newton-Raphson
    while (true) {
        if (DM->Residual()->Test()) {
            return;
        }
        if (this->NewJacobian()) {
            DM->Jacobian();
        }
        LS->Solve();
        DM->Update();
    }
}
SS::Advance() {
    SS->Predict();
    NLS->Solve();
}
```

Fig. 2. Newton–Raphson nonlinear solver inner loop.

- a "linear solver" (LS), which takes care of the linear algebra involved in either the iterative solution of the direct nonlinear solver, or in the preconditioning for the iterative nonlinear solvers.

Figure 2 illustrates the main inner loop in pseudo-code when the Newton–Raphson nonlinear solver is used. During its evolution, the need to improve the performances in terms of problem size and complexity, and of solution time led to the development of different components for each layer; the different step solvers are not addressed in this work.

3.1 Linear solution strategies

An abstract layer for the solution of linear equations has been designed from the beginning. The interface, specified by the `LS` abstract class illustrated in Figure 3, only defines the methods that are required to access and, possibly, modify, the matrix and the right hand side and solution vectors.

The reference problems required sparse linear solvers to efficiently handle the very sparse problems from a few hundreds to a few thousands of equations that typically result from models of deformable mechanical and multidisciplinary systems formulated with a *Redundant Coordinate Set* approach, leading to a differential algebraic set of equations (DAE).

Eight different sparse solvers (ten considering the parallel version of two of them) are currently supported, and the neutral, generic programming interface allows to easily add new ones. Some have only historical value, and may only be useful as guidelines for further software integration. The "workhorse" is Umfpack, from the University of Florida [7], which is the standard sparse solver used by Matlab and other numerical packages.

```
class LS {
public:
    // constructor
    LS(void);
    // destructor
    virtual ~LS(void);
    // reset the matrix
    virtual void MatrReset(void) = 0;
    // solve the linear system
    virtual void Solve(void) = 0;
    // access to the matrix handler
    virtual MatrixHandler* pMatHdl(void) const = 0;
    // access to the right hand side
    virtual VectorHandler* pResHdl(void) const = 0;
    // access to the solution (may return the pResHdl())
    virtual VectorHandler* pSolHdl(void) const = 0;
};
```

Fig. 3. Linear solution manager basic interface.

Each specialized solver uses its dedicated `MatrixHandler` and `VectorHandler` classes; for example, the `Lapack` solver uses the `FullMatrixHandler`, while most of the sparse solvers may use any of the sparse matrix representations presented in Section 3.2.

The need to address very specific problems, significantly very small ones (100÷200 equations) within the very tight scheduling required by real-time simulation forced into looking for as much efficiency as possible. For this purpose, support for the LAPACK dense LU solver has been added to the suite of linear solvers supported by MBDyn. In most of the applications, this outperforms Umfpack when addressing problems up to 100 equations with a fill-in of 25% and above.

Nonetheless, since the typical problem fill-in was limited enough to make sparse matrix handling appealing, and the need to address slightly larger problems (up to 200 equations) with the same stringent requirements, suggested the possibility to design a very specific solver, indicated as the "naïve" solver, which combines the minimal access cost of dense matrices with the minimal computational cost of sparse matrix factorization and substitution [17]. This specialized solver outperforms all the linear solvers that were considered in most cases, based on the sparsity of the matrix and on the sparsity pattern, up to 1500÷2000 equations, as illustrated in [17, 19] and discussed later in the applications.

3.2 Assembly strategies

3.2.1 Sparse matrix handling

The use of sparse matrices may dramatically reduce the factorization and substitution time when efficient linear solvers are used. However, they introduce a cost when the coefficients of the matrix are accessed to add or modify a coefficient during the problem assembly.

```
typedef std::map<int, double> row_cont_type;
std::vector<row_cont_type> col_indices;

double& operator()(int i_row, int i_col)
{
    row_cont_type::iterator i;
    row_cont_type& row = col_indices[i_col];

    i = row.find(i_row);
    if (i == row.end()) {
        return row[i_row] = 0.;
    }
    return i->second;
}
```

Fig. 4. Sparse map data structure for sparse matrices.

Sparse matrix handling in MBDyn, for use with Umfpack and other sparse solvers, is based on C++ STL containers. The matrix, called "sparse map", is designed as a vector of associative containers (maps) that contain the real-typed value with the related row index, as illustrated[1] in Figure 4. The sparse map matrix provides a method that packs the matrix in column-compressed form when it is passed to the sparse linear solver. The packing must be repeated before each matrix factorization.

3.2.2 Column compressed (CC) storage

An interesting evolution is represented by the "column compressed" sparse matrix handling. It consists in directly accessing the matrix in the column compressed form obtained prior to factorization, under the assumption that the shape of the matrix non-zeroes does not change between assemblies. The column-compressed form is made of a vector of reals containing the non-zeroes, a vector of integers containing the row indices of the non-zeroes, sorted throughout each column, and another vector of integers pointing to the beginning of the columns. A generic non-zero can be accessed by directly accessing a column with the column index, and performing a binary search across it for the row index. If it is not present, the symbolic structure of the sparse matrix changed; this is handled by throwing an exception that causes the matrix to return into the sparse map form, and the assembly to restart.

In typical problems, the sparsity pattern of the matrix is usually preserved for most of the iterations; as a consequence, the use of the column-compressed sparse matrix allows to entirely save the packing phase, while the cost of the binary search required to access the generic coefficient is comparable to that of accessing the sparse map.

[1] Note that the (pseudo-)code of Figure 4 actually fills the matrix of zeroes if, for example, the matrix is accessed by means of the operator ( ) for read; a different operator must be designed for read operations.

```
std::vector<double>& values;
const std::vector<int>& row_indices;
const std::vector<int>& column_start;

double& operator()(int i_row, int i_col)
{
    int row_begin = column_start[i_col - 1];
    int row_end = column_start[i_col] - 1;
    int idx;
    int row;

    if (OutOfRange(i_row)) {
        // out of range: rebuild
        throw ErrRebuildMatrix();
    }

    // binary search
    while (row_end >= row_begin) {
        idx = (row_begin + row_end)/2;
        row = row_indices[idx];
        if (i_row < row) {
            row_end = idx - 1;
        } else if (i_row > row) {
            row_begin = idx + 1;
        } else {
            return values[idx];
        }
    }

    // not found: rebuild
    throw ErrRebuildMatrix();
}
```

Fig. 5. Compressed-column data structure for sparse matrices.

3.2.3 Specialized sparse + dense storage ("naïve")

The specialized structure of the "naïve" solver [17] represents yet another radical change in matrix structure. In this latter case, a dense matrix is used to store the values, while the row and column indices to non-zeroes are stacked in other two dense matrices. This approach is quite memory intensive, and quickly loses efficiency when the problem size grows not only because of physical memory constraints, but also, and essentially, because large memory buffers quickly destroy CPU cache locality.

3.3 Nonlinear solution strategies

As for the linear solvers, different nonlinear solvers can be cast under a common framework. Basically, a nonlinear solver must be able to solve a nonlinear problem, provided in form of a pointer to a `NonlinearProblem` data type.

3.3.1 Direct methods: Newton–Raphson

The original solution method used by MBDyn is based on Newton–Raphson iterations. This method requires the factorization of the Jacobian matrix and, for large problems, it can be computationally intensive. In the current implementation, the modified Newton iteration reuses the same Jacobian matrix factorization for a fixed number of iterations, and at least one new Jacobian matrix is computed for each time step. This approach is recommended, for instance, when performing simulations in real-time, because a fresh Jacobian matrix factorization is assumed to yield a faster convergence, so the cost of each time step is almost constant without the dramatic increase resulting when the worst case of a Jacobian matrix re-computation and re-factorization occurs.

3.3.2 Indirect methods: Iterative methods

It is well known that the linear equation resulting from the Newton method can be solved in an approximate manner. The resulting methods are usually called *Inexact Newton* and see broad applications in computational mechanics and other fields [12]. Among them, iterative Newton methods represent a very interesting class, where basically the linear equation is solved for each step using an approximate iterative solution. Typically, the nonlinear iteration generates a sequence, called *outer iteration*, and the linear iteration that generates the approximation for each step, called *inner iteration*. Krylov methods are usually adopted as iterative inner solvers, like the Generalized Minimum RESidual (GMRES) or the BiCGStab [24]. They basically require, during each inner iteration, a simple multiplication of the Jacobian tangent matrix of the problem with a vector, plus the adoption of a preconditioner. The product of the Jacobian matrix $\mathbf{J}$ times a generic vector $\mathbf{w}$ can be approximated by a finite difference formula using the residual vector $\mathbf{r}$ as

$$\mathbf{J}(\mathbf{x})\mathbf{w} = \|\mathbf{w}\| \frac{\mathbf{r}\left(\mathbf{x} + h\mathbf{w}\dfrac{\|\mathbf{x}\|}{\|\mathbf{w}\|}\right) - \mathbf{r}(\mathbf{x})}{h\|\mathbf{x}\|}. \tag{1}$$

In this expression h is the amplitude of the finite difference step, which must be always as small as possible, although compatible with the sensitivity of the problem and the round-off errors. For this reason the selected step is usually scaled keeping into account the differences between the scale of the solution vector $\mathbf{x}$ and the test vector $\mathbf{w}$. This attenuates the need to have a precise approximation of the Jacobian matrix for the equations that are solved; it is only required to be able to assemble the residual vector $\mathbf{r}$ for perturbations of the state $\mathbf{x}$ in different testing directions. The resulting methods are denominated *matrix free* since no tangent matrix is required. The preconditioner is used to accelerate the convergence of the inner iterative solution. Ideally, if the preconditioner is the exact approximation of the inverse of the Jacobian matrix, convergence will be reached in just one step; so it is necessary to build an approximation of this matrix, for example adopting the inverse of the matrix assembled using the elements for which the tangent matrix can be easily derived. To make the problem less computationally intensive, the preconditioner may be retained, until the number of inner iteration to converge grows past a specific threshold.

Finally, the inner iteration does not need to converge with high accuracy, so the solution can be approached in a faster manner by relaxing the convergence criteria. The application of this method can be crucial for all those complex elements for which the generation of the analytical tangent matrix is either extremely complex or impossible (see for example [9]).

3.4 Parallelization strategies

3.4.1 Domain decomposition: Schur complement

A coarse-grained parallelization of the analysis can be obtained by partitioning the problem into subdomains that are solved separately on different CPUs, while the interface problem is solved at the end, based on the contributions from each subproblem. It is apparent that a small interface is key to the effectiveness of this "divide et impera" approach, because the size of the interface problem governs the amount of data that needs be transmitted and, since it is performed after the solution of the subproblems, it represents a bottleneck and inevitably reduces to a dense problem, with cubic solution cost.

The computational domain Ω is first split into the s subdomains Ω_i by means of an element-based partition (Figure 6). This means that no element must be split between two subdomains, i.e. all the information related to a given element is mapped to the same processor. As a result, there is no need for information exchange while the assembly phases are performed. Anyway, it is worth noting that while the elements can belong to one subdomain only, nodes may belong to multiple subdomains. By reordering the unknowns to label the interface nodes at the end, the linear system associated with the problem assumes the structure:

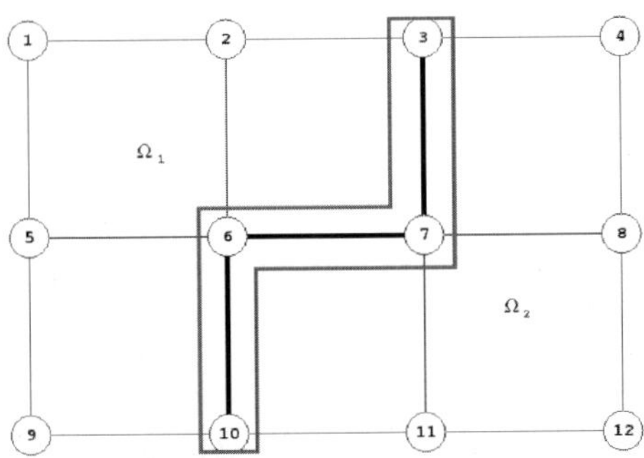

Fig. 6. Element-based partition.

$$\begin{pmatrix} \mathbf{B}_1 & 0 & \cdots & 0 & \mathbf{E}_1 \\ 0 & \mathbf{B}_2 & & \vdots & \mathbf{E}_2 \\ \vdots & & \ddots & & \vdots \\ 0 & \cdots & & \mathbf{B}_s & \mathbf{E}_s \\ \mathbf{F}_1 & \mathbf{F}_2 & \cdots & \mathbf{F}_s & \mathbf{C} \end{pmatrix} \begin{pmatrix} \mathbf{x}_1 \\ \mathbf{x}_2 \\ \vdots \\ \mathbf{x}_s \\ \mathbf{y} \end{pmatrix} = \begin{pmatrix} \mathbf{f}_1 \\ \mathbf{f}_2 \\ \vdots \\ \mathbf{f}_s \\ \mathbf{g} \end{pmatrix}, \tag{2}$$

where the $\mathbf{B}_i$ are the local subdomain matrices, $\mathbf{C}$ is the interface matrix and $\mathbf{E}_i$, $\mathbf{F}_i$ are the coupling matrices. Each $\mathbf{x}_i$ is a vector containing the internal unknowns of the subdomain Ω_i, and $\mathbf{y}$ is the vector of the interface unknowns; $\mathbf{f}_i$ and $\mathbf{g}$ are the corresponding right-hand terms. In a more compact form the system can be expressed as

$$\begin{pmatrix} \mathbf{B} & \mathbf{E} \\ \mathbf{F} & \mathbf{C} \end{pmatrix} \begin{pmatrix} \mathbf{x} \\ \mathbf{y} \end{pmatrix} = \begin{pmatrix} \mathbf{f} \\ \mathbf{g} \end{pmatrix}. \tag{3}$$

Provided $\mathbf{B}$ is non-singular, the unknown $\mathbf{x}$ can be expressed as

$$\mathbf{x} = \mathbf{B}^{-1}(\mathbf{f} - \mathbf{E}\mathbf{y}). \tag{4}$$

By substituting Equation (4) into the second block-row of equation (3), the following reduced system is obtained:

$$\mathbf{S}\mathbf{y} = \mathbf{g} - \mathbf{F}\mathbf{B}^{-1}\mathbf{f}. \tag{5}$$

where the matrix $\mathbf{S}$ is called the *Schur complement matrix*, also known as *substructuring* in structural analysis software; it assumes the form:

$$\mathbf{S} = \mathbf{C} - \mathbf{F}\mathbf{B}^{-1}\mathbf{E}. \tag{6}$$

After assembling all the elements that belong to subdomain Ω_i, a local sub-matrix results with the structure:

$$\begin{pmatrix} \mathbf{B}_i & \mathbf{E}_i \\ \mathbf{F}_i & \mathbf{C}_i \end{pmatrix}. \tag{7}$$

By calling $\mathbf{R}_i$ the restriction operator to the local interface values, so that $\mathbf{R}_i\mathbf{y} = \mathbf{y}_i$, the Schur matrix is obtained as

$$\mathbf{S} = \sum_{i=1}^{s} [\mathbf{R}_i\mathbf{C}_i\mathbf{R}_i^T - \mathbf{R}_i\mathbf{F}_i\mathbf{B}_i^{-1}\mathbf{E}_i\mathbf{R}_i^T] = \sum_{i=1}^{s} \mathbf{R}_i\mathbf{S}_i\mathbf{R}_i^T. \tag{8}$$

A solution method based on this approach involves five steps:

1. The local matrices $\mathbf{B}_i$ are factored, fully exploiting any natural sparsity.
2. The local parts of the right-hand side of the reduced system Equation (5) are assembled and transmitted to a "master" processor that will deal with the interface problem.
3. The local parts of the Schur complement matrix are assembled and transmitted as well.

4. The reduced system is solved.
5. The other unknowns are computed by the back-substitution shown in Equation (4).

Only the 4th step cannot be performed concurrently in a parallel environment, so it must be considered the bottleneck phase. Furthermore, when a modified Newton–Raphson method is used, since the Jacobian matrix is not updated at each iteration, steps (1) and (3) must be performed only at the beginning of the iterative solution, and many of the operations required during the assembly phase (2) need to be performed only once as well.

Usually, the direct substructuring method described here is not considered feasible for large structural problems, because the size of the interface problem grows rapidly; moreover, the Schur matrix presents a lower grade of sparsity than the original system, so an iterative inner solver should be considered, which does not require the explicit assembly of matrix S. On the contrary, this strategy can be very effective when some special conditions are met. This is the case of complex systems with a peculiar topology of the computational domain that allows the generation of a partition with very small interfaces [5]. Many common mechanical problems show a topology that meets this requirement; a clear example is represented by the rotorcraft analysis models described later in the results section.

It is not easy to define an appropriate dimension for the computational domain that is analyzed with a multibody multidisciplinary simulator. The multidisciplinarity requirement obviously does not allow any structure in the domain, because structurable, i.e. physical space related, and non-structurable, i.e. abstract, unknowns coexist in the solution space. Anyway a typical problem can be usually thought as quasi-monodimensional, with some multiple paths or closed-circuits. This is basically true because the underlying structure is usually made of rigid bodies connected by algebraic constraints, which are the irreducible parts of the computational domain. As a consequence, the computational grid can be subdivided into parts with an optimal ratio between internal and interface nodes. Clearly, the search for a minimal-interface partition is crucial, so this task cannot be performed manually; it is delegated to standard partitioning tools (for example, METIS [11]).

3.4.2 Multithread assembly/factorization

Another parallelization approach has been explored in view of its application to real-time simulation. In essence, the availability of sparse matrix storage schemes that result in very compact mapping of the non-zeroes, like the previously discussed column-compressed form, suggested to address the parallelization of the assembly on SMP architectures by allowing multiple concurrent assembly threads to access the elements in a competitive manner, without any prior model partitioning, by way of a specifically designed concurrency-aware element iterator. Each thread assembles its portion of problem on its local compressed buffer, accessing in read-only mode

```
for (int row = thr; row < N; row += Nthr) {
        for (int t = 1; t < Nthr; t++ ) {
                A[0][row] += A[t][row];
        }
}
```

Fig. 7. Contribution of thread `thr` to parallel assembly of buffers, where thread #0 buffer is used as the main storage.

the common structure of the indices as described in Figure 5. At the end, the threads sum their buffers in a single storage. This operation can be performed concurrently as well, because the buffers exactly map on the same coefficients, so, for instance, each thread may safely add a specific portion of the coefficients of all buffers, as illustrated in Figure 7.

This approach relies on the fact that assembly may account for up to 30% of the execution time, especially when multiple connections exist between a relatively limited set of nodes. This is the case, for instance, of aeroelastic problems, where elastic, inertial and aerodynamic elements contribute to the same equations and some of these elements may require intensive operations (e.g. table lookups for aerodynamic forces). Unfortunately the parallel assembly did not bring all the expected advantages, because the impact of assembly on the overall simulation cost reduced dramatically when the column-compressed form, which is essential for the multithread assembly, was introduced. Only very limited additional speedups were obtained, and the overhead related to task scheduling makes the parallel assembly ineffective or even adverse for very small models, so it is not applicable to real-time simulation.

Among the directions for future development, there remain few combinations of the above described improvements that are currently unsupported. Some of them may be of interest; for instance, the possibility to exploit multithreaded parallelization of assembly on SMP machines in conjunction with Schur parallelization on remote machines in a cluster may allow to exploit both performance gains simultaneously when addressing very large problems, in a staggered distributed solution layout.

4 Applications

4.1 Landing gear simulation

The problem addresses the simulation of the gear-walk phenomenon, as described in [9]: an instability related to the interaction of the Automatic Braking System (ABS) with the deformability of the landing gear.

The modeling of the forces that the tire exchanges with the ground represents a crucial point in this type of problems. The tire and shock absorber models implemented in MBDyn were inherited from an already validated software program, but they did not provide any Jacobian matrix, and some of the empirical functions they were based on could not be easily differentiated analytically. Furthermore, during

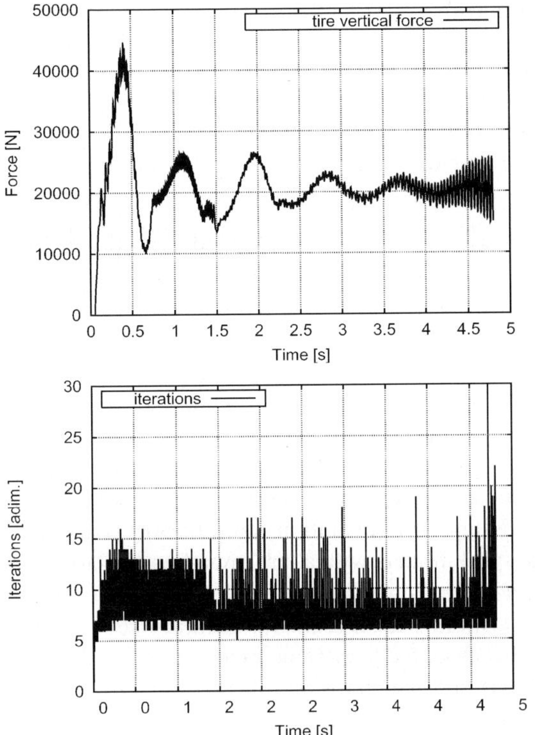

Fig. 8. Tire normal force and iterations during landing and braking maneuver.

the landing phase, the wheel, the shock absorber and the gear deformable structure are subjected to very high, impulsive loads, and the wheel and the shock absorber showed a very high stiffness that increased with gear compression. As a result, the default Newton–Raphson solver is not able to successfully integrate the equations of motion, even with the smallest time step that would appear reasonable for the problem at hand ($1e{-}3 \div 1e{-}4$ s).

The iterative matrix-free solver proved to be a valid alternative to the numerical differentiation of that specific user-defined elements, and allowed to efficiently simulate the whole landing and braking phases with fairly reasonable time steps. Figure 8 shows the iterations count as opposed to the tire normal force during a landing impact followed by braking in a manner that initiates the gear walk instability phenomenon. Note that the number of iterations remains limited and fairly independent from the changes in ground forces. The time step is $0.5e{-}3$ s.

4.2 Real-time simulation

The real-time simulation of robots and rotorcraft wind-tunnel models has been addressed by means of MBDyn [2, 13, 18, 19]. In all cases, the modeling needs could not be satisfied by models with less than 120–150 equations; thus the performances of the analysis were essentially dictated by the need to find an acceptable trade-off between the highest possible sample rate, requested by accuracy needs (e.g. a minimal number of steps per rotor revolution) or by interaction requirements with other tasks participating in the simulation (e.g. the controller process). Remember that, as already mentioned, the needs of the real-time simulation were the actual drivers of many of the performance-oriented speculations described in this paper.

Significant achievements, illustrated in [18], were:

a. rigid robots with up to 120 equations (6 DoF manipulator), with friction in the joints, have been simulated in real-time at up to 2 kHz sampling rate;
b. rigid-blade wind tunnel rotorcraft models with up to 180 equations have been simulated in real-time at realistic rotational speeds with a realistic, accuracy dictated number of steps per revolution at up to 900 Hz; issues remain for small scale (i.e. high rpm) models with a large number of blades;
c. rigid-blade, non-optimized helicopter models with up to 280 equations have been simulated in real-time with a realistic, accuracy dictated number of steps per revolution; in this case, the very same model used for fluid-structure interaction investigations was run, which exemplifies the versatility of the software and the potential for code and model reuse.

All the results illustrated above have been obtained with off-the-shelf, relatively inexpensive hardware: the multibody software was running on an Athlon 64 3000+ (2 GHz) single CPU, 64 bit architecture, with a 512 kB L2 cache.

One of the main points of speculation was the capability to exploit Symmetric Multi Processor (SMP) architectures to split the easily parallelizable portions of the solution into multiple threads of execution. Unfortunately, no appreciable success was obtained by spreading the Jacobian matrix and residual vector assembly, or by parallelizing the factorization and the back substitution with the "naïve" solver described in [17]. The most significant improvements, apart from an overall "squeezing" of the code, came from the "naïve" solver in scalar form.

4.3 Rotorcraft simulation

The parallelization of Jacobian matrix and residual vector assembly partially gave the expected improvements when addressing much larger models. For rotorcraft dynamics analysis, typical models [16, 22, 23] consisting in deformable rotor blades, detailed rotor hub and control system kinematics and, for tiltrotors, deformable wing dynamics, result in 600-2000 equations, with hundreds of beam, inertia and aerodynamics elements, and dozens of joints. The problem is very sparse, so the solution

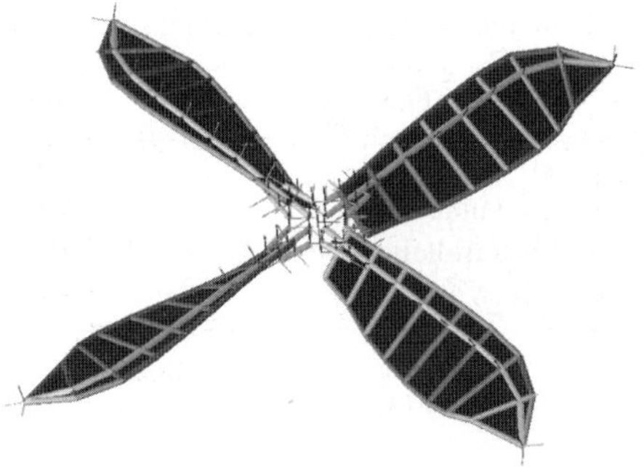

Fig. 9. Deformable rotor model.

by means of efficient sparse solvers, ranging from Umfpack to the built-in "naïve" solver, result in very fast analysis. Nonetheless, the factorization and back substitution, namely the linear solution, still accounts for a vast majority of the analysis time, while assembly reduces to at most 30% of the overall solution time, despite the high elements versus equations ratio.

The use of the column-compressed form of the sparse matrix storage showed dramatic reductions in computational time; this indicates that the cost of creating the sparse maps and then packing them into the form required by the solver accounts for a significant amount of the simulation time.

The parallelization of the assembly phase, on dual Athlon CPU systems, showed further reductions in the solution time of about 3÷4%, thus indicating that the actual cost of the elements assembly lies between 6 and 10% of the overall computational time; this is consistent with results from standard profiling software.

Tables 1–3 show some figures about computational time reductions for the tilt-rotor model described in [22]. The model of the rotor is illustrated in Figure 9; in the two cases, it consists in respectively about 900 and 1500 equations that describe the dynamics of the deformable rotor of a tiltrotor aircraft, supported by a deformable wing. The integration occurs with a fixed time step of 2.5e−4 that is required to capture the dynamics of the blades with the desired accuracy.

As a comparison, the performances related to two benchmarks, respectively made of about 1450 and 2900 equations, representing a string of beams clamped at one end and excited by an impulsive force at the other end, are illustrated in Tables 4 and 5. The assembly is required only every 5 residuals; as each step converges in 2–3 iterations, only one Jacobian matrix per time step is assembled. The figures above change dramatically if the Jacobian matrix is re-factored during each iteration, as shown in Table 6; note that in that case, the reference time is slightly more than 2.5 times that of Table 4. The advantages of the parallel assembly are essentially eaten up by the time spent in factorization; on the contrary, the enormous savings in creating and packing the sparse matrix are preserved almost untouched. The "naïve" solver

Table 1. Rotorcraft dynamics analysis: 890 equations (coarse model, realistic).

Feature	Time
Baseline	1.00
Column-compressed (CC)	0.81
Assembly parallelization + CC (2 CPU)	0.78
Naïve solver	0.71

Table 2. Rotorcraft dynamics analysis: 1455 equations (refined model, realistic).

Feature	Time
Baseline	1.00
Column-compressed (CC)	0.83
Assembly parallelization + CC (2 CPU)	0.80
Naïve solver	0.78

Table 3. Rotorcraft dynamics analysis: 2415 equations (over-refined model, unrealistic).

Feature	Time
Baseline	1.00
Column-compressed (CC)	0.83
Assembly parallelization + CC (2 CPU)	0.79
Naïve solver	0.75

magnifies its performances, thanks to its great factoring efficiency, while the Schur solver does not improve as well, because the transmissions related to Jacobian matrix assembly and partial factorization impact on the performance reductions.

This benchmark may not be considered representative of a realistic analysis for many reasons. First of all, it appears that the "naïve" solver performs better than Umfpack with an even larger problem; however, realistic applications show the opposite. Also, the Schur parallelization appears to behave quite well, since it gives a quasi-linear scaling; this behavior is not confirmed by realistic applications, where the interface is larger (it is exactly 1 node in the above example) and the partitioning is not exactly symmetric. The above results have been obtained using Umfpack as local subproblem linear solver and LAPACK as interface linear solver. Better performances are expected when using the "naïve" solver for the subproblems, although some more refinement is needed. Unfortunately, no significant scalings appear for the rotor essentially because no optimal partitioning on two CPUs can be determined. However, earlier analysis on a 8 CPU SMP architecture showed interesting performances also on realistic cases, as discussed in [21]; note however that the optimistic results obtained in that work were partially related to less than ideal scalar perform-

Table 4. Structural dynamics of beam benchmark: 1452 equations; modified Newton.

Feature	Time
Baseline	1.00
Column-compressed (CC)	0.80
Assembly parallelization + CC (2 CPU)	0.77
Naïve solver	0.66
Solution parallelization (Schur; 2 CPU)	0.58

Table 5. Structural dynamics of beam benchmark: 2892 equations; modified Newton.

Feature	Time
Baseline	1.00
Column-compressed (CC)	0.82
Assembly parallelization + CC (2 CPU)	0.78
Naïve solver	0.69
Solution parallelization (Schur; 2 CPU)	0.61

Table 6. Structural dynamics of beam benchmark: 1452 equations; full Newton.

Feature	Time
Baseline	1.00
Column-compressed (CC)	0.77
Assembly parallelization + CC (2 CPU)	0.77
Naïve solver	0.50
Solution parallelization (Schur; 2 CPU)	0.53

ances. However, the above figures should give an idea of the capabilities brought in by all the approaches considered in order to speed-up the analysis of models that, although "small" from the point of view of linear solution, can be considered "large" from the point of view of multibody analysis.

5 Conclusions

The paper discussed some of the strategies recently applied to the multibody software MBDyn to further reduce its execution time. Generally beneficial improvements have been sought, although mainly focusing on the most promising for the types of problems commonly addressed by this software.

Apart from the dramatic improvements obtained by designing a specialized linear solver for small and very small sparse systems, described in a companion paper, significant improvements have been obtained by addressing the assembly of sparse matrices, which is now performed in parallel on SMP architectures, and reusing as much as possible the sparsity patterns across different assembly iterations and time steps.

The introduction of inexact, iterative matrix-free solvers allowed to efficiently solve very stiff problems with incomplete Jacobian matrix.

A coarse-grained topological parallelization scheme for the solution of the non-linear problem has been revitalized for the very specific class of almost tree-like problems that are typical of rotorcraft dynamics simulation.

Only some of the developments investigated in this work were beneficial for the purpose of improving the performances when running simulations in real-time. Nevertheless, the versatility and the overall efficiency of the software as a whole was improved, and very promising real-time simulation capabilities have been obtained, along with better overall performances in batch simulations.

The enhancements discussed in this paper may serve as guidelines for the improvement and optimization of general-purpose software of the same class, and in general for numerical software with similar requirements.

References

1. E. Arge, A. M. Bruaset, and H. P. Langtangen (eds). *Modern Software Tools for Scientific Computing*. Birkhäuser, 1997.
2. M. Attolico and P. Masarati. A multibody user-space hard real-time environment for the simulation of space robots. In *Fifth Real-Time Linux Workshop*, Valencia, Spain, November 9–11 2003.
3. M. Attolico, P. Masarati, and P. Mantegazza. Trajectory optimization and real-time simulation for robotics applications. In *Multibody Dynamics 2005, ECCOMAS Thematic Conference*, Madrid, Spain, June 21–24 2005.
4. W. Bangerth, R. Hartmann, and G. Kanschat. `deal.II` *Differential Equations Analysis Library, Technical Reference*. `http://www.dealii.org`.
5. P. E. Bjørstad and A. Hvidsten. Iterative methods for substructured elasticity problems in structural analysis. In G. Meurant R. Glowinski, G. Golub and J. Périaux (eds), *Proceedings of the first international symposium on Domain Decomposition Methods for Partial Differential Equations*, SIAM, Philadelphia, PA, 1988.
6. M. Dæhlen and A. Tveito (eds). *Numerical Methods and Software Tools in Industrial Mathematics*. Birkhäuser, 1997.
7. T. A. Davis. Algorithm 832: Umfpack, an unsymmetric-pattern multifrontal method. *ACM Transactions on Mathematical Software*, 30(2):196–199, June 2004.
8. L. Goldthwaite. Technical report on C++ performance. Technical Report WG21/N1666 J16/04-0106, ISO/IEC, May 2004. Downloadable from `http://www.open-std.org/jtc1/sc22/wg21/docs/papers/2004/n1666.pdf`.

9. S. Gualdi, M. Morandini, P. Masarati, and G. L. Ghiringhelli. Numerical simulation of gear walk instability in an aircraft landing gear. In *IFASD 2005*, Muenchen, Germany, June 28–July 1 2005.
10. W. D. Henshaw. Overture: An object-oriented framework for overlapping grid applications. In *2002 AIAA conference on Applied Aerodynamics*, St Louis MO, September 2–6 2002.
11. G. Karypis and V. Kumar. A fast and high quality multilevel scheme for partitioning irregular graphs. *SIAM J. Sci. Comput.*, 20(1):359–392, 1998.
12. C. T. Kelley. *Iterative methods for linear and nonlinear equations.* SIAM, Philadelphia, PA, 1995.
13. P. Masarati, M. Attolico, M. W. Nixon, and P. Mantegazza. Real-time multibody analysis of wind-tunnel rotorcraft models for virtual experiment purposes. In *AHS 4th Decennial Specialists' Conference on Aeromechanics*, Fisherman's Wharf, San Francisco, CA, January 21–23 2004.
14. P. Masarati, M. Morandini, G. Quaranta, and P. Mantegazza. Open-source multibody analysis software. In *Multibody Dynamics 2003, International Conference on Advances in Computational Multibody Dynamics*, Lisboa, Portugal, July 1–4 2003.
15. P. Masarati, M. Morandini, G. Quaranta, and P. Mantegazza. Computational aspects and recent improvements in the open-source multibody analysis software "MBDyn". In *Multibody Dynamics 2005, ECCOMAS Thematic Conference*, Madrid, Spain, June 21–24 2005.
16. P. Masarati, D. J. Piatak, G. Quaranta, and J. D. Singleton. Further results of soft-inplane tiltrotor aeromechanics investigation using two multibody analyses. In *American Helicopter Society 60th Annual Forum*, Baltimore, MD, June 7–10 2004.
17. M. Morandini and P. Mantegazza. Using dense storage to solve small sparse linear systems. Submitted to *ACM Transactions on Mathematical Software* (ACM TOMS).
18. M. Morandini, P. Masarati, and P. Mantegazza. Performance improvements in real-time general-purpose multibody virtual experiment on rotorcraft systems. In 31st *European Rotorcraft Forum*, pages 18.1–14, Firenze, Italy, September 13–15 2005.
19. M. Morandini, P. Masarati, and P. Mantegazza. A real-time hardware-in-the-loop simulator for robotics applications. In *Multibody Dynamics 2005, ECCOMAS Thematic Conference*, Madrid, Spain, June 21–24 2005.
20. G. Quaranta, Giampiero Bindolino, P. Masarati, and P. Mantegazza. Toward a computational framework for rotorcraft multi-physics analysis: Adding computational aerodynamics to multibody rotor models. In *30th European Rotorcraft Forum*, pages 18.1–14, Marseille, France, 14–16 September 2004.
21. G. Quaranta, P. Masarati, and P. Mantegazza. Multibody analysis of controlled aeroelastic systems on parallel computers. *Multibody System Dynamics*, 8(1):71–102, 2002.
22. G. Quaranta, P. Masarati, and P. Mantegazza. Dynamic characterization and stability of a large size multibody tiltrotor model by pod analysis. In *ASME 19th Biennial Conference on Mechanical Vibration and Noise (VIB)*, Chicago Il, September 2–6 2003.
23. G. Quaranta, P. Masarati, and P. Mantegazza. Assessing the local stability of periodic motions for large multibody nonlinear systems using POD. *Journal of Sound and Vibration*, 271(3–5):1015–1038, 2004.
24. Y. Saad. *Iterative methods for sparse linear systems.* PWS Publishing Company, Boston, MA, 1996.

25. T. L. Veldhuizen and M. E. Jernigan. Will C++ be faster than Fortran? In *Proceedings of the 1st International Scientific Computing in Object-Oriented Parallel Environments (ISCOPE'97)*, Lecture Notes in Computer Science. Springer-Verlag, 1997.
26. T. Veldhuizen. Using C++ template metaprograms. *C++ Report*, 7(4):36–43, May 1995. Reprinted in C++ Gems, ed. Stanley Lippman.

Engineering Education in Multibody Dynamics

Paul Fisette and Jean-Claude Samin

Department of Mechanical Engineering, Université Catholique de Louvain, 2, Place du Levant, 1348 Louvain-la-Neuve, Belgium; E-mail: {fisette,samin}@prm.ucl.ac.be

Abstract. This chapter of the book is devoted to teaching the discipline of Multibody Dynamics. This discipline concerns young students (baccalaurean) as well as more senior students (master's degree, and even PhD degree). The first part of this chapter is strongly inspired by various opinions, which are also those of the authors of this chapter. In particular, some of the comments already expressed in [1] are rigorously reproduced in this part, as well as various comments which have also already been published [2, 3]. The content of this first part of the chapter also greatly benefits from the opinions expressed at the occasion of the round-table during the session on "Education of multibody dynamics" of the International ECCOMAS Thematic Conference on Advances in Computational Multibody Dynamics (Universidad Politecnica de Madrid, Spain, June 21–24, 2005). The continuation of the chapter gives an example of the experiment lived at the Université Catholique de Louvain (Belgium) [3, 4] with undergraduate students, experience which has also been successively reproduced in Equator (University of Quito) [5] with graduate students.

1 Introduction

Even at the present time, Multibody Dynamics is often still considered as a research topic rather than an established subject for mandatory university courses in the education curriculum of mechanical engineers. The current situation is somewhat similar to the pre-finite elements period. During the sixties, structural dynamics was mainly a theoretical subject and students were trained to solve problems by means of analytical methods. But thanks to the evolution of computer facilities, today no university offers a degree in mechanical or civil engineering without a mandatory course in finite elements and various projects based on commercial finite elements computer codes. This matter obviously responds to an industrial demand.

Similarly today, multibody dynamics simulation packages are more and more used in industries. Most of these software are distributed by commercial companies and issue from various research teams who previously published scientific papers

J.C. García Orden et al. (eds), Multibody Dynamics. Computational Methods and Applications, 159–178.

describing the details of their methods. However, software implementation packages without a detailed theoretical manual are unfortunately not rare and the training courses organized by these companies focus on the working features of their own codes and not on the theoretical bases. As a consequence, there is a real danger of practical engineers misusing a multibody code because they are not aware of the underlying hypothesis and/or limitations included in the code. Facing this situation, and since the number of engineers which are potential users of such software is growing, there is a real need for including the field of Multibody Dynamics (MBD) in a standard engineering education curriculum.

As already said, teaching multibody dynamics to engineering students responds to a specific industrial demand. But in addition to this, and thanks to the very large class of practical applications covered by this field, MBD can also be very useful in the global engineering education context. In particular, two basic domains are immediately concerned: classical mechanics and mathematics.

At the undergraduate level, courses of "classical mechanics" generally treat the kinematics of rigid bodies, Newton–Euler dynamical equations, etc. Unfortunately, these mechanical concepts are generally illustrated by pure academic examples because of the relatively poor means at student's disposal: paper and pencil. For the same reason, student's exercises are strictly limited: for instance a ball rolling on a plane, a simple or double pendulum, this kind of application is obviously not very attractive and students may naturally ask "equations of motion, what for in a practical and realistic situation"?

Concerning various courses of mathematics (linear algebra, numerical analysis, and computational methods), all teachers face the same problem, i.e. the difficulty of finding attractive and realistic applications of their fields. As for the domain of classical mechanics, the MBD domain offers a large class of potential illustrations.

2 General Considerations

2.1 Computer programming

MBD is based on different theoretical bases, like for instance: choice of generalized coordinates (e.g. Euler angles, Cardan angles, Euler parameters, Denavit–Hartenberg, dual numbers) and methodologies for the systematic description of kinematic constraints (e.g. method of constraints, loop-closure equations) in mechanical systems. Moreover, equations of dynamics may be deduced from different approaches (Newton–Euler, Lagrange, Gibbs–Appell, d'Alembert principle, Gauss, graph theory). As a consequence, engineers who deal with MBD models must build their skills on firm physical, mathematical and numerical foundations. For this reason, lessons learned must be applied to realistic models to reach a deep understanding of these concepts.

At this point, we think that it is necessary to distinguish between undergraduate and graduate levels of engineering education.

2.1.1 Undergraduate level

The learning of MBD principles does not require any specific programming skill, but the development of even simple MBD software implies non-trivial programming knowledge, which may not be strictly related to the wider curriculum of mechanical engineers. For educational purposes at the undergraduate level, it is thus unnecessary and unrealistic to expect students to write even elementary MBD software from scratch. On the other hand, when inspecting the contents of equations of motion (whatever the underlying formalism used to produce them), a major part (from 50% for a few d.o.f. systems to more than 95% for large systems) is of a pure calculating nature and thus represents more student perspiration than inspiration. Thus, conferring this part (but only this part!) to an MBD generator allows students to concentrate on far more relevant tasks (e.g., implementation of a specific wheel/ground model, use of a kinematic Jacobian) and also to dealing with realistic applications.

Using "computer aided" MBD, even at the undergraduate level, raises several questions. Indeed the interest in using a multibody program to model realistic applications (car, truck, motorbike, bicycle, walking robot, etc.) might be questionable with respect to educational objectives. Let us think about programs for which user intervention is restricted to "clicking and dragging the mouse" to selecting user-friendly menus and entering data. Such "black-box" programs, while quite suitable and powerful for research projects and industrial applications, may be inappropriate or even "anti-pedagogical" for teaching fundamental mechanics to beginning students. Thus even at this stage, students must have normal skills of computer programming in some higher level language of their choice (MATLAB, C, etc.) and they should be required to develop software for assembling different modules: data input and output, automatic build up of equations and numerical solutions. Through the computer assembling of a "methodology", the students usually reach a deep understanding of the subject. For this reason computer programming is strongly recommended in basic multibody dynamics classes, but pieces of code that students may use during the development of their own software should be made available. This will help students avoid getting lost in coding all the software, instead of concentrating on the overall structure of the program. Students should also be encouraged, as in other fundamental courses, to use professionally tailored linear algebra and numerical integration subroutines available in packages such as MATLAB, MAPLE, and MATHEMATICA, etc.

Depending on the strategy adopted by the teacher, different levels of MBD basic tools can be put at the student's disposal. The "lowest level" consists in a library of elementary functions allowing them to construct the elementary kinematic properties of the system and in defining, in great detail, all the forces and moments acting on the system. This approach was for instance adopted early on by Kane with his AUTOLEV software [6, 7] and by Nikravesh in his book [8] and its companion library of sub-routines. Both were based on the assembly of elementary FORTRAN modules dealing with vector manipulations, elementary kinematics, transformation matrices, etc. Presently this approach is still adopted by some educators, but using progresses in computer languages. A good example is provided by the EASYDYN software [9] developed at FPMs (Faculté Polytechnique de Mons, Belgium): it is a symbolic software for which students only define on their own the relative positions and orientations of the bodies constituting the system in terms of independent (relative) generalized coordinates. If they are, for instance, facing a closed-loop mechanism, they also have to give the explicit equations relating the dependent coordinates in term of the independent ones. Students must then describe all interacting forces and torques, without forgetting fundamental principles like Newton's third principle of action and reaction. Then, by symbolic manipulations, EASYDYN automatically produces all the requested expressions for the linear and angular velocities and accelerations, thus saving all student perspiration over tedious computations. The software is based on the d'Alembert variation principle and thus also automatically produces the dynamical residual equations needed by the implicit Newmark integration scheme which, in turn, is also provided in the code as a sub-routine.

Without any doubt, this type of "low level" MBD approach presents several advantages for undergraduate students: in particular, it is obvious that to use this computer tool students must first dominate the fundamentals of classical mechanics before being able to tackle practical and realistic problems. One disadvantage comes from the fact that the simulation code generated by students cannot be computationally optimized, but everybody should accept that this is not an important educational matter at the undergraduate level: the unique practical consequence is a limit to the size and complexity of the mechanical systems which can be simulated within a reasonable computer time.

Also using the symbolic computation capabilities of recent computer languages like (C and C++), ROBOTRAN [10–12], which will be illustrated in Sections 4 and 5 is to be considered as a "medium level" MBD software from the educational point of view. Based on the same philosophy as EASYDYN [9], the main difference lies in the fact that it also offers a suitable tool for training graduate students in more sophisticated problems in terms of numbers of bodies, loop non-linearities, etc., as well as complicated geometry (for instance, the wheel-rail contact in railway dynamics, the contact patch in cam-follower systems).

2.1.2 Graduate level

In multibody dynamics there is a wide variety of theoretical approaches. Which kinematic formulation and dynamical principle is the most effective for teaching at the graduate level? Although this is a key question, a definite answer cannot be given and the final choice depends on the instructor's personal judgment and preferences. However, teaching the dynamical formulations adopted in commercial software is recommended in at least a few introductory lectures. The student will thus have a better understanding of the theoretical bases, performances and possible limitations of these software.

Students may also gain more confidence in any MBD program if they first use it on academic examples for which they have previously written the equations "by hand". They then accept using a "toolbox" MBD optimized program for more sophisticated systems, whose equations' complexity only comes from a large amount of "trivial" computation.

2.2 Learning physical "modeling" and "analysis"

To learn how to effectively model a mechanical system, and then to correctly analyze the results constitutes an objective impossible to reach by means of classical lectures. In particular, choosing reasonable hypotheses (not too restrictive but still retaining enough detail as to be able to tackle the physical phenomena one is interested in) is really difficult. Nevertheless, it is the teacher's duty to transmit his own experience to young students intent on becoming engineers as best he can. In the academic world, it is unanimously recognized that this transmission can be made only by the channel of projects and the personal contacts between student and teacher those imply.

3 Project-Based Learning

The Project-based Learning approach (PBL) is now more and more recognized to be very effective and profitable for student education in various disciplines, such as mechanical design, system modeling, robotics, mechatronics [13]. In addition to the fact that student projects allow them to develop their capability in working together, in managing a group and in scheduling work over a rather long period (i.e. several weeks), such projects exhibit two interesting characteristics:

- Contrary to a problem or an exercise, a project starts from realistic applications or phenomena, such as those students will regularly encounter during their professional career. Moreover, the project is formulated under the form of open questions to answer, which reflect a real engineering problem to solve (for example, the design of a small robot to automatically clean windows or the modeling and understanding of "truck jacknifing" phenomena).
- The above-mentioned scope and range of student projects unavoidably make them multidisciplinary and this is of course a very feature of enrichment and appeal in the modern and up-to-date thinking of engineering education.

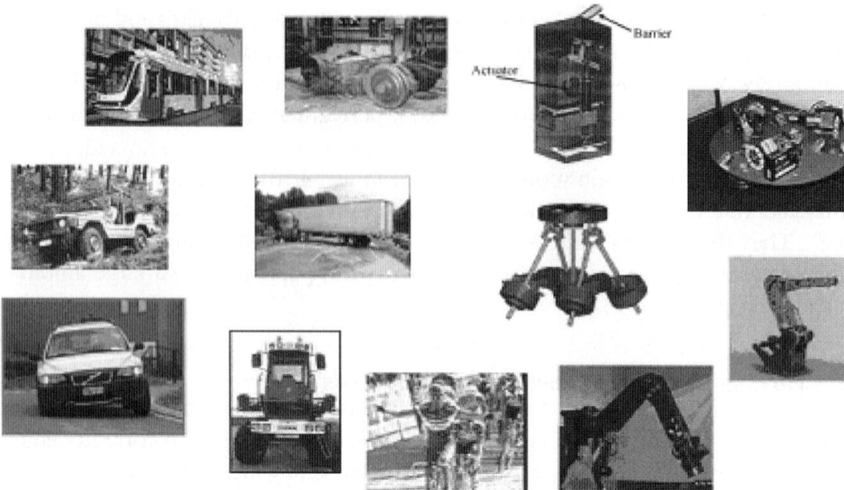

Fig. 1. Multibody families.

In view of the previous considerations, it is obvious that the MBD discipline is a perfect candidate for fulfilling the above requirements of the PBL approach in terms of scope, range, application and multi-disciplinarity.

Regarding the applications, we can assert that the domain is outstandingly rich and very varied. We need only think about the multibody families of applications: cars and trucks, motorbikes, trains, cranes, manipulators, walking and rolling robots, mechanisms, machine tools, carnival equipment (see, for instance, Figure 1). Going one step further, even inside a given family there exist numerous problems which can lead to interesting project definition. For instance in cases of cars and trucks, one can study suspension kinematics, the stability of a truck with trailer, the comfort of a car or the handling of a sport car, etc. In some sense, it is really a "gift of God" for a teacher to be involved in the field of MBD, when he has to set up projects in the field of virtual engineering, with no problem whatsoever in "imagining" a new theme every year!

In terms of *multi-disciplinarity*, the same positive conclusion can be drawn. Indeed, as we shall see in more detail in the next sections, the modeling and analysis of a multibody system involve various topics (and thus, practically, various courses), namely: 3-D geometry, body kinematics and dynamics, linear algebra and numerical methods. Depending on the project scale and ambition, one can also resort to programming methods and Computer-Aided-Drawing.

Thus, for setting up a project, one can easily conclude from this that the room for maneuver is large in terms of applications to propose, problems to solve and disciplines to involve. From our point of view, this has a direct consequence on the type of student such projects can be organized for. After more than ten years of experience, we can assert that a multibody project can be proposed not only at the graduate level but also at the undergraduate one, just by "scaling" the project's scope and ambition accordingly: this will be the subject of Section 4. In the past two years, we have even set up a multibody project for "technical-commercial engineering" graduate students, within the framework of their course in mechanics, to improve their understanding of Newton–Euler laws via modeling and analysis of real examples.

4 Multibody Projects at Undergraduate Level

From our teaching experience, we realize that students in engineering require successive learning "layers" (e.g., lectures, projects, student theses) to acquire sufficient expertise in a given field. In particular, as a sequel to the undergraduate lecture in classical mechanics (which covers, as far as we are concerned, rigid body motion: theory and exercises), student capabilities are limited to applications proposed within the lecture framework (planar pendulum, small cart, winch, etc.). Further to these observations, we decided to set up a project in multitbody dynamics (length: 11 weeks, groups of 6–8 students) which takes place after the main course.

After a few years of experience, we can claim that this multi-disciplinary project really improves the student's skills in the field of multibody system modeling, including a timid but existing engineer attitude with respect to the work they produce. This attitude will be of course more deeply developed during subsequent learning "layers" at graduate level, and later on.

Let us highlight the students learning problems on the basis of the academic example of Figure 2, issuing from a final examination in classical mechanics.

The system consists of a planar one-degree of freedom pendulum sustained by rollers constrained to roll without slip on a circular ring. The unique equation of motion (to be found) is:

$$(Ml^2 + 3mb^2)\ddot{\varphi} + (Ml + 2mb\cos\alpha)g\sin\varphi = 0. \tag{1}$$

The following observations deserve to be emphasized.

1. Although the system appears to be quite simple and somewhat "artificial" (that is, with respect to realistic applications), obtaining the correct equation of motion is not a trivial task for undergraduate students because frequent mistakes appear at various steps of the modeling process, such as:
   - wrong definition of frames and/or vectors
   - wrong sign of sine/cosine functions
   - non-exhaustive inventory of forces (often!)
   - wrong inertia computation
   - hazardous use of the Newton–Euler equations or d'Alembert principle
   - etc.

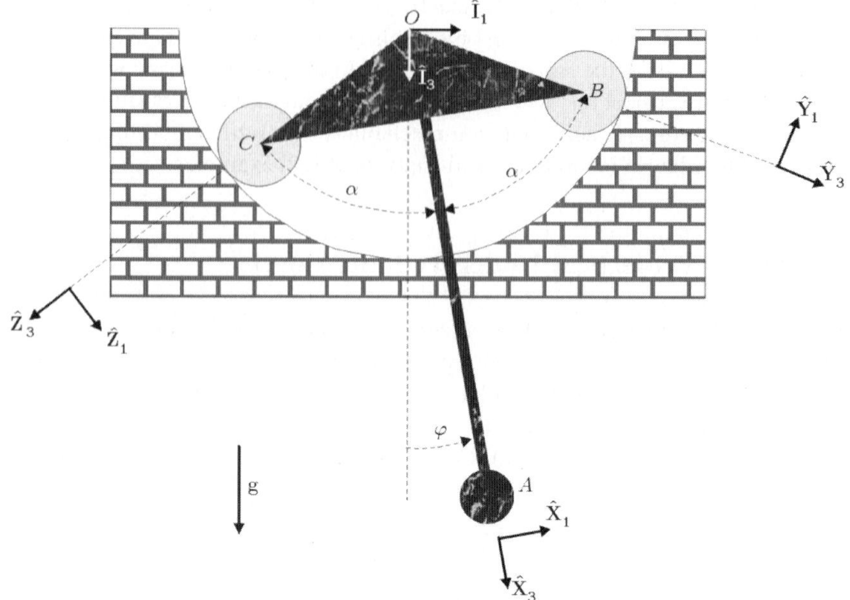

Fig. 2. Classical mechanics course: example of examination problem.

2. The artificial character of the system is not only a question of number of bodies involved but also comes from the rather restrictive nature of the modeling hypotheses with respect to reality: perfect joints (no backlash, no friction), perfect rolling, perfect bodies, unlimited motion, etc. With this respect, let us distinguish two kinds of student attitude. The first one "feels" the restrictive nature of the hypotheses and is generally – and logically – frustrated thinking to real applications that he will encounter as a future engineer. The other one, being too much "immersed" into the perfect word of system modeling could naïvely liken real applications to academic ones and this is a rather poor – or even dangerous – engineering attitude.
3. The final comment relates to the equation of motion itself: once obtained (correct or not) – what to do with this equation and how to do it? If students are already a bit familiar with numerical methods (for example, time integrators) and if teachers think about establishing a connection between "equations of motion" and "numerical time integration" for instance, then students can probably accept – and even appreciate – their investment when writing differential equations of motion.

In the above considerations, there is no attempt to discredit the contents of the course in classical mechanics: the difficulty in learning it, requires a rather slow and long process based on academic and progressive applications, starting from spring-mass pendulum to systems like that of Figure 2. Precisely, the major conclusion is that this long – but indispensable – learning process deserves to be "crowned" with an interesting activity which will exploit fresh students skills in mechanics, numerical methods and – to a lesser degree – in computer science: this is the purpose of the modeling project, whose methodology is the subject of the next section.

4.1 Student background and pedagogical objectives

Before to take up the project, the students have a theoretical background which is given by two courses in classical mechanics covering the following topics respectively:

- Fundamental physics and dynamics of particles
  - Kinematics: position, velocity and acceleration vectors
  - Frames, inertial frames
  - Linear and angular momentum vectors
  - Newton's three laws of fundamental mechanics
  - Energy, power, conservative motions.
- Dynamics of rigid bodies (based on the first two chapters of [14])
  - Frames, transformation matrices, Euler and Tait–Bryan angles
  - Angular velocity vector
  - Geometry of masses (center of mass, inertia matrix)
  - Newton–Euler equations for rigid bodies
  - Generalized coordinates, holonomic and non-holonomic constraints, degrees of freedom
  - Principle of virtual power and Lagrange multipliers technique for constrained systems.

The new learning layer of the project aims to answer the following questions and expectations:

- At the close of the course in classical mechanics, both the nature of the exercises (often 2D "academic" problems) and the associated results (analytical equations of motion) are often perceived by students as a weak "return on investment". Roughly speaking: equations for what purpose?
- A very important point relates to the modeling hypotheses: which ones and how to formulate them? This aspect is generally not covered by standard exercises since all the modeling hypotheses are often included in the problem formulation itself.

Obtaining equations of motion is an important and crucial phase but analytical equations are intrinsically useless: a numerical post-process (for example, equilibrium solution, modal analysis, time integration), which is generally tackled in another undergraduate course, is necessary to study the system motion and – this is not negligible – to promote an actual multi-disciplinary activity. Additionally in our case, a CAD course also participates actively, as a third partner, to ensure the 3D animation of the mechanical system under consideration.

In Figure 3, these disciplines are illustrated for the modeling and analysis of a sidecar skidding (project of the academic year 2002–2003), successively:

Mechanics for the modeling phase:

- Estimation and/or computation of dynamical data (centers of mass, inertia matrices)
- Suspension characteristics – constitutive laws
- Tire/ground force – constitutive laws

Computer science:

- Elaboration of a well-structured algorithmic process
- Development of a numerical simulation program

Numerical methods:

- Use or implementation of the Newton–Raphson method (equilibrium search)
- Use or implementation of a Runge–Kutta method (time integration)
- Implementation of interpolation techniques (for data fitting, results, etc.)

Computer Aided Drawing:

- 3D Drawing of mechanical components
- Implementation of a camera and a target in a 3D scene for animation.

4.2 Project methodology

The envisaged project consists of the understanding, modeling, simulation, 3D animation and – elementary – analysis of a multibody system in a given situation and/or environment. Recent projects are illustrated in figure 4; they dealt with the dimensioning of a merry go round, the stability of a careless cyclist, the guidance capabilities of a railway bogie, the lateral skid of a sidecar, the motorbike wheeling, etc.

The subjects are carefully chosen to match the above-mentioned objectives: they represent *realistic* applications where the necessary data must be collected and preprocessed by the students; they are suitable for formulating *relevant hypotheses* and for observing some *numerical method* limits and performances (for example, the "stiff" mathematical character of a lateral wheel/ground slip model); they absolutely require a *computer program* to be developed by the students, as well as a *3D animation* which is helpful for interpreting the simulation results.

Here below the main steps of the project are briefly explained in a chronological way (project length: 11 weeks).

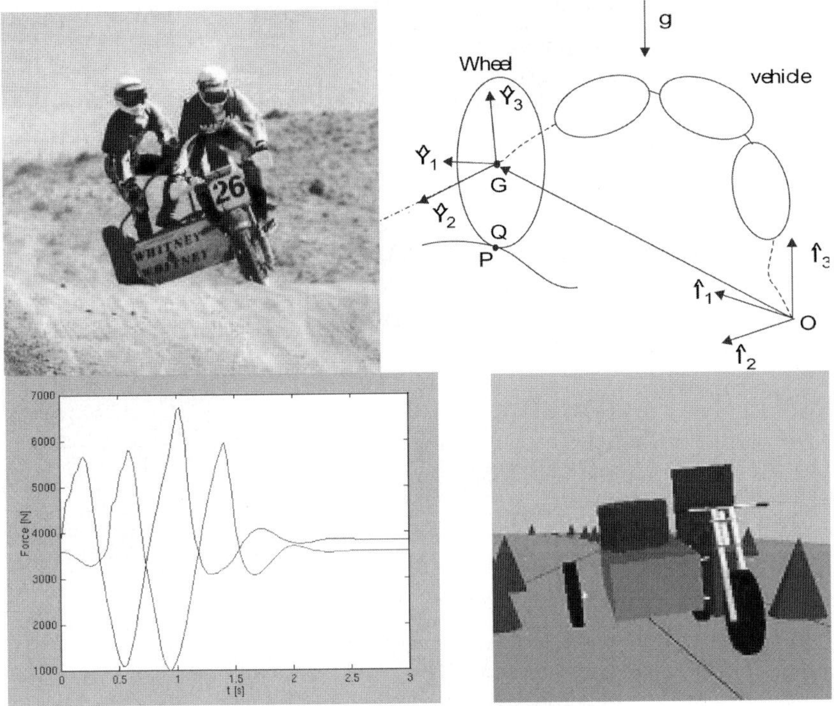

Fig. 3. Multibody dynamics, a real multi-disciplinary activity.

4.2.1 Understanding of the system to analyze

Obviously, the first task asked to the students is to find a practical example of the proposed system (the motorbike of a cousin, the ATV of a sports association, a bogie of a railway company, etc.) and to understand it from a geometrical, topological (set of bodies and joints) and dynamical (masses, compliant elements, etc.) point of view. Indeed, they need to match the real system with the multibody modeling requirements.

4.2.2 Getting familiar with the computer tools

Since the project will involve various computer tools, i.e. a symbolic program (see below), a numerical environment (for example, MATLAB) and a CAD program for 3D animation, a pre-project based on the single example shown in Figure 5 is proposed to the students, for which hypotheses, data and results are given: they – only – have to make the multibody model, the simulation program and a simplified animation.

Fig. 4. Examples of recent project applications.

Fig. 5. A simplified vehicle suspension (pre-project).

4.2.3 Project planning

Being aware of the problem to analyze, of the full-project objectives and of its final deadline(!), each group of students must properly schedule their 11 weeks project, in a dynamic manner, using a PERT (Program Evaluation and Research Task) diagram in which all the tasks are listed in advance (but not allotted). An example, illustrated in Table 1, is delivered to the students for the first three weeks of the "pre-project" in which they familiarize with the computer tools for simple examples (i.e., spring-masses and simplified suspension systems).

Table 1.

Code	Tasks	Actors	w1	w2	w3
10	Getting started with the problem	all			
20	Kick off meeting with the tutor	all			
30	Spring masses : equations of motion (analytically)				
40	Spring masses : equilibrium search (analytically)				
50	Getting started with Robotran software				
60	Three "spring/masses" models	all			
70	Spring masses and suspension : Matlab model				
80	Spring masses and suspension : simulation				
90	Spring masses and suspension : analysis				
100	3D drawing and animation				
110	Redaction of a report				
120	Report presentation	all			

Fig. 6. Double wishbone suspension (ATV suspension).

4.2.4 Formulation of relevant modeling hypotheses

In this task, the students must select and justify the modeling hypotheses they will take for their model. Sometimes, the justification can only be done a posteriori, i.e. at the end of the project, but this is not a real problem pedagogically speaking. From our point of view, selecting and justifying the hypotheses is the most relevant and important tasks in an engineering context. Let us give some examples.

Neglecting the internal engine motion is a valid hypothesis for a "motorbike wheeling" and it is rather clear for most of the students. But replacing a double wishbone suspension (Figure 6) by the simplified one of Figure 5 for a pure dynamical analysis is more difficult to accept, while being valid too in most situations.

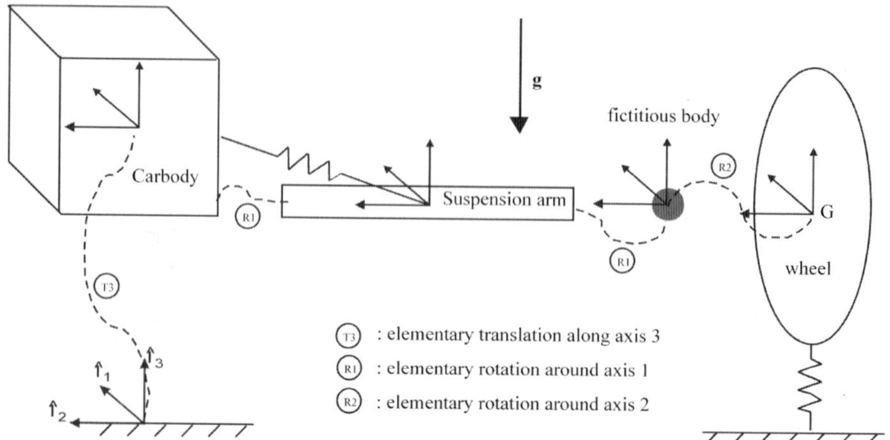

Fig. 7. Multibody model representation.

4.2.5 Symbolic multibody model

Once the hypotheses have been taken and unanimously accepted by the group, the multibody model can be drawn and described in terms of bodies, joints, degrees of freedom, position vectors, existence of internal/external forces and torques. Figure 7 shows the so-called "reference configuration" of a multibody system as required by ROBOTRAN [12], precisely for the simplified suspension depicted in Figure 5.

Since the model that ROBOTRAN will generate on the basis of this configuration is fully symbolic, it is not yet vital to have all the numerical data at one's disposal at this level. Moreover, as regards the internal/external forces whose constitutive law will represent a key task in the project (see below), only their existence must be indicated at this level (i.e., on which points of which bodies they must be applied).

4.2.6 Data acquisition and/or computation

Numerical data are of different natures and must also be collected (or computed) with an engineering "feeling" to avoid unnecessary work (see model hypotheses). Beside the geometrical data that students can obtain quite easily (data sheet, direct measurements, etc.), they must compute the center of mass and the inertia matrix of the various bodies. Once again, more than applying formulae for computing those elements (involving complex integrals), an important aspect of the project relates to the way students "feel" which are the mechanical devices which have a significant contribution in computing the center of mass and the inertia matrix of a body (for example, for a motorbike frame: the engine, the tubular chassis, etc.) and which elementary shape and density to confer to them, for sake of simplicity.

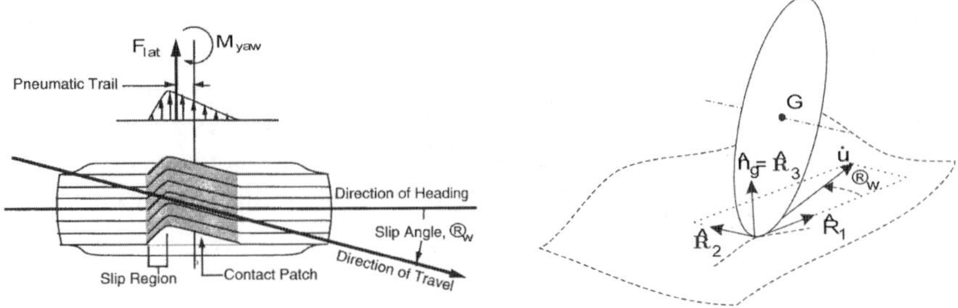

Fig. 8. A specific constitutive law: the lateral tire/ground force.

4.2.7 Understanding and computation of environment forces

This is also a crucial point for which the symbolic generation process (see below) is really helpful since it provides the students with most of the kinematic ingredients required for the introduction of these forces: absolute position, orientation, velocity of a body, etc., whose computation really require more perspiration than inspiration and are thus not really interesting, pedagogically speaking.[1]

Thus, having those kinematic ingredients at their disposal, students have to understand a specific constitutive law (for example, the transverse tire/ground slip force shown in Figure 8), to adapt it to the symbolic model and finally to introduce it in their simulation program.

4.2.8 Numerical program and first simulations

This indispensable task – performed in the MATLAB environment – can start as soon as the previous steps related to system modeling are completed, i.e. with equations "ready to be programmed". Let us point out that, thanks to the symbolic ingredients (i.e. full kinematics and full dynamics of the multibody system, but with no force description neither kinematic constraints resolution) provided by ROBOTRAN under the form of MATLAB functions with suitable i/o arguments, the number of statements of the simulation program (including data and post-processing) is rather low (between 50 and 100) whatever the size of the system (a bicycle, a bogie, etc.). This avoids that the project be too much programming-oriented. In fact, it is just the opposite since the most part of the work is to assemble a few blocks: some symbolic ROBOTRAN outputs, a couple of user functions, an algebraic system solver, a Runge–Kutta integration method, etc. In spite of that, unfortunately, undergraduate students skills in

[1] On two accounts: first because this has already be done during the previous course, secondly because the size of the proposed systems is too large for manual kinematic computations.

computer science often lack for rigor and/or practice, sometimes leading to a rather long debugging phase.

Close of this work, simulation can take place, the first step being generally to find the equilibrium of the system (for example, the vehicle at rest) with a Newton–Raphson (N–R) type method. This, apparently simple, task could also be detailed here for various reasons. Let us only mention some interesting features, mainly related to vehicle applications:

- Non-linearties such as the normal intermittent tire/ground contact can be simply disregarded during the N–R iterative process, by artificially considering a ground that is able to pull a wheel vertically. Using this trick, the correct equilibrium will be iteratively found far more straightforwardly.
- Wheel spin rotation to which a generalized variable corresponds is ignorable for vehicle equilibrium; if one does not care, the N–R iterative matrix will be singular being insensitive to these variables.
- N–R algorithm requires a valuable initial guess to converge: in case of vehicle oscillating suspensions with high stiffness spring, good estimates are indispensable to avoid unrealistic (often funny!) equilibrium of the system.

4.2.9 System analysis and parameterization

From a pure engineering point of view, we could say that the interesting part of the project only starts[2] at this level, i.e. to answer the original questions by performing the appropriate simulations (e.g., ATV side slipping, motorbike wheeling) and parameterizations. Depending on the quality of the group and of their tasks planning during weeks 8 to 10, the following analyses can be performed and discussed in the final report:

- Influence of the suspension parameters (stiffness, damping) on the vehicle dynamical performance;
- Comparison of different integration schemes in terms of accuracy, CPU time, etc;
- Modeling of a more realistic model by disregarding some of the hypotheses (related to locked motions, to constitutive laws, to the environment, etc.);
- Comparison of systems issuing from the same family: e.g. a bogie with rigid axles and the same with independent wheels;
- etc.

4.2.10 Project report, presentation and evaluation

Without entering into details, a final report is asked in which all the previous points must be detailed. Emphasize is made during the oral presentation on the analysis phase for which students, sometimes a bit clumsily, give their own interpretation of

[2] This is hopefully not true from the pedagogical aspect.

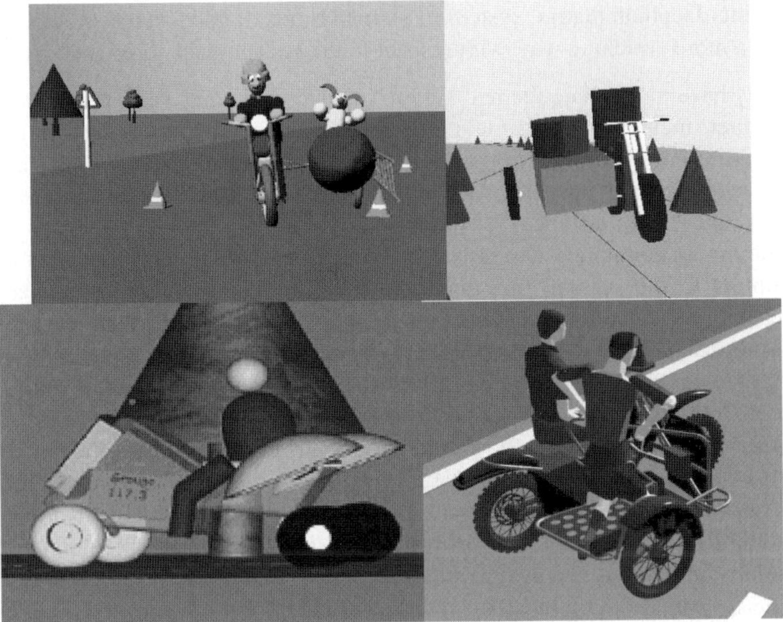

Fig. 9. Snapshots of 3D animation.

a specific phenomenon (e.g., a peak or a high frequency in a variable time history). 3D animations (see snapshots in Figure 9) help them to illustrate their simulation observations.

The project is then evaluated taking into account all the previous aspects, but without weighting the analysis phase too much, recalling that undergraduate students are not yet mature engineers!

5 Symbolic Generation

For such a project, an important question relates to the multibody model: how to avoid the trap of a "press-button" black-box multibody program which would be completely useless for undergraduate students, pedagogically speaking? From the modeling point of view, the answer relies on our conviction that extrapolating kinematic and dynamical computations of 2–3 interconnected bodies (as done in the course) to a larger multibody system (10–15 bodies) requires, as already mentioned, more "perspiration than inspiration" to develop rotation matrices, absolute velocities, accelerations, etc. Entrusting these "low-level" tasks to a multibody symbolic program like ROBOTRAN [12] for various ingredients such as sub-chain kinematics, constraints Jacobian matrix, system mass matrix, etc., allow students to spend their time in a more enriching way, with respect to the pedagogical objectives. Namely:

- They have to translate the real system in terms of bodies and joints in a correct manner, including the formulation of pertinent hypotheses regarding the proposed topology and environment. Of course, they must end up with a model description which must be in accordance with the symbolic program conventions (see Figure 7 for instance).
- They must compute – and introduce – the internal/external forces and torques applied to the system: this range from a single spring/dashpot element to a wheel/ground model. Since 90% of the necessary kinematics can be generated symbolically from the ROBOTRAN program; the students can focus on more fundamental points: interpretation of the wheel slip angle, geometrical contact on a profiled ground, etc.
- Once the necessary symbolic ingredients have been collected and the model completed (forces, driven joints, sensors, ...), students can implement an algorithm for the numerical simulation phase. For that purpose, they directly manipulate the symbolic output and the user functions they developed in completing the model.
- Finally, once their model, their simulation and animation are ready, students have time and thus the opportunity to tackle a more "noble" objective, namely the critical analysis of the model and the system behavior: validity or limit of some hypotheses, influence of given parameters, time integrator performance, etc.

6 Conclusions

A few decades after the advent of the finite elements method, the MBD field arrives today at a stage where it also becomes necessary to include it in the engineering educational curriculum. This not only meets a request of an industrial world using MBD software more and more. Indeed we illustrated in this chapter that MBD is extremely rich from the teaching point of view. Indeed, the room of manoeuver is large in terms of applications, problems to solve and disciplines to involve. This is true of course for the graduate level of study, but also for the undergraduate level. After more than 10 years of experience, we can claim that a multi-disciplinary project in multibody dynamics can really improve the *undergraduate* student's skills in the field of system modeling, including a timid but existing engineering attitude with respect to the work they produce. Of course, this behavior must be more deeply developed during subsequent learning "layers" at the graduate level for students in mechanical engineering, eventually on the occasion of their master's theses and, later on, in their professional activities. From this point of view, the potentialities offered by an open symbolic MBD software like ROBOTRAN make it possible to take students by the hand from the beginning of their studies to their PhD or professional degree.

Acknowledgements

This research, related to the development of ROBOTRAN, has been sponsored by the Belgian Program on Interuniversity Attraction Poles initiated by the Belgian State – Prime Minister's Office – Science Policy Programming (IAP V/6). The scientific responsibility is assumed by the authors. The authors also thank all the speakers for the opinions expressed at the round table on "Education of multibody dynamics" of the International ECCOMAS Thematic Conference on Advances in Computational Multibody Dynamics (Universidad Politecnica de Madrid, Spain, June 21–24, 2005).

References

1. Pennestri E., Vita L. (2005) Multibody dynamics in advanced education. In: Advances in Computational Multibody Systems, Ambrósio, JA.C. (ed.), Springer-Verlag, Berlin.
2. Quaranta G., Masarati P., Morandini M. (2005) Multibody free software for teaching purposes. In: Multibody Dynamics 2005, ECCOMAS Thematic Conference, Goicolea J.M., Cuadrado J., Garcia Orden J.C. (eds.), Madrid, Spain, 21–24 June, 2005.
3. Fisette P., Samin J.-C. (2005) Teaching Multibody Dynamics from Modeling to Animation, Multibody System Dynamics, Vol. 13, Number 3, Dordrecht, pp. 339–351.
4. Fisette P., Samin J.-C. (2005) Multibody model symbolic generation: a fruitful approach to teaching and applying fundamental mechanics. In: Multibody Dynamics 2005, ECCOMAS Thematic Conference, Goicolea J.M., Cuadrado J., Garcia Orden J.C. (eds.), Madrid, Spain, 21–24 June, 2005.
5. Sass L., Fisette P., Samin J.-C. (2005) Simulation projects with a symbolic multibody modeling software for students in Belgium and Ecuador. In: Proc. of IDETC/CIE (ASME International Design Engineering Technical Conferences & Computers and Information in Engineering Conference), Long Beach, California, USA, September 24–28, 2005, DETC2005-84323.
6. Kane Th.-R., Levinson D.-A. (1985) Dynamics: Theory and Applications, McGraw-Hill, New York.
7. Kane Th.-R., Levinson D.-A. (1990) AUTOLEV: A New Approach to Multibody Dynamics. In: Multibody Systems Handboook, Schiehlen W. (ed.), Springer-Verlag, Berlin, pp. 81–102.
8. Nikravesh P.-E. (1988) Computer Aided Analysis of Mechanical Systems, Prentice-Hall, London.
9. Verlinden O., Kouroussis G., Conti C. (2005) EASYDYN: a framework based on free symbolic and numerical tools for teaching multibody systems. In: Multibody Dynamics 2005, ECCOMAS Thematic Conference, Goicolea J.M., Cuadrado J., Garcia Orden J.C. (eds.), Madrid, Spain, 21–24 June, 2005.
10. Maes P., Samin J.-C., Willems P.-Y. (1990) AUTODYN and ROBOTRAN. In: Multibody Systems Handboook, Schiehlen W. (ed.), Springer-Verlag, Berlin, pp. 225–264.
11. Fisette P. and Samin J.C. (1993) ROBOTRAN: Symbolic generation of multibody system dynamic equations. In: Advanced Multibody System Dynamics: Simulation and Software Tools, Schiehlen W. (ed.), Kluwer Academic Publishers, Dordrecht, pp. 373–378.

12. www.robotran.be
13. Raucent B. (2001) Introducing PBL in a Machine Design Curriculum: Result of an Experiment, Journal of Engineering Design, Vol. 12, No. 4, 2001, pp. 293–308.
14. Samin J.C. and Fisette P. (2003) Symbolic Modeling of Multibody Systems, Kluwer Academic Publishers, Dordrecht.

Rollover Tendency in Embankment and Ramp Maneuvers with Ground Contact

Simulation and Experiment

Karina Hirsch and Manfred Hiller

Chair of Mechatronics, Faculty of Engineering, University Duisburg-Essen, Lotharstraße 1, 47057 Duisburg, Germany; E-mail: {hirsch,hiller}@imech.de

Abstract. Passive safety systems in vehicles, which are directly responsible for the optimal passenger protection in the case of an accident, are critical factors in contemporary vehicle development. Rollover maneuvers, e.g. rides over an embankment or a ramp, are considered to be especially hazardous, as they always involve a high risk of injury or even death and of vehicle damage. Today, for reasons of cost and development time, computer simulations play an important role in the development of such safety systems.

For simulating the vehicle dynamics, the vehicle is realized as a complex multibody system, as a base for the mechanical part, and which is combined with additional non-mechanical components like hydraulics, sensors, driver and environment into an overall mechatronical system. This concept is realized within the three-dimensional vehicle dynamic simulation environment FASIM_C++. The central issue in this paper is the modeling of the vehicle ground contact, which may occur between the underbody and the ground during embankment and ramp maneuvers. Finally, the results from test drives for a cabriolet, which has been selected by the car manufacturer from simulations with FASIM_C++, show a good coincidence between simulations and experiment.

1 Introduction

In order to increase the safety in road traffic, more and more active and passive safety systems, as typical examples of mechatronic systems, are integrated into the vehicles. While the active safety systems, such as antilock braking system (ABS), traction control system (TCS) and electronic stability program (ESP), support the driver to prevent an accident, the passive safety systems are activated only when an accident is occurring. In this connection, the occupant protection systems, such as airbags and seat-belt tighteners, should reduce the risk of injury to the passengers, which are especially high in accidents with rollover tendency. In the case of cabriolets, the passengers should be further protected using rollover bars during a rollover maneuver.

J.C. García Orden et al. (eds), Multibody Dynamics. Computational Methods and Applications, 179–200.

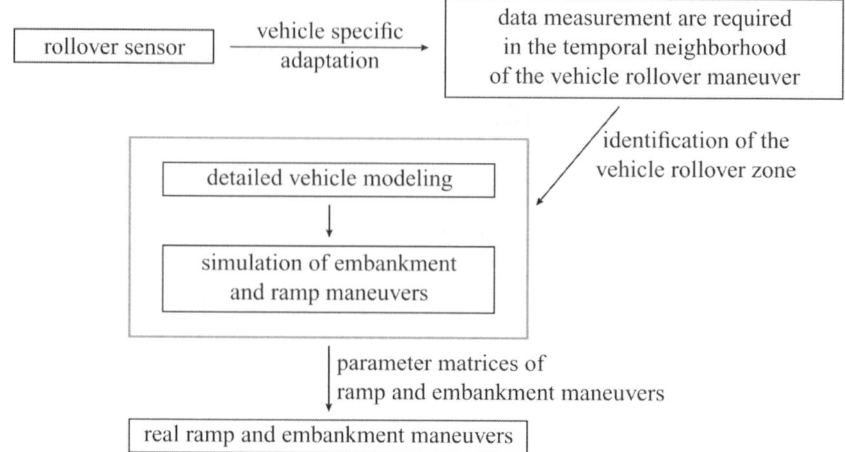

Fig. 1. Vehicle dynamic simulations to support the development of controllers for restraint systems.

Before activating the occupant protection systems, the rollover tendency of the vehicle has to be detected (Figure 1) [8]. Hence, the different rollover behaviors of vehicles have to be considered, as this plays a critical role in developing the controller. For these vehicle specific controller designs, measurements such as the rolling and vertical accelerations, at the time of vehicle rollover are required. For investigating the rollover behavior, in the past, embankment and ramp maneuvers were well-proven methods. Before performing the actual test drive, which are technically complex and expensive, it is thus suggested to exhaustively simulate and analyze the rollover behavior of vehicles on different embankment and ramp maneuvers in the computer [3]. Based on these results specific driving maneuvers can be chosen, limiting to only a few the number of experiments.

To simulate the vehicle dynamics, the mechanics of the vehicle is modeled as a multibody system and together with the non-mechanic components, such as hydraulics, sensors, driver and environment combined to a complete mechatronic vehicle model, inside the three-dimensional vehicle dynamics simulation environment FASIM_C++ [5]. The concept behind FASIM_C++ is presented in Section 2.

During the simulation of embankment and ramp maneuvers of sport cabriolets, hitting the ground of the vehicle underbody has to be considered due to the low ground clearance of the chassis. Hence, the vehicle dynamics simulation environment FASIM_C++ has to be extended (Section 3).

Modeling the ground contact by means of a mathematical substitution model for the vehicle underbody and for the embankment or ramp, their relative position to each other is investigated. The surface load acting on the undercarriage contact is

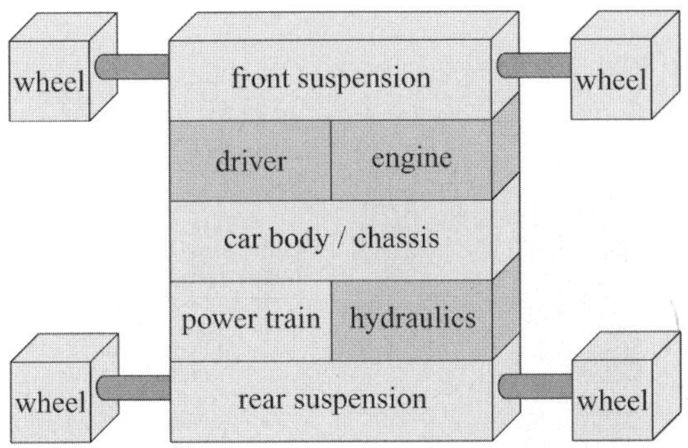

Fig. 2. Modular structure of a passenger car in FASIM_C++.

modeled as a concentrated force, and the mathematical algorithm to calculate the point of application of the force is explained. Finally, the determination of the external forces and moments that act on the vehicle in-between the vehicle carriage and embankment or ramp contact, as well as, the integration of the equations of motion are presented.

In Section 4, the simulation results for a sports cabriolet are presented, which show the influence of modeling the ground contact to the simulation results. A comparison of the simulation results with the measured data from actual test-drives illustrate the influence of the undercarriage contact to the dynamic behavior of the vehicle.

2 Vehicle Dynamic Simulation with FASIM_C++

Development of vehicle controllers requires an appropriate model of the vehicle dynamics built into a versatile simulation environment [2]. This simulation environment has to be able to simulate different vehicle types or models without any recompilation. The vehicle model has to have a modular form so that single component of the vehicle may be exchanged, depending on the simulation task. Thus, models of the vehicle dynamics with differing levels of complexity can be defined covering correspondent physical effects with the desired accuracy. The modular structure of a vehicle model in FASIM_C++ is shown in Figure 2 using the example of a passenger car.

The structure presented does not show the construction details of the modules, e.g. which kind of front suspension is used. During initialization this is not important, because the required information for generating the equations of motion is part of the modules and only at the beginning of simulation it is evaluated. As an example, the topology of an all-wheel driven upper class passenger car, showing the

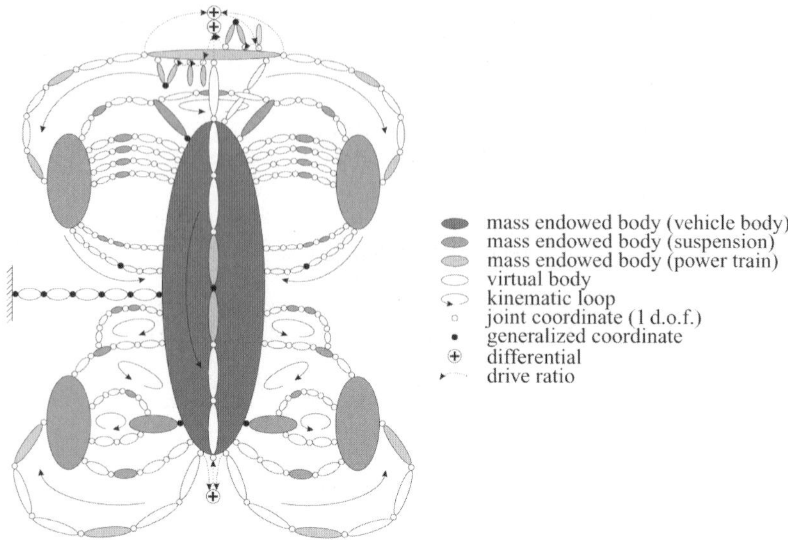

Fig. 3. Kinematic topology of an all-wheel driven upper class passenger car.

kinematic coupling of the individual modules, is shown in Figure 3. For reasons of clarity the modules engine hydraulics (braking system), driver and environment are not shown. Using this modeling technique it is possible to decide during runtime which configuration of a vehicle is used without any recompilation of the program.

FASIM_C++ contains a large library of different vehicle modules such as suspensions, tire models, drive trains, engines, engine mounts, controllers, sensors, elasticities, a rigid or flexible car body, several hydraulic braking systems, a driver and an environment model. The structure of the modules makes it easy to expand the library by adding new modules. The equations of motion are based on d'Alembert's principle:

$$\sum_{i=1}^{n_B}(m_i\ddot{\mathbf{r}}_{S_i} - \mathbf{f}_{S_i}) \cdot \delta\mathbf{r}_{S_i} + (\boldsymbol{\Theta}_{S_i}\dot{\boldsymbol{\omega}}_i + \boldsymbol{\omega}_i \times \boldsymbol{\Theta}_{S_i}\dot{\boldsymbol{\omega}}_i - \boldsymbol{\tau}_{S_i}) \cdot \delta\boldsymbol{\varphi}_i = 0, \tag{1}$$

where n_B is the number of mass-endowed bodies, m_i, $\boldsymbol{\Theta}_{S_i}$ are the mass and inertia tensor of body i, $\ddot{\mathbf{r}}_{S_i}$ the acceleration of center of gravitation, $\mathbf{f}_{S_i}$, $\boldsymbol{\tau}_{S_i}$ the applied force and torque, and $\delta\mathbf{r}_{S_i}$, $\delta\boldsymbol{\varphi}_i$ are the virtual linear and angular displacement.

Due to the constraints in the system, the virtual displacements are not independent. To generate the equations of motion in minimal coordinates the choice of n independent generalized coordinates $\mathbf{q} = [q_1, q_2, \ldots, q_n]^T$, is necessary, corresponding to the number of degrees of freedom (d.o.f.) in the system. The equations of motion of the mechanical system in minimal coordinates can then be written as:

$$\mathbf{M}(\mathbf{q})\ddot{\mathbf{q}} + \mathbf{b}(\mathbf{q}, \dot{\mathbf{q}}) = \mathbf{Q}(\mathbf{q}, \dot{\mathbf{q}}, t), \tag{2}$$

with the notations and dimensions:

n : number of d.o.f.,
$\mathbf{q}$: $(n \times 1)$-vector of generalized coordinates,
$\mathbf{M}$: $(n \times n)$-mass matrix,
$\mathbf{b}$: $(n \times 1)$-vector of generalized Coriolis, centrifugal and gyroscopic forces,
$\mathbf{Q}$: $(n \times 1)$-vector of generalized applied forces.

With help of the method of kinematical differentials developed by Kecskeméthy, the elements of the equations of motion can be calculated using the pseudo velocities $\hat{\dot{\mathbf{r}}}_i^{(j)}$, $\hat{\boldsymbol{\omega}}_i^{(j)}$ and the pseudo accelerations $\hat{\ddot{\mathbf{r}}}_i$, $\hat{\dot{\boldsymbol{\omega}}}_i$ [6]. The elements gets the form

$$\mathbf{M}_{j,k} = \sum_{i=1}^{n_B} \left[m_i \hat{\dot{\mathbf{r}}}_i^{(j)} \cdot \hat{\dot{\mathbf{r}}}_i^{(k)} + \hat{\boldsymbol{\omega}}_i^{(j)} \cdot \left(\boldsymbol{\Theta}_{Si} \hat{\boldsymbol{\omega}}_i^{(k)} \right) \right], \tag{3}$$

$$\mathbf{b}_j = \sum_{i=1}^{n_B} \left[m_i \hat{\dot{\mathbf{r}}}_i^{(j)} \cdot \hat{\ddot{\mathbf{r}}}_i + \hat{\boldsymbol{\omega}}_i^{(j)} \cdot \left(\boldsymbol{\Theta}_{Si} \hat{\dot{\boldsymbol{\omega}}}_i + \boldsymbol{\omega}_i \times \boldsymbol{\Theta}_{Si} \boldsymbol{\omega}_i \right) \right], \tag{4}$$

$$\mathbf{Q}_j = \sum_{i=1}^{n_B} [\hat{\dot{\mathbf{r}}}_i^{(j)} \cdot \mathbf{f}_{S_i}^{e} + \hat{\boldsymbol{\omega}}_i^{(j)} \cdot \boldsymbol{\tau}_{S_i}^{e}]. \tag{5}$$

The vectors, which are bookmarked with the symbol "$\hat{\ }$", are called pseudo velocities and pseudo accelerations, respectively. They result from the analysis of the global kinematics with the pseudo input velocities

$$\hat{\dot{q}}_i^{(j)} = \begin{cases} 1 & \text{for} \quad i = j \\ 0 & \text{for} \quad i \neq j \end{cases}, \qquad i, j = 1, \ldots, n, \tag{6}$$

and are defined by

$$\left.\begin{aligned}
\hat{\dot{\mathbf{r}}}_i^{(j)} &= \dot{\mathbf{r}}_i \Big|_{\dot{\mathbf{q}} = \hat{\dot{\mathbf{q}}}^{(j)}} &&= \frac{\partial \mathbf{r}_i}{\partial q_j}, \\
\hat{\ddot{\mathbf{r}}}_i &= \ddot{\mathbf{r}}_i \Big|_{\ddot{\mathbf{q}} = \hat{\ddot{\mathbf{q}}} = \mathbf{0}} &&= \sum_{j=1}^{f} \sum_{k=1}^{f} \frac{\partial^2 \mathbf{r}_i}{\partial q_j \partial q_k} \dot{q}_j \dot{q}_k, \\
\hat{\boldsymbol{\omega}}_i^{(j)} &= \boldsymbol{\omega}_i \Big|_{\dot{\mathbf{q}} = \hat{\dot{\mathbf{q}}}^{(j)}} &&= \frac{\partial \boldsymbol{\omega}_i}{\partial \dot{q}_j}, \\
\hat{\dot{\boldsymbol{\omega}}}_i &= \dot{\boldsymbol{\omega}}_i \Big|_{\ddot{\mathbf{q}} = \hat{\ddot{\mathbf{q}}} = \mathbf{0}} &&= \sum_{j=1}^{f} \sum_{k=1}^{f} \frac{\partial^2 \boldsymbol{\omega}_i}{\partial \dot{q}_j \partial \dot{q}_k} \dot{q}_j \dot{q}_k.
\end{aligned}\right\} \tag{7}$$

Based on the modular structure of the vehicle model (Figure 2), the equation of motion can be written for every module separately. So, the matrices and vectors of the equations of motion show also a modular structure. Their elements can easily be calculated from the corresponding modules. For this reason the terms for the mass matrix **M**, the generalized Coriolis, centrifugal and gyroscopic forces **b**, and generalized applied forces **Q** have to be rewritten [7]. They are now subdivided into an inner sum, inside the module l considering all its bodies n_{Bl} and in an outer sum considering all modules n_M:

$$\mathbf{M}_{j,k} = \sum_{l=1}^{n_B} \sum_{i=1}^{n_{Bl}} \left[m_i \hat{\dot{\mathbf{r}}}_i^{(j)} \cdot \hat{\dot{\mathbf{r}}}_i^{(k)} + \hat{\boldsymbol{\omega}}_i^{(j)} \cdot \left(\boldsymbol{\Theta}_{Si} \hat{\boldsymbol{\omega}}_i^{(k)} \right) \right], \tag{8}$$

$$\mathbf{b}_j = \sum_{l=1}^{n_B} \sum_{i=1}^{n_{Bl}} \left[m_i \hat{\dot{\mathbf{r}}}_i^{(j)} \cdot \hat{\ddot{\mathbf{r}}}_i + \hat{\boldsymbol{\omega}}_i^{(j)} \cdot \left(\boldsymbol{\Theta}_{Si} \hat{\dot{\boldsymbol{\omega}}}_i + \boldsymbol{\omega}_i \times \boldsymbol{\Theta}_{Si} \boldsymbol{\omega}_i \right) \right], \tag{9}$$

$$\mathbf{Q}_j = \sum_{l=1}^{n_B} \sum_{i=1}^{n_{Bl}} \left[\hat{\dot{\mathbf{r}}}_i^{(j)} \cdot \mathbf{f}_{Si}^{e} + \hat{\boldsymbol{\omega}}_i^{(j)} \cdot \boldsymbol{\tau}_{Si}^{e} \right] \tag{10}$$

with the pseudo velocities $\hat{\dot{\mathbf{r}}}_i^{(j)}$, $\hat{\boldsymbol{\omega}}_i^{(j)}$ and the pseudo accelerations $\hat{\ddot{\mathbf{r}}}_i$, $\hat{\dot{\boldsymbol{\omega}}}_i$ from Equation (7).

3 Modeling of the Ground Contact

In vehicles with low ground clearance, a further difficulty may arise during embankment and ramp maneuvers, as the vehicle underbody may hit the ground, which significantly influences the vehicle dynamics. This effect will be considered by modeling the vehicle dynamics, especially for sports cabriolets.

Before calculating the contact forces and moments at the time of ground contact, it should first be detected, if there is any contact between the vehicle and the road, in this case with the embankment or the ramp. Hence, the actual position of the vehicle underbody relative to the embankment or the ramp is investigated for every discrete time-interval of the simulation. The mathematical substitution model of the vehicle underbody, the embankment and the ramp, respectively, required for the investigation, are introduced next.

3.1 Substitute model for the vehicle underbody

The underbody of the vehicle is simplified as a rectangular planar surface, whose position in the inertial system $\mathcal{K}_I$ is calculated from the current position and orientation of the vehicle system $\mathcal{K}_V$ (Figure 4). The vehicle system is placed in the

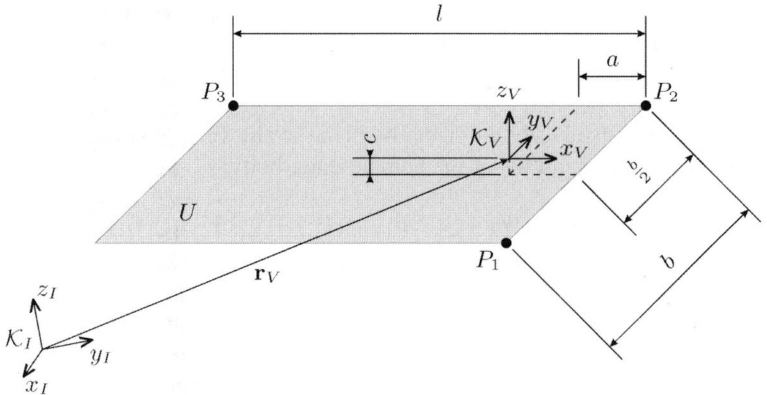

Fig. 4. Geometry of the vehicle underbody.

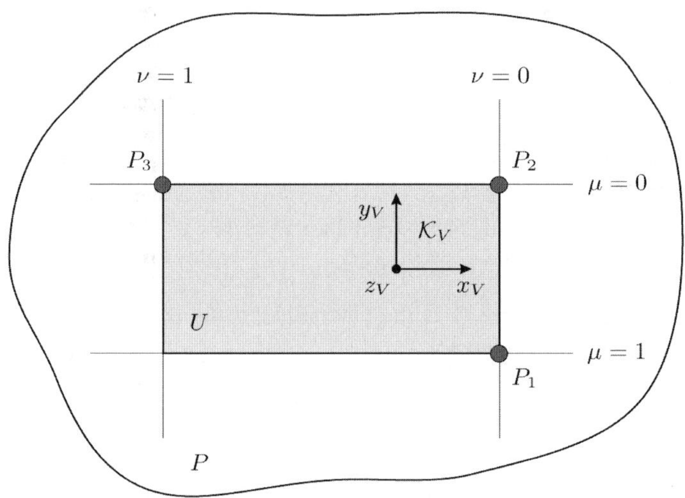

Fig. 5. Plane of the vehicle underbody.

mid-point of the front axle. The position of the vehicle underbody relative to the vehicle system and its parameters are fixed by the geometric measures a, b, c and l, where l describes the total length of the vehicle, a refers to the distance of the front axle from the front end of the vehicle, along the length of the vehicle, b refers to the width of the vehicle and c the distance of the vehicle underbody from the front axle along the vertical of the vehicle system.

With the help of these parameters, the position vectors of the corners $\mathbf{P}_1$, $\mathbf{P}_2$ and $\mathbf{P}_3$ (Figure 5), in the vehicle system, can be calculated:

$$ {}^V\mathbf{P}_1 = \begin{bmatrix} a \\ -\frac{b}{2} \\ c \end{bmatrix}, \quad {}^V\mathbf{P}_2 = \begin{bmatrix} a \\ \frac{b}{2} \\ c \end{bmatrix}, \quad {}^V\mathbf{P}_3 = \begin{bmatrix} a-l \\ \frac{b}{2} \\ c \end{bmatrix}. \tag{11} $$

These three points clearly describe the plane of the vehicle underbody, and one can set up the planar equations of the vehicle underbody:

$$ {}^VP: \quad {}^V\mathbf{x}_P = {}^V\mathbf{P}_2 + \mu({}^V\mathbf{P}_1 - {}^V\mathbf{P}_2) + \nu({}^V\mathbf{P}_3 - {}^V\mathbf{P}_2) \tag{12} $$

$$ = \begin{bmatrix} a \\ \frac{b}{2} \\ c \end{bmatrix} + \mu \begin{bmatrix} 0 \\ -b \\ 0 \end{bmatrix} + \nu \begin{bmatrix} -l \\ 0 \\ 0 \end{bmatrix}, \quad \mu, \nu \in \mathbb{R}. \tag{13} $$

For $\mu, \nu \in [0, 1]$ all the points ${}^V\mathbf{x}_P = {}^V\mathbf{x}_P(\mu, \nu)$ lie within the vehicle underbody, and for all other values μ, ν they lie outside the vehicle underbody:

$$ {}^VU: \quad {}^V\mathbf{x}_U = {}^V\mathbf{x}_P(\mu, \nu), \quad \mu, \nu \in [0, 1]. \tag{14} $$

To describe the plane of the vehicle underbody in the inertial system, the position and orientation of the vehicle system and that of the inertial frame have to be known. With the transformation matrix

$$ {}^I\mathbf{T}_V = \begin{bmatrix} c\theta\, c\psi & s\varphi\, s\theta\, c\psi - c\varphi\, s\psi & c\varphi\, s\theta\, c\psi + s\varphi\, s\psi \\ c\theta\, s\psi & s\varphi\, s\theta\, s\psi + c\varphi\, c\psi & c\varphi\, s\theta\, s\psi - s\varphi\, c\psi \\ -s\theta & s\varphi\, c\theta & c\varphi\, c\theta \end{bmatrix} \tag{15} $$

($c :=$ cos, $s :=$ sin), where ψ is the yaw angle, θ the pitch angle and φ the roll angle, and with the position vector

$$ \mathbf{r}_V = {}^I\mathbf{r}_V = \begin{bmatrix} {}^Ix_V \\ {}^Iy_V \\ {}^Iz_V \end{bmatrix} \tag{16} $$

the equation for a plane is set up in the inertial system as given below:

$$ {}^IP: \quad {}^I\mathbf{x}_P = {}^I\mathbf{T}_V\, {}^V\mathbf{x}_P + {}^I\mathbf{r}_V \tag{17} $$

$$ = {}^I\mathbf{T}_V \begin{bmatrix} a - \nu l \\ \frac{b}{2} - \mu b \\ c \end{bmatrix} + \begin{bmatrix} {}^Ix_V \\ {}^Iy_V \\ {}^Iz_V \end{bmatrix}, \quad \mu, \nu \in \mathbb{R}. \tag{18} $$

Analogous to Equation (14) it applies in the inertial system:

$$ {}^IU: \quad {}^I\mathbf{x}_U = {}^I\mathbf{x}_P(\mu, \nu), \quad \mu, \nu \in [0, 1]. \tag{19} $$

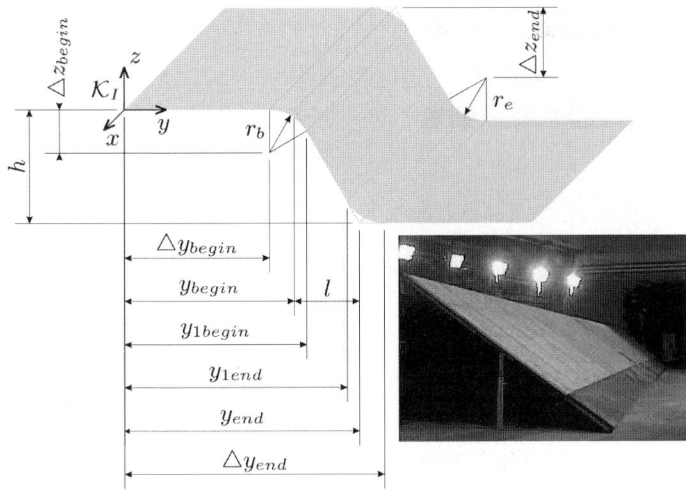

Fig. 6. Nomenclature of embankment geometry.

3.2 Embankment and ramp geometry

Besides the plane equation of the vehicle underbody, the geometric descriptions of the embankment and the ramp, respectively, are required for the contact calculations. For simulation purposes, the sharp edges of the embankment and the ramp, as they occur on the test track, are replaced by radii, to guarantee that each wheel has only one contact point on the track. This deviation in the model is also valid for the calculation of the contact points between the vehicle underbody and the track.

The upper and lower embankment edges run parallel to the x-axis of the inertial system and the embankment shall stretch the entire length of the track. Thus the height of the embankment z- is only dependent on the y-coordinate:

$$z_{embankment}(x, y) = z_{embankment}(y) \tag{20}$$

$$= \begin{cases} 0 & , \quad y \leq \Delta y_{begin} \\ \Delta z_{begin} + \sqrt{r_b^2 - (y - \Delta y_{begin})^2} & , \quad \Delta y_{begin} < y < y_{1begin} \\ \frac{h}{l}(y_{begin} - y) & , \quad y_{1begin} \leq y \leq y_{1end} \\ \Delta z_{end} - \sqrt{r_e^2 - (y - \Delta y_{end})^2} & , \quad y_{1end} < y < \Delta y_{end} \\ -h & , \quad \Delta y_{end} \leq y \end{cases} . \tag{21}$$

In contrast, the ramp has limited dimensions. The track height, depending on x and y, is thus given by:

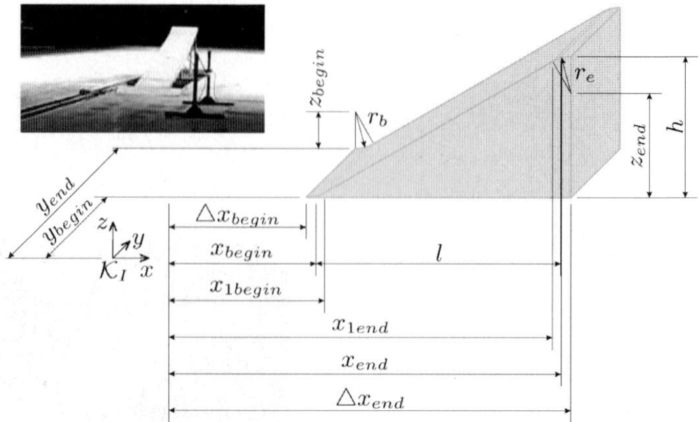

Fig. 7. Nomenclature of ramp geometry.

$$z_{ramp}(x,y) = \begin{cases} \Delta z_{begin} - \sqrt{r_b^2 - (x - \Delta x_{begin})^2}, & \begin{cases} \Delta x_{begin} < x < x_{1begin} \\ y_{begin} \leq y \leq y_{end} \end{cases} \\ \frac{h}{l}(x - x_{begin}), & \begin{cases} x_{1begin} \leq x \leq x_{1end} \\ y_{begin} \leq y \leq y_{end} \end{cases} \\ \Delta z_{end} + \sqrt{r_e^2 - (x - \Delta x_{end})^2}, & \begin{cases} x_{1end} < x < \Delta x_{end} \\ y_{begin} \leq y \leq y_{end} \end{cases} \\ 0, & \text{otherwise} \end{cases} . \quad (22)$$

3.3 Center point of the contact surface

The surface loads, acting on the contact surface during the ground contact, should be modeled as a concentrated force. The point of application of the force is determined from the center point, which is here the center of the contact surface.

3.3.1 Contact during embankment maneuvers

Before determining the actual point of application of the force in the vehicle chassis, the position of the vehicle underbody relative to the embankment top surface has to be examined. The following four cases are possible:

1. the vehicle underbody does not touch the embankment at all (Figure 8a),
2. the vehicle underbody touches the embankment surface at one point (Figure 8b),
3. the vehicle underbody touches the embankment surface along a straight line (Figure 8c) or

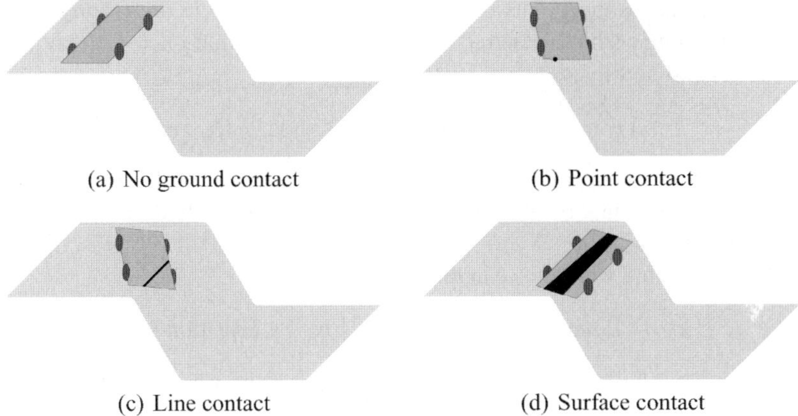

(a) No ground contact (b) Point contact

(c) Line contact (d) Surface contact

Fig. 8. Comparison of the positions of the vehicle underbody relative to the embankment.

4. the vehicle underbody penetrates the embankment (Figure 8d), i.e. the vehicle underbody rectangle cuts the embankment top surface, while in the real vehicle, the vehicle underbody deforms elastically due to the embankment contact. It is stretched over the embankment edge. The depth of penetration can be used as a measure for the elastic deformation.

To summarize, in the first case there is no contact between the vehicle underbody and the embankment, whereas there exists a contact in the remaining three cases.

In order to generalize the contact definition, the different positions of the vehicle underbody relative to the embankment will not be considered. On the basis of the points of intersection between the vehicle underbody and the embankment top surface, as well as the corner points that lie under the embankment top surface, the shape of the contact is approximately described. Hence there exists no need to differentiate between a point, a line or a surface contact.

The position of the corner points of the vehicle underbody with respect to the inertial system results from the Equation (19) for the following pairs of parameters

$$(\mu_C, \nu_C) \in \{(0,0), (0,1), (1,0), (1,1)\}. \tag{23}$$

The corner point

$${}^I\mathbf{x}_{UC} = {}^I\mathbf{x}_U(\mu_C, \nu_C) := [x_U(\mu_C, \nu_C), y_U(\mu_C, \nu_C), z_U(\mu_C, \nu_C)]^T$$

lies exactly on or already inside the embankment, if its components satisfy the following inequality

$$z_U(\mu_C, \nu_C) \leq z_{embankment}(x_U(\mu_C, \nu_C), y_U(\mu_C, \nu_C)). \tag{24}$$

In order to define the points of intersection of the vehicle underbody with the embankment, using each of these $\mu_I = 0$, $\mu_I = 1$, $\nu_I = 0$ and $\nu_I = 1$, the following equations need to be solved

$$z_U(\mu_I, \nu_I) = z_{embankment}(x_U(\mu_I, \nu_I), y_U(\mu_I, \nu_I)). \tag{25}$$

If the calculated μ_I or rather ν_I lie inside the interval [0, 1], then the point ${}^I\mathbf{x}_{U\,I} = {}^I\mathbf{x}_U(\mu_I, \nu_I)$ lies on the border of the vehicle underbody and on the embankment top surface.

All the points that were thus calculated

$$\{\mathbf{S}_i\}_{i\in\mathbb{N}} := \left\{{}^I\mathbf{x}_{U\,C} : \text{Equation (24) applies}\right\} \cup \left\{{}^I\mathbf{x}_{U\,I} : \text{Equation (25) applies}\right\} \tag{26}$$

represent the boundary points of the contact surface of the vehicle underbody. Here it should be observed, that because of the separate examination of the corner points, and because of rounding errors, a few points are duplicated. These should later be eliminated.

3.3.2 Contact during ramp maneuvers

Analogous to defining the points of contact between a vehicle underbody and the embankment (Section 3.3.1), the contact between the vehicle underbody and the ramp can, during ramp maneuvers, also be divided into four different cases:

1. the vehicle underbody does not touch the ramp at all (Figure 9a),
2. the vehicle underbody touches the ramp-edge at one point (Figure 9b),
3. the vehicle underbody touches the ramp-edge in a straight line (Figure 9c) or
4. the vehicle underbody penetrates the ramp (Figure 9d), which means, the rectangle of the vehicle underbody cuts the ramp top surface, whereas the real vehicle underbody will be elastically deformed by the ramp. It will stretch itself over the ramp edge. The penetration depth defines the strength of the elastic deformation.

Hence, in the first case there is no contact between the vehicle underbody and the ramp, while in the other three cases there exists a ground contact.

When hitting the ramp, the chassis always compresses its lateral edges of the vehicle underbody the most. Therefore, it is sufficient to determine the points of intersection between the side edges of the underbody and the side surface of the ramp. For which, using each of the following $\mu_I = 0$, $\mu_I = 1$, $\nu_I = 0$ und $\nu_I = 1$, the subsequent two equations have to be solved:

$$y_U(\mu_I, \nu_I) = \begin{cases} y_{begin} \\ y_{end} \end{cases} . \tag{27}$$

If the calculated μ_I or rather ν_I lie inside the interval [0, 1], then the point ${}^I\mathbf{x}_{U\,I} = {}^I\mathbf{x}_U(\mu_I, \nu_I)$ lies on the border of the vehicle underbody. It also applies

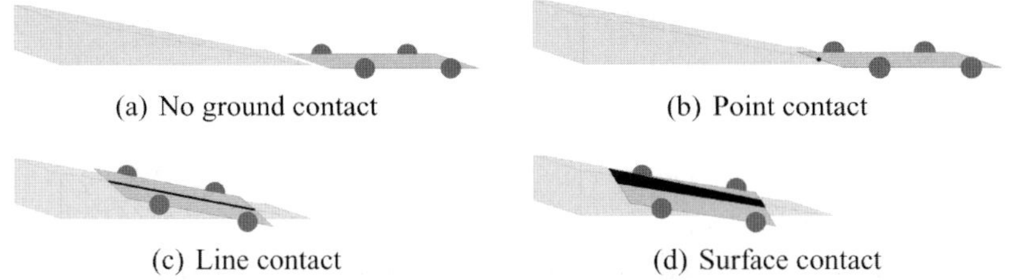

(a) No ground contact (b) Point contact

(c) Line contact (d) Surface contact

Fig. 9. Comparison of the different positions between vehicle underbody and the ramp.

$$z_U(\mu_I, \nu_I) \leq z_{ramp}(x_U(\mu_C, \nu_C), y_U(\mu_C, \nu_C)), \tag{28}$$

hence, ${}^I\mathbf{x}_{U\,I}$ is a point of intersection of the vehicle underbody edge with the ramp side. All the calculated points of intersection

$$\{\mathbf{S}_i\}_{i \in \mathbb{N}} := \left\{ {}^I\mathbf{x}_{U\,I} : \text{Equation (27) and (28) apply} \right\} \tag{29}$$

represent the corner points of the ground contact surface. Duplicated points are not considered.

3.3.3 Center point of contact surface

Out of the multitude of the pair-wise different points of intersection $\mathbf{S}_1, \ldots, \mathbf{S}_n$ (Equation (26) or (29)) the center point or the surface center of mass $\mathbf{M}$ (Figure 10) is determined using

$$\mathbf{M} = \frac{1}{n} \sum_{i=1}^{n} \mathbf{S}_i. \tag{30}$$

3.4 Point of application of the force

The center point $\mathbf{M}$, which was calculated in Sections 3.3.1 and 3.3.2, respectively, is also the point $\mathbf{A}$ of application of the force of the reaction forces on the vehicle underbody, if it lies exactly on top of the embankment top surface

$${}^I\mathbf{M}_z = z_{embankment}({}^I\mathbf{M}_x, {}^I\mathbf{M}_y), \tag{31}$$

or on the side of the ramp edge, as the case may be,

$${}^I\mathbf{M}_z = z_{ramp}({}^I\mathbf{M}_x, {}^I\mathbf{M}_y). \tag{32}$$

The vehicle underbody touches the embankment, or the ramp, at only one point or a straight line.

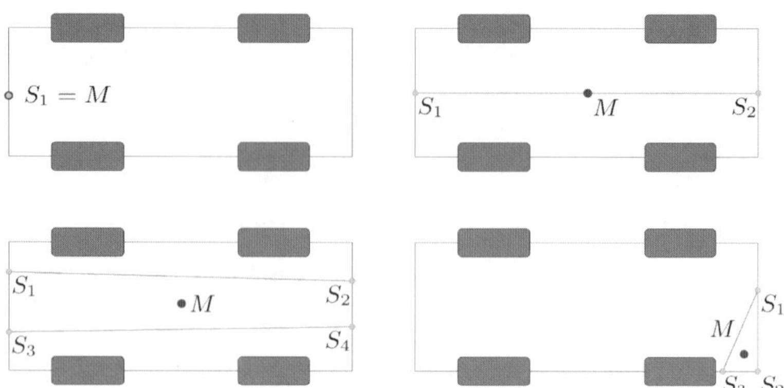

Fig. 10. Center point of the contact surface.

If the center point **M** is lying under the embankment top surface, or under the side wall of the ramp, it is no longer the required point **A** of application of the force. Based on the assumption that the vehicle underbody is elastic, the distance between the calculated midpoint and the embankment top surface, or for that matter the top ramp edge, gives the deformation of the vehicle underbody (Figures 11 and 12, respectively).

3.4.1 Point of application of the force on the embankment

The midpoint **M**, which is presently under the embankment top surface, and the calculated point **A** of application of the force shall both lie on a line, parallel to the z_V-axis:

$$^{V}g: \quad {}^{V}\mathbf{x}_g = {}^{V}\mathbf{M} + \lambda \begin{bmatrix} 0 \\ 0 \\ 1 \end{bmatrix}, \quad \lambda \in \mathbb{R}. \tag{33}$$

Here, the parameter λ_A is chosen in such a way that the point **A** of application of the force lies on the top surface of the embankment. With the help of the numerical integration routine Regula Falsi, the parameter λ_A is approximated. With this, the point of application of the force can finally be determined from

$$^{V}\mathbf{A} = {}^{V}\mathbf{x}_g(\lambda_A)\,. \tag{34}$$

3.4.2 Point of application of the force on the ramp

If the midpoint **M** of the contact surface lies within the side of the ramp, it must then be moved to the top edge of the ramp. Thus a point **A** of application of the force is obtained by a translation along the z_I-axis:

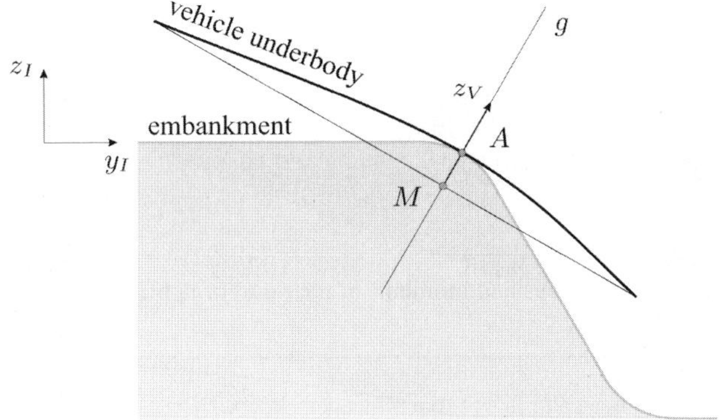

Fig. 11. Point of application of the force on the embankment.

$$
{}^{I}\mathbf{A} = \begin{bmatrix} {}^{I}\mathbf{M}_x \\ {}^{I}\mathbf{M}_y \\ z_{ramp}({}^{I}\mathbf{M}_x, {}^{I}\mathbf{M}_x) \end{bmatrix} . \tag{35}
$$

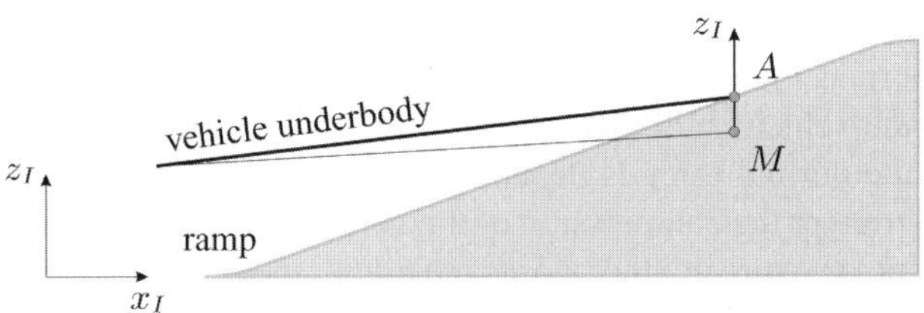

Fig. 12. Point of application of the force on the ramp.

3.5 Reaction forces in the ground contact

When the vehicle underbody hits the embankment or the ramp, it slides over the embankment top surface and ramp top surface, respectively. As a consequence, there is friction in the contact surface. The surface loads are summarized into one force acting on the surface center of mass. The point of application of the force was already calculated in Section 3.4. In the following, the reaction forces, which act on the ground contact, are determined.

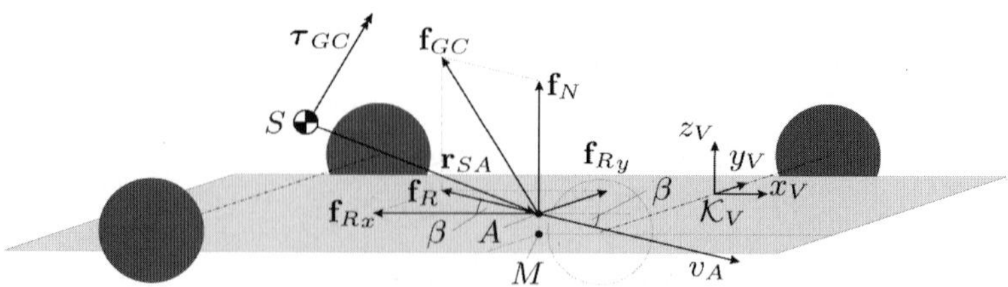

Fig. 13. Reaction forces and moments at the point of application of the force.

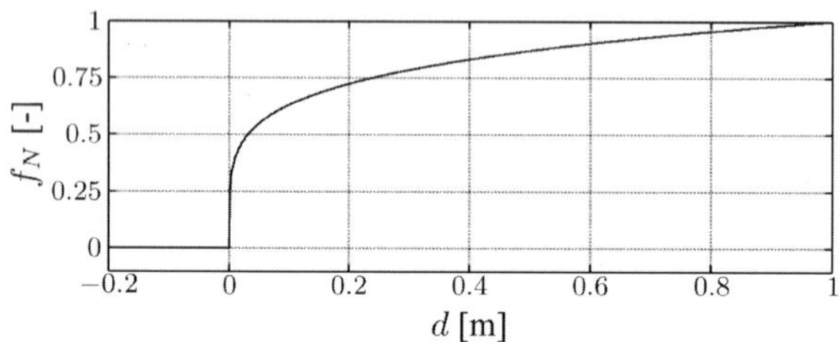

Fig. 14. Normal force on the ground contact (normalized).

3.5.1 Normal force

The applied normal force of the vehicle chassis, at the time of ground contact, is orthogonal to the vehicle underbody (Figure 13). It is positive in z_V-direction. The magnitude of the normal force is dependent on the deformation of the vehicle underbody. The stronger the compression of the vehicle underbody during contact with the embankment, or ramp, in other words the higher the point **A** of application of the force lies inside the vehicle, the higher must be the normal force to generate this deformation. To model the quick increase in the normal force from the time of contact, so that the vehicle can react quickly to the situation, a declining normal variation is chosen (Figure 14). If there is no contact between the vehicle underbody and the embankment, or the ramp, there exists no normal force, i.e. it is zero. This yields for the normal force:

$$f_N = \begin{cases} 0 & , \quad d < 0 \\ f_{deg}(d) \, , & \quad d \geq 0 \end{cases} , \tag{36}$$

where f_{deg} is a degressive function, depending on the distance $d = {}^V\mathbf{A}_z - c$.

3.5.2 Friction force

The friction force is calculated, with the help of the Coulomb friction law, from the above determined normal force:

$$f_R = \mu_R f_N , \tag{37}$$

where μ_R is the friction coefficient between the vehicle underbody and the embankment, or the ramp as the case may be.

The friction force acts tangentially to the normal force, hence it lies on the x_V, y_V-plane, and acts opposite to the direction of motion (Figure 13). Based on the relative kinematics, the velocity of the point **A** of application of the force can be determined:

$$^V\mathbf{v}_A = {}^V\mathbf{v}_V + {}^V\boldsymbol{\omega}_V \times {}^V\mathbf{A}. \tag{38}$$

With this, the components x_V und y_V of the friction force can be set up as:

$$\mathbf{f}_R = -f_R \begin{bmatrix} \cos\beta \\ \sin\beta \end{bmatrix} \quad \text{with} \quad \beta = \arctan(\mathbf{v}_{Ay}, \mathbf{v}_{Ax}) \,. \tag{39}$$

3.6 Integration in the equations of motion

The above calculated reaction force on the ground contact

$$\mathbf{f}_{GC} = \begin{bmatrix} \mathbf{f}_{Rx} \\ \mathbf{f}_{Ry} \\ f_N \end{bmatrix} \tag{40}$$

induces a moment with respect to the vehicle center of mass S:

$$\boldsymbol{\tau}_{GC} = \mathbf{r}_{SA} \times \mathbf{f}_{GC} = (\mathbf{r}_A - \mathbf{r}_S) \times \mathbf{f}_{GC}. \tag{41}$$

The reaction force and the resulting moment will be considered as acting on the center of mass:

$$\mathbf{Q}_{GCj} = \hat{\dot{\mathbf{r}}}_S^{(j)} \cdot \mathbf{f}_{GC} + \hat{\boldsymbol{\omega}}_S^{(j)} \cdot \boldsymbol{\tau}_{GC} \,, \quad j = 1, \ldots, n_{ch}, \tag{42}$$

with the pseudo velocities $\hat{\dot{\mathbf{r}}}_S^{(j)}, \hat{\boldsymbol{\omega}}_S^{(j)}$ and the pseudo accelerations $\hat{\ddot{\mathbf{r}}}_S, \hat{\dot{\boldsymbol{\omega}}}_S$ in accordance with Equation (7). This vector of the generalized ground contact forces completes the right hand side of the equations of motion of the vehicle chassis (according to Equation (2)):

$$\mathbf{M}_{ch}(\mathbf{q}_{ch})\ddot{\mathbf{q}}_{ch} + \mathbf{b}_{ch}(\mathbf{q}_{ch}, \dot{\mathbf{q}}_{ch}) = \mathbf{Q}_{ch}(\mathbf{q}_{ch}, \dot{\mathbf{q}}_{ch}) + \mathbf{Q}_{GC}, \tag{43}$$

where $\mathbf{q}_{ch} \in \mathbb{R}^{n_{ch}}$ is the vector of generalized coordinates, $\mathbf{M}_{ch} \in \mathbb{R}^{n_{ch} \times n_{ch}}$ the generalized mass matrix, $\mathbf{b}_{ch} \in \mathbb{R}^{n_{ch}}$ is the vector of generalized Coriolis, centrifugal and gyroscopic forces, and $\mathbf{Q}_{ch} \in \mathbb{R}^{n_{ch}}$ is the vector of generalized applied forces (without the generalized ground contact forces).

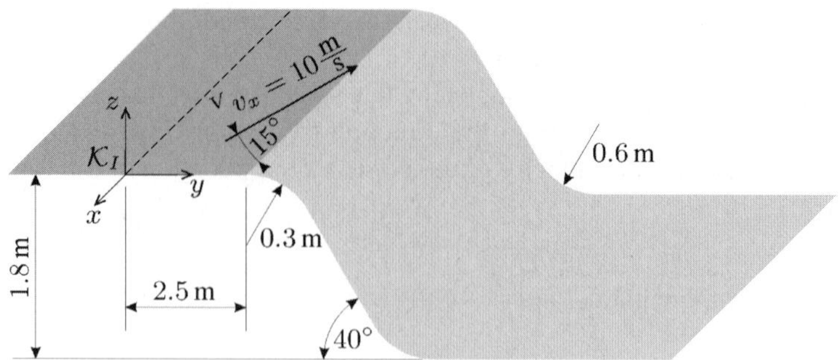

Fig. 15. Construction of a test drive.

Fig. 16. Measurements of the test vehicle.

4 Simulation Results

The results from Sections 2 and 3 shall be tested by an embankment maneuver with a sports cabriolet. The simulation results obtained, with and without ground contact, will be compared with the actual measurements from test drive.

The embankment begins at $y_{begin} = 2.5$ m and is $h = 1.8$ m high. Based on an embankment angle of 40°, an embankment length of $l = 2.2$ m results. At the begin of the embankment the radius is chosen as $r_b = 0.3$ m, and $r_e = 0.6$ m for the end radius. The vehicle shall drive onto the embankment with an initial velocity of $^V v_x = 10$ m/s and an approach angle of 15°. Thus for the vehicle coordinate system in the front axle of the vehicle at start time $t = 0$ s, its position is given by

$$^I x_{V0} = 50.000\,\text{m}, \quad ^I y_{V0} = 0.999\,\text{m} \quad \text{and} \quad ^I z_{V0} = 0.298\,\text{m}, \tag{44}$$

and its orientation is given by a yaw angle ψ, a pitch angle θ and a roll angle φ:

$$\psi_0 = 165.00°, \quad \theta_0 = -0.17° \quad \text{and} \quad \varphi_0 = 0.00°. \tag{45}$$

For the investigation, whether a contact takes place between the vehicle underbody and the embankment, the geometric measurements of the vehicle underbody are required:

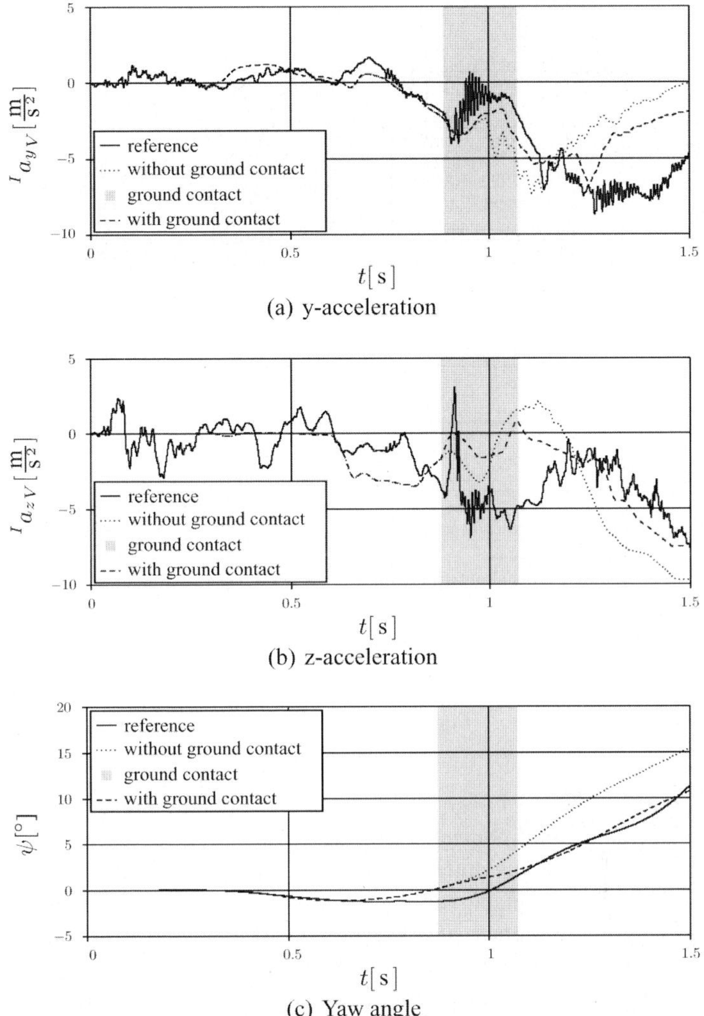

(a) y-acceleration

(b) z-acceleration

(c) Yaw angle

Fig. 17. Comparison of simulation results and measurements.

$$l = 4.547\,\text{m}, \quad a = 0.921\,\text{m}, \quad b = 1.766\,\text{m} \quad \text{and} \quad c = -0.120\,\text{m}. \qquad (46)$$

The normal force that exists during the ground contact is modeled as a declining force depending on the elastic vehicle underbody deformation, as explained in Section 3.5.1 (Figure 14). To calculate the friction force, the friction coefficient is given by $\mu_R = 0.3$.

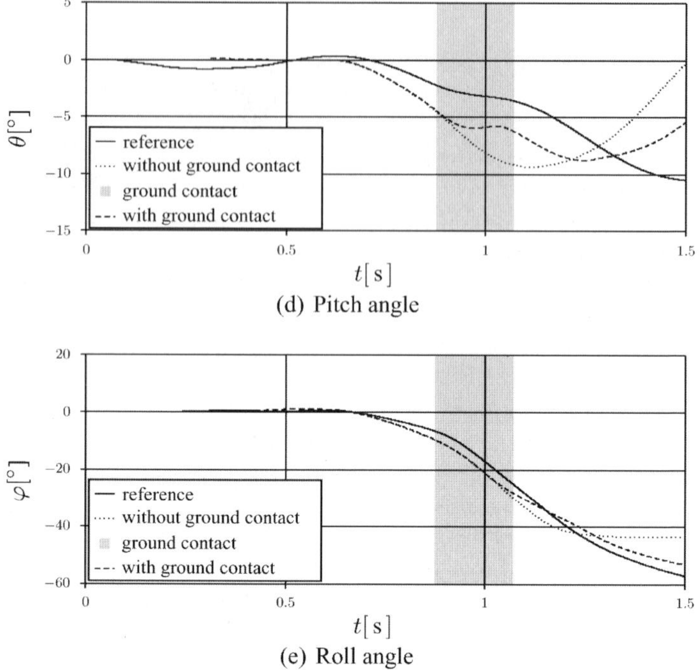

Fig. 17. (Continuation)

The diagrams in Figure 17 show how much the simulation results are influenced by considering the ground contact. Here, the solid graph represents the measurements from the test drive, the dotted graph represents the simulation results without considering the ground contact, and the dashed graph shows the simulation results considering the ground contact. The function has calculated a contact between the vehicle underbody and the embankment in the gray time interval.

Compared to the simulation results without considering the ground contact, a relevant improvement towards the measured values of the test drive is obtained with considering the ground contact. It shows, that the computed ground contact lies exactly in the time interval where vibrations of the measured y- und z-accelerations occur, which are caused by the contact between vehicle underbody and embankment. Furthermore, it can be noted from the time history that the simulated pitch angle shows a reaction similar to that of the test vehicle. The vehicle bounces back a little when hitting the ground. During the simulation without consideration of ground contact, the roll angle does not show a rollover tendency at the end of the maneuver. But with consideration of ground contact, the gradient of the simulated roll angle agrees with the one from the measurement. In analogy to the experiment, the simulation with consideration of ground contact yields a rollover maneuver.

5 Conclusions and Outlook

The simulation of vehicle dynamics provides a very cost efficient tool, by which different test maneuvers can be exactly duplicated with minimum effort in changes and exact boundary conditions. Thus, one can eliminate the necessity for dangerous, complex and vehicle sacrificing testing maneuvers. Hence, simulation of vehicle dynamics is introduced in the development of sensors or electronic control units (ECU), illustrated here by the rollover sensor that is responsible for the activation of occupant protection systems, like airbags, seat-belt tighteners and rollover bars.

In this connection, the three dimensional vehicle dynamics simulation environment FASIM_C++ has been presented, in which the mechanics of the vehicle is modeled as a multibody system, together with the non-mechanic components, such as hydraulics, sensors, driver and environment assembled to a complete mechatronic vehicle model. Special consideration was given to the modeling of the ground contact, which has a significant influence on the vehicle dynamics in vehicles with a low ground clearance. Finally some of the simulation results were presented and compared with results from actual test drives.

Future experiments should optimize and finally validate the contact model with the help of measured data from different test drives, a calculation of the point of application of the force dependant on the position of the vehicle center of mass, and to include a modeling of the impact behavior that occurs while the vehicle underbody hits the ground.

References

1. Bardini R, Hiller M (1999) The contribution of occupant and vehicle dynamics simulation to testing occupant safety in passenger cars during rollover. SAE Technical Paper Series 1999-01-0431, Warrendale, PA, USA
2. Bertram T, Bekes F, Greul R, Hanke O, Haß C, Hilgert J, Hiller M, Öttgen O, Opgen-Rhein P, Torlo M, Ward D (2003) Modelling and simulation for mechatronic design in automotive systems. Control Engineering Practice, 11(2):177–188
3. Hiller M, Bardini R (1998) Vehicle and occupant dynamics simulation - important tools in development of sensor concepts for control of restraint systems. In: Proceedings of the 8th German-Japanese-Seminar on "Nonlinear Problems in Dynamical Systems – Theory and Applications", Kobe, Japan
4. Hiller M, Kecskeméthy A (1989) Equations of motion of complex multibody systems using kinematical differentials. Transactions of the Canadian Society of Mechanical Engineers, 13(4):113–121
5. Hiller M, Schuster C, Adamski D (1997) FASIM_C++ – A versatile developing environment for vehicle dynamics simulation. International Journal of Crashworthiness, Vol. 2, No. 1
6. Kecskeméthy A (1993) Objektorientierte Modellierung der Dynamik von Mehrköpersystemen mit Hilfe von Übertragungselementen. Fortschritt-Berichte VDI, Reihe 20, Nr. 88, VDI Verlag GmbH, Düsseldorf

7. Pichler V (1999) Modellbildung der Dynamik von Kraftfahrzeugen unter Anwendung objektorientierter Konzepte. Fortschritt-Berichte VDI, Reihe 12, Nr. 382, VDI Verlag GmbH, Düsseldorf
8. Wottreng W, Mehler G, Mattes B, Henne M, Lang H-P (1998) Rollover sensing (ROSE). In: Proceedings of the 2nd International Conference on Advanced Microsystems for Automotive Applications, Berlin, Germany

Aircraft Subsystems Modelling Using Different MBS Formalisms

Krzysztof Arczewski and Janusz Frączek

Institute of Aeronautics and Applied Mechanics, Warsaw University of Technology, Nowowiejska 24, 00-665 Warsaw, Poland;
E-mail: {krisarcz,jfraczek}@meil.pw.edu.pl

Abstract. The paper presents three case studies of dynamic analysis of aircraft during landing manoeuvre using two basic formalisms encountered in rigid and flexible multibody system (MBS) modelling. In the first case a formulation in natural coordinates has been used to analyze the dynamics of a medium size aircraft. Equations of motion have been formulated and solved using velocity transformation method. The aircraft has been modelled as consisting of rigid bodies connected by universal joints with springs. Aerodynamic forces have been taken into account by applying the Vortex Lattice Method (VLM) to the calculations performed. The effect of ground proximity on the results (ground effect) has been analyzed. In the second case, a dynamic analysis of a glider during the landing manoeuvre has been carried out from the point of view of stress recovery by means of various methods. Body positions and orientations have been written in absolute coordinates with floating frame approach for flexible bodies. Finite element method (FEM) and component mode synthesis has been used to model the flexibility of the bodies. A comparison of stress results obtained for different computation methods has been carried out. In the third analysis a MBS model of the Su-22 military airplane main landing gear has been presented. The absolute coordinates and the differential algebraic equations (DAE) formulations were used in all calculations. The whole landing gear model includes individual models of hydraulic actuators, shock absorber, flexible tire and contacts between some landing gear parts. Several types of simulations like landing gear extension and selected ground manoeuvres were performed. On that basis values of the forces which will allow to assess fatigue and durability of landing gear in future experiments were obtained. The received results were compared to the experimental measurements which were carried out on a real military airplane. The key issues of that comparison and general remarks were formulated. In the final part of the paper general conclusions regarding application of various computation MBS methods to dynamical analyses of aircrafts have been presented.

1 Introduction

In the early stage of the airplane design the static and dynamic analyses of the whole airplane and many subsystems play a significant role. Many of these analyses are usually devoted to forces and stresses evaluation appearing in subsystems of the airplane. Particularly forces which come from ground in manoeuvres like landing, taxiing or taking off [1] in the presence of aerodynamic forces must be determined. There exist many publications with respect to the simulation of air-

J.C. García Orden et al. (eds), Multibody Dynamics. Computational Methods and Applications, 201–220.

craft ground manoeuvres and whole aircraft modelling. For example an early overview of computer simulation of aircraft and landing gear is given by Doyle [2]. Shepherd, Catt and Cowling [3] describe a program founded by British Aerospace for the analysis of aircraft landing-gear interaction with a high level of details with flight tests involving ground contact. Two publications of International Association for Vehicle System Dynamics, Hitch in 1981 [4] and Kruger et al. [5] in 1997 and one of NASA Langley Research Center [6] provide state of the art of aircraft ground simulations.

One of common techniques used in some airplane subsystems modelling is MBS formalism. Several MBS formulations exist. Two common ones are based on the body kinematics written in natural coordinates [7] and in absolute coordinates [8]. Also, the flexibility of aircraft elements is being modelled in various ways [9, 10]. In some cases, flexible parts are modelled as rigid bodies interconnected by kinematic pairs which include elastic elements whereas in others their flexibility effects are modelled by means of a mixed formulation – MBS and FEM [11].

This work presents three cases of aircraft dynamics analysis during the landing manoeuvre. In the first case, a medium-size aircraft is modelled in natural coordinates. It has been assumed that the system consists only of rigid bodies interconnected by universal joints. Element flexibility is taken into account by introducing elastic (springs) elements. Aerodynamic forces are calculated by means of the VLM [12]. Equations of motion have been formulated in the dependent variable system. They have been integrated afterwards by means of the velocity transformation method. As a result, a set of ordinary differential equations (ODE) has been achieved. The presented model has been used to illustrate the dynamics of an aircraft during its landing with the ground effect taken into account.

In the second case, the dynamics analysis has been carried out for a glider during landing. The presented model has been used to perform a comparison of stress results obtained with different computation methods. This time, the aerodynamic forces have been neglected and the element flexibility has been taken into account by means of the FEM. Motion equations in the form of DAE [13] have been built using absolute coordinates. The system can be reduced to a system of a smaller size using the Component Mode Synthesis.

In the third case the virtual model of the Su-22 main landing gear [14] has been built which could be used to conduct variety of dynamic analysis. The main reason of investigation of existing construction (usually landing gear simulations and laboratory tests are being performed in the early design stage) is an attempt to predict forces, fatigue effects and reliability of the landing gear due to the lack of these data. In the second step simulation results were compared to experimental measurements carried out on board of the real airplanes during various maneuvers on the ground. These measurements have been carried out in the project by courtesy and with cooperation of the Polish Air Force Institute of Technology [14]. The aim of the measurements was to obtain stress, strains and forces results between the landing gear parts during specific manoeuvres like taxiing, landing and taking off.

2 Dynamic Analysis of Aircraft in Natural and Joint Coordinates

A three dimensional medium size aircraft model is used in this investigation [15]. The aircraft MBS under consideration consists of three main components, fuselage, wing and landing gear. In the present analysis, the aircraft MBS consists of rigid bodies interconnected by flexible joints. Natural coordinates will be used to describe each body, the interconnection between different bodies and their motions.

2.1 Multibody model of the aircraft

The three dimensional aircraft model used in this research is shown in Figure 1. The system has fifteen rigid bodies and two identical suspensions which represent the landing gear. Body 1 is the aircraft fuselage, and bodies 2-7 and 9-14 represent both wings of the aircraft while bodies 8 and 15 are the landing gear. The suspension mass is assumed to be concentrated at the wheel centre. The total aircraft weight is ($72x10^3$ kg) which represents the aircraft weight during landing.

The fuselage of the aircraft is considered to be a single body (body 1) with its coordinate frame fixed to the mass centre. The mass of the body represents the total weight of the fuselage, the aircraft engines, passengers, fuel etc.

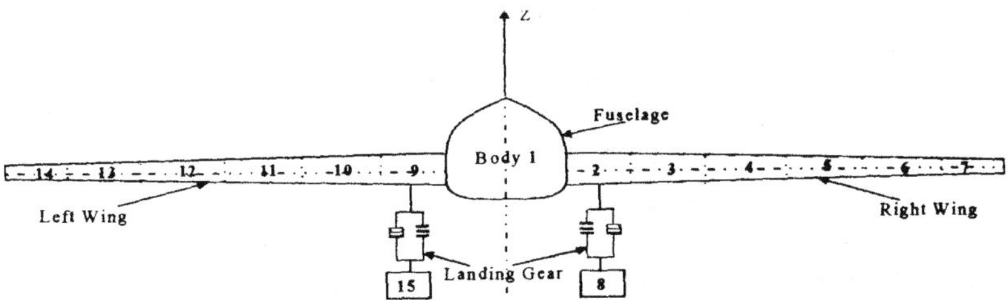

Fig. 1. Three dimensional aircraft model.

The wings of the aircraft, which represent the lifting surface of the system, have been modelled as two identical flexible cantilever beams discretized by a series of rigid bodies joined by elastic joints. Each joint is of a universal type and its flexibility is represented by springs (the type of the joint chosen allows for bending and torsion of each body). It is assumed that each body has the same cross section but with different dimensions.

The landing gear is modelled by two identical suspensions. The masses of the two suspensions are assumed to be concentrated at the wheel centres. Elasticity of the landing gear is introduced by means of a nonlinear spring and damper elements. The total mass of landing gear components is lumped at bodies 8 and 15 (Figure 1).

2.2 External and aerodynamic forces

The forces and moments acting on the MBS model can result from air pressure, gravity or others such as those caused by springs, dampers or contact with the ground. The calculation of the aerodynamic force seems to be most interesting.
The aerodynamic force is a result of motion of air around the wing which produces pressure and velocity differences. The elementary concepts of how these forces and moments are produced are presented in many texts dealing with aerodynamics (e.g. [12]). We make basic assumptions about symmetric flow action on the aircraft wing, and only the lift and drag forces together with the pitching moments are taken into account. For such assumptions, several methods have been developed to compute the flow about a wing which is operating at a small angle of attack, so that the resultant flow may be assumed to be inviscid, irrotational and incompressible. A lifting surface theory known as the Panel Method has been used in this paper for aerodynamic force calculations [14].

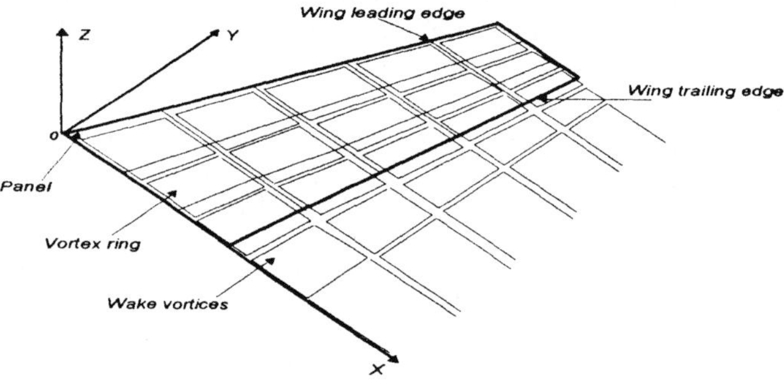

Fig. 2. Panels and vortex rings model for a thin lifting surface

In this method the surface of the body is covered by a finite number of small areas called panels, each of which has distributed singularities of a certain kind that have an undetermined uniform density. The distributed singularities are used to deflect the oncoming stream so that it will flow around the body. In general, source panels are used on the nonlifting surfaces, and vortex panels are used on the lifting surfaces of a body. The requirement for the oncoming flow to be tangent to every panel at a particular location gives a set of equations, which is used to compute the singularity densities on the panels. In this approach, the differential equation is converted to an integral over the configuration surface by means of Green's theorem. This integral equation is then solved by the discretization process. Thus, the overall flow consisting of a uniform flow and a flow induced by the singularities on a finite number of panels, becomes determined and the velocity and pressure at any point in the flow field can be calculated.

The evaluations of the aerodynamic lift, drag and moment coefficients are all based on the proper integration of the pressure coefficient on the lifting surface.

The Panel Method is solved numerically by means of a very common *modified Vortex Lattice Method* [12] which is an extension to the classical VLM used for the calculation of aerodynamic forces on lifting surfaces undergoing complex unsteady 3D motion. It is assumed that all considerations are restricted to low angles of attack.

From a theoretical point of view it is assumed that the problem to be solved is described by the Laplace equation for the velocity potential:

$$\nabla^2\Phi = 0 \tag{1}$$

where Φ is a velocity potential.

For an irrotational and inviscid flow a velocity potential can be defined such that:

$$\mathbf{V} = \nabla\Phi \tag{2}$$

where ∇ is the gradient operator. If the free stream Mach number is significantly small, the flow may also be considered incompressible. The principle of mass conservation for an incompressible flow has the form:

$$\nabla \cdot \mathbf{V} = 0 \tag{3}$$

Equation (1) follows immediately from equations (2) and (3).

In order to complete the problem proper boundary conditions on the body surface, its image at the trailing edge and at infinity must be defined. At first, two conditions should be satisfied. One is for the velocity normal to the wing surface and to the ground to be equal to zero, and the other is for the velocity to have a finite value:

$$\nabla\Phi\mathbf{n} = 0, \nabla\Phi < \infty \tag{4}$$

where $\mathbf{n}$ is a unit vector normal to the body and ground plane surfaces respectively).

The third condition requires that the influence of the wing on the flow field should vanish at large distances from the wing. The fourth requires the Kelvin condition to be fulfilled for both the body and the wake.

It should be pointed out that aerodynamic characteristics of an aircraft are influenced by ground proximity during takeoff and landing. The interaction of the aircraft with the ground is known in the literature as the "ground effect" [15]. The flow during takeoff and landing is inherently unsteady even if the aircraft is moving at constant velocity. These phases are among the most dangerous phases of flight and they are included in the aerodynamic modelling presented in this paper.

The VLM is described in detail from a theoretical and computational point of view in a majority of books dealing with aerodynamics [12]. Therefore only the main steps of computations of unsteady aerodynamic loads are mentioned in this paper:

1. Define geometry,
2. Calculate panel coordinates, vortex rings coordinates, etc.
3. Define ground effect,
4. Select flight parameters
5. Calculate influence coefficients and wake influence,
6. Calculate momentary right hand side vector
7. Solve matrix equation
8. Wake roll up,
9. Calculate pressure and load.

Steps from 5 to the 9 are repeated in a time loop. As mentioned before, the wing of the aircraft is modelled as consisting of small bodies. The aerodynamic forces and moments are calculated for each body. The point of action of these forces and moments is placed in the intersection point between the quarter of the mean aerodynamic chord line of each body and the line dividing the body into two equal halves.

2.3 Equations of motion

Natural coordinates corresponding to a three-dimensional MBS are used to describe the position of each body by means of the Cartesian coordinates of the basic points distributed throughout the elements and by means of the Cartesian components of several unit vectors. The body motion is defined through the motion of its points and vectors. Each body of the system should have a sufficient number of points and vectors linked to it, so that their motions completely define the motion of the body.

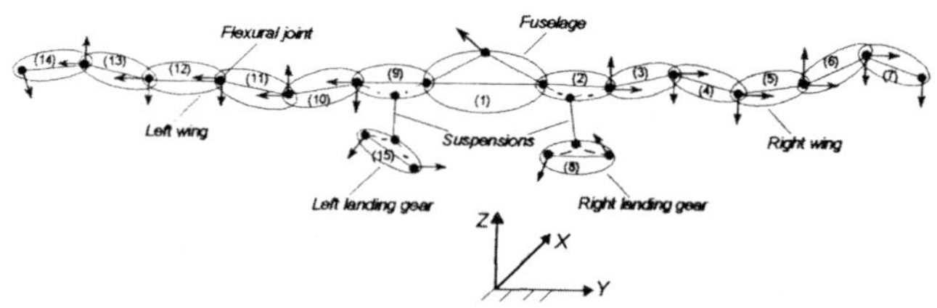

Fig. 3. Panels and vortex rings model for a thin lifting surface

The details of calculations and constraint equations formulation in natural coordinates are given in [7]. Figure 3 shows the complete system and the set of points and vectors used to define the MBS model of the aircraft.

The equations of motion of the MBS can be derived and expressed in a variety of forms depending on the type of formalism used. Probably the most commonly used with MBS are Newton-Euler methods and Lagrangian methods [10]. Also the principle of virtual power [7] which leads to a direct formulation of the inertial forces and avoids the differentiation process inherent in Lagrange's equations can be encountered.

One of the final forms of the equation of motion of the MBS in natural coordinates (which are in general dependent coordinates) obtained for instance by using the principle of virtual power can be written in a general, well known DAE [7] form:

$$\begin{cases} \mathbf{M}\ddot{\mathbf{q}} = -\boldsymbol{\Psi}_{\mathbf{q}}^{T}(\mathbf{q},t)\boldsymbol{\lambda} + \mathbf{Q}(\mathbf{q},\dot{\mathbf{q}},t) \\ \boldsymbol{\Psi}(\mathbf{q},t) = \mathbf{0} \end{cases} \tag{5}$$

where $\mathbf{q}$ is a vector of generalized natural coordinates, $\mathbf{Q}$ is a vector of external (including aerodynamic) and velocity-dependent (Coriolis and centrifugal) forces, $\boldsymbol{\lambda}$ is a vector of Lagrange multipliers responsible for magnitude of constraints reaction forces, $\boldsymbol{\Psi}$ is a vector of constraints equations (usually vector $\boldsymbol{\Psi}$ represents scleronomic constraints).

2.4 Numerical approach – velocity transformation.

It is well known that from a numerical point of view equation (5) is in general a index-3 DAE equation [13]. It can be numerically integrated directly or in a modified form in dependent coordinates (DAE equation) or after transformation to independent coordinates (ODE equation). Introduction to these methods is given in detail in [13].

In case the equations of motion are integrated in dependent coordinates, then BDF (Gear) algorithms can be applied directly to index-3 formulation or to a formulation having a lower index (2 or 1). Usually in integration of DAE equations having an index 2 or 1 formulation constraints stabilization methods have to be applied in order to avoid drifting effects and problems with numerical stability (i.e. projection methods, Baumgarte or GGL algorithms). Independently, augmented Lagrangian methods are proposed as well as algorithms based on promising IRK methods [13].

In this paper the method of integration, which actually belongs to the second group of integration algorithms based on transformation equation (5) to formulation in independent coordinates has been applied. The method known as velocity transformation has also been applied [8]. In general, velocity transformation follows from the fact that dependent velocities can be represented by independent (or simply alternative) ones:

$$\dot{\mathbf{q}} = \mathbf{R}\dot{\mathbf{z}} \tag{6}$$

In case the constraints are scleronomic the matrix $\mathbf{R}$ is orthogonal to the Jacobian matrix, i.e.

$$\boldsymbol{\Psi}_{\mathbf{q}}\mathbf{R} = \mathbf{0} \tag{7}$$

Substituting (6) (and its time derivative) into equation (5) (assuming scleronomic constraints), multiplying by $\mathbf{R}^T$, then using (7) yields:

$$\mathbf{R}^T\mathbf{MR}\ddot{\mathbf{z}} = \mathbf{R}^T(\mathbf{Q} - \mathbf{M}\dot{\mathbf{R}}\dot{\mathbf{z}}) \tag{8}$$

Equation (8) is ODE and can be integrated using classical predictor-corrector, BDF or RK methods [13]. It should be pointed out that numerical efficiency of the mentioned integration method is strongly influenced by the proper choice of matrix **R**. In the general case, there exist various methods for matrix **R** calculation based on projections algorithms, SVD or QR algorithms.

In the velocity transformation method the vector of joint coordinates is taken as the vector **z**. In such case, for a kinematical loop having a tree structure and typical joints, matrix **R** can be calculated very efficiently. Details of calculations used in this paper are given e.g. in [8].

2.5 Simulation results

The aircraft model developed in this paper is solved numerically. At first, the motion equations of the wing MBS are integrated in time using a numerical integration method. To validate this model, a case study is performed by investigating the dynamic characteristics of a wing flying in subsonic flow near and far from the ground. Finally, the whole aircraft MBS is introduced through the investigation of the effect of touchdown impact on the aircraft's response during its landing operation for different study cases.

In one of the first simulation experiment the aircraft wing shown in Figure 3 is rigidly fixed to the ground, thus simulating a root fixed condition representative of the wing-fuselage connection on the actual aircraft. This wing is divided into six small bodies. These bodies are interconnected by elastic joints consisting of universal joints and springs. The system has ten degrees of freedom. Natural coordinates are used to describe the position of each body in the system. The Unsteady VLM (vortex rings singularities) is used to calculate the aerodynamic loads acting on the wing with ground effect taken into account. These calculated loads are finally determined in terms of natural coordinates. The constraint equations and equations of motion are first determined in natural coordinates (dependent coordinates), and then transformed into independent coordinates (the system's degrees of freedom) using the velocity transformation process. To validate this model, a case study to predict the dynamic characteristics of real aircraft wing flying in subsonic flow near and far from the ground was performed.

The wing MBS resembles a cantilever beam, its different shape modes are well known in dynamics literature. In this case, the wing behaviour has been investigated for different flow airspeeds with the angle of attack kept constant (5 degrees). Different speeds of the airflow are used as input - external excitation of the wing structure. Vertical displacements of each body were determined at each airspeed. Afterwards, these vertical displacements were plotted against the wing half span. The wing MBS was excited in a velocity range between 200 and 600 m/s, each excitation case data was acquired for a period of one minute and several cases were performed.

In Figure 4a the exemplary first mode shape of the discretized wing near and far from the ground has been shown. As can be seen, the ground has a great effect on the response of the wing MBS.

In the second simulation, a case study of investigating the effect of the touchdown impact on the aircraft response during its landing is chosen. The formulations presented are used to simulate the impact between the landing gear and the ground.

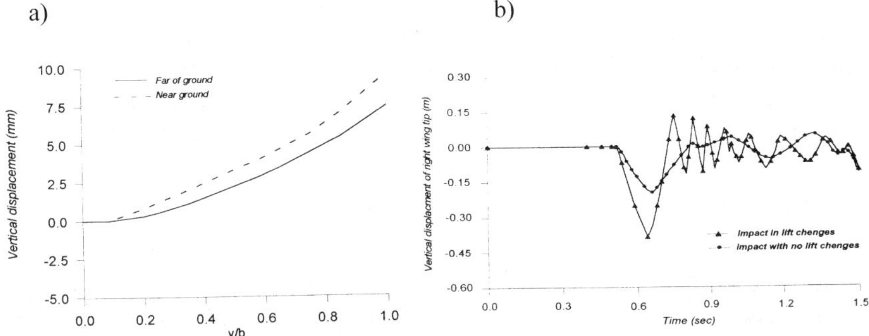

Fig. 4. a) First mode shape of the discretized wing near and far from ground b) Vertical displacement of the right wing tip point with and without lift changes.

The response of the aircraft was investigated for two different situations. In the first one, it has been assumed that the aircraft landed in normal conditions, i.e. no wake, wind shear (calm atmosphere) or any external disturbances were present. The lift coefficient has been set equal to the ground steady-state value (C_{Ls}=1.84) determined in earlier simulation, representing the landing configuration with the ground effect included. In the second situation, it has been assumed that the aircraft landed in the presence of sudden changes in its lift coefficient (caused for instance by "wake vortices" or due to a wind-shear caused side gust). In this case the steady-state value of the lift coefficient with the ground effect (C_{La}=1.84) included is assumed to drop to a value of 0.34 ($C_L \approx 0.2C_{La}$).

In Figure 4b the vertical response of the tip deflection of the right wing relative to the coordinate system of the two joints connecting the wing to the fuselage (rigid joints) has been presented. The results are plotted from the time of release of the model (t=0s) through impact (t=0.51s), rebound of the landing gear (t=0.86s) and secondary impact (t = 1.41s).

Figure 4b shows that significant differences which can be observed in tip vibration between two cases of landing – with lift changes and without lift changes. It is clear that the force induced on the suspension system of the aircraft landing gear at the time of impact significantly affects the dynamic response of the wing and the entire aircraft. If these forces are high, they will be transmitted to the wing through the suspension. Therefore, these forces excite the deformation modes of the wing, thus producing oscillations that can significantly affect the dynamic response and passenger's ride comfort of the aircraft. This is a clear indication that the sudden changes (drop) in the lift coefficient are very dangerous to the aircraft.

It should be pointed out that the presented model built using MBS proved also to be useful for simulation of other dangerous situation arising during landing such as landing on one wheel or rolling motion.

3 Dynamical Analysis of a Glider Using FEM and MBS Methods

In this section a dynamic analysis of a glider during the landing manoeuvre is presented and discussed from the point of view of stress estimation by means of various methods. A comparison of results obtained for different computation methods has been carried out. For the glider dynamics calculations body positions and orientations have been written in absolute coordinates. FEM and component mode synthesis has been used to model the flexibility of the bodies. The most frequent landing variant was tested - landing on the main wheel with the front wheel just above the ground. During the landing approach phase, the glider should have the proper rate of descent, which is defined in international regulations concerning the design and operation of JAR-22 gliders - the velocity amounts to 1.5 m/s. In the study, the rate of descent was established at 1.955m/s in order to test the behaviour of the glider. The total simulation time of the glider's landing manoeuvre equals 2s. The gear-ground contact phase lasts approximately 1.8s.

3.1 The MBS and FEM model of the PW-6 glider

To perform the analysis of the glider's dynamics during the landing manoeuvre, a calculation model was developed based on the MBS method. The floating frame approach algorithm and absolute coordinates formulation were used. In the FEM models employed here, the tacit assumption has been made which idealized that single grid connections between system components can be made. Within the different calculation variants, a comparison of different methods of stress estimation was performed.

The glider's fuselage [16] (Figure 5a) was modelled in an FEM software [20] environment using quad shell elements (2125 elements) and rigid body elements (RBE) to model external fixing points [17]. The model of the front and main (rear) suspension was prepared using the MBS method.

The calculations of the glider's dynamics (kinematical parameters) were performed using both the MBS method and the floating frame approach [10]. FEM was used to estimate stresses using various methods.

The calculations were performed with the simplifying assumption that the material of the fuselage is duralumin. In fact, the fuselage of the PW-6 glider is made of glass/epoxy composite. For this material the floating frame approach methods could not be applied.

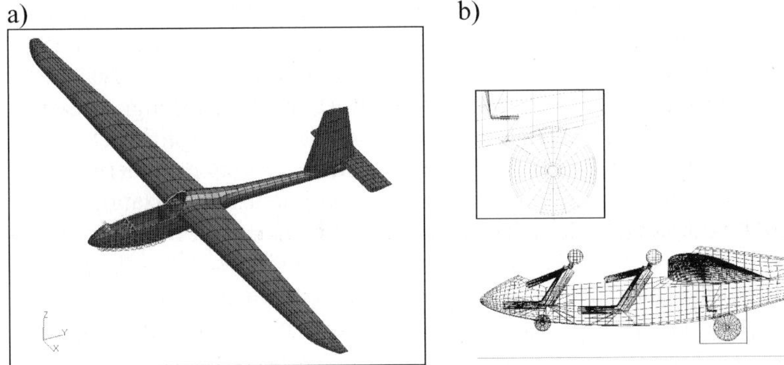

Fig. 5. a) FEM model of the PW-6 glider fuselage b) Model of rear suspension and simplified rigid model of pilots.

The front suspension consists only of the front wheel, which is directly fixed to the glider's fuselage. The rear suspension consists of the main wheel, bellcrank and the shock absorber, which are connected with each other and with the fuselage by means of kinematical pairs (Figure 5b). Simplified models of the two pilots are added to the glider's model with front and rear suspensions (Figure 5b). It was assumed that during the flight and landing manoeuvre the pilots are not in motion.

3.2 Equation of motion. Numerical integration

If a system of rigid bodies is modelled by means of the floating frame approach in the range of small deformations, the equations of motion in absolute nodal coordinates can be written in the form analogous to equation (5) [10]:

$$\begin{bmatrix} \mathbf{m}_{SS} & \mathbf{m}_{Sf} \\ \mathbf{m}_{fS} & \mathbf{m}_{ff} \end{bmatrix} \begin{bmatrix} \ddot{\mathbf{q}}_S \\ \ddot{\mathbf{q}}_f \end{bmatrix} + \begin{bmatrix} \mathbf{0} & \mathbf{0} \\ \mathbf{0} & \mathbf{D}_{ff} \end{bmatrix} \begin{bmatrix} \dot{\mathbf{q}}_S \\ \dot{\mathbf{q}}_f \end{bmatrix} + \begin{bmatrix} \mathbf{0} & \mathbf{0} \\ \mathbf{0} & \mathbf{K}_{ff} \end{bmatrix} \begin{bmatrix} \mathbf{q}_S \\ \mathbf{q}_f \end{bmatrix} = \begin{bmatrix} (\mathbf{Q}_z)_S \\ (\mathbf{Q}_z)_f \end{bmatrix} + \begin{bmatrix} (\mathbf{Q}_v)_S \\ (\mathbf{Q}_v)_f \end{bmatrix} - \mathbf{\Psi}_{\mathbf{q}}^T \boldsymbol{\lambda} \tag{9}$$

where: $\mathbf{Q}_z$ – vector of external forces applied to a flexible body, $\mathbf{Q}_v$ - vector of centrifugal, Coriolis and other forces resulting from differentiation of the kinetic energy with respect to time and each of the coordinates. The above matrices can be obtained by means of classical FEM algorithms used in linear range.

In order to reduce the number of degrees of freedom, the system of equations (9) is usually written in the form:

$$\begin{bmatrix} \mathbf{m}_{SS} & \mathbf{m}_{Sf}\mathbf{\Lambda} \\ \mathbf{\Lambda}^T\mathbf{m}_{fS} & \mathbf{I} \end{bmatrix} \begin{bmatrix} \ddot{\mathbf{q}}_S \\ \ddot{\mathbf{p}} \end{bmatrix} + \begin{bmatrix} \mathbf{0} & \mathbf{0} \\ \mathbf{0} & \mathbf{d} \end{bmatrix} \begin{bmatrix} \dot{\mathbf{q}}_S \\ \dot{\mathbf{p}} \end{bmatrix} + \begin{bmatrix} \mathbf{0} & \mathbf{0} \\ \mathbf{0} & \mathbf{\Omega} \end{bmatrix} \begin{bmatrix} \mathbf{q}_S \\ \mathbf{p} \end{bmatrix} = \begin{bmatrix} (\mathbf{Q}_z)_S \\ \mathbf{\Lambda}^T(\mathbf{Q}_z)_f \end{bmatrix} + \begin{bmatrix} (\mathbf{Q}_v)_S \\ \mathbf{\Lambda}^T(\mathbf{Q}_v)_f \end{bmatrix} - \begin{bmatrix} \mathbf{\Psi}_{\mathbf{q}_S}^T \\ \mathbf{\Lambda}^T\mathbf{\Psi}_{\mathbf{q}_f}^T \end{bmatrix} \boldsymbol{\lambda} \tag{10}$$

The matrix $\mathbf{\Lambda}$ can be obtained through different modal synthesis techniques. The elastic deformations of all degrees of freedom are approximated by a linear combination of suitable modes. The component modes contain the static and dynamic behaviour of the structure and consist of two families of modes: constraints modes and normal modes. In order to get a decoupled set of modes the constraint modes and normal modes are often transformed into a set of orthogonalised component modes [7], in consequence it is not possible to distinguish between pure static and pure dynamic modes. Equation (10) was integrated numerically using direct integration method in index-1 formulation with index stabilization.

It should be pointed out, that the inertia matrix in equation (10) is a function of a few invariant matrices [7]. Through omitting or including some of them it is possible to control the analysis type and the numerical complexity.

According to the size of displacements or forces obtained after integration of the equations (6), the stresses in flexible bodies can be estimated. The values of the obtained stresses can vary significantly depending on the calculation method. The issue of stress evaluation for different estimation methods will be discussed on the basis of the PW-6 glider.

3.3 Stress recovery methods

In order to estimate stresses in selected parts of the glider's structure, the following evaluation methods were used [17, 18]:

A. Rigid body stress recovery – force based. Neglecting body deformations, it is possible to calculate forces and accelerations of the studied system, and then the stresses using FEM software, with a statically determinate support.
B. Flexible body stress recovery – force based. The estimation of stresses is performed as above but all the bodies are treated as flexible. Then, using FEM software, static calculations of the system, with statically determinate supports, were performed. Calculated reaction forces are used to evaluate the correctness of computations (force values should be numerically close to zero).
C. Flexible body stress recovery – force based – inertia relief. Basing on the inertia and gravity forces calculated in a MBS analysis, the system is kept in kinetostatic equilibrium.
D. Modal stress recovery. During the modal basis generation phase in the FEM code, additional information can be also precomputed in order to later combine the modal coordinates to the FE stresses. This so-called modal stress tensor identifies the stress component associated with each orthogonalized mode shape. All calculations for stress evaluations are carried out in MBS software.
E. Deformation based stress recovery. Stresses are calculated based on node displacements obtained from calculations in MBS software. This can be done using FEM analysis by stress estimation based on node displacement in consecutive time steps.

The methods mentioned above were used for stress evaluation of different elements of the glider in many different places of the fuselage.

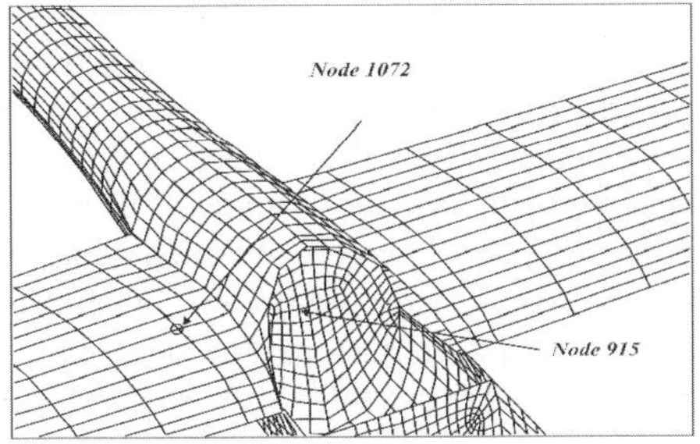

Fig. 6. Location of nodes 1072 (high stresses area) and 915 (moderate stresses area).

The further part of this study presents selected results of the stress analysis carried out for a chosen time interval in the neighbourhood of two selected nodes - node 1072 and node 915 (Figure 6). Node 1072 is located in the place on the wing of the glider where high levels of stresses are observed whereas node 915 is placed on the rear wall of the cabin of the glider in the area of moderate stresses.

3.4 Stress recovery methods. Results

The comparisons of different stress recovery methods in the form of stress trajectories calculated in the time interval <0.19s,0.5s> in nodes 1072 and 915 with methods A through E are presented in Figures 7a and 7b respectively.

In the area of stress concentration (Figure 7a) in the first phase of landing (approximately from 0.2s to 0.3s) the representation of the flexible structure as rigid (method A) as well as the force methods for stress recovery (B,C) lead to stress overestimation in comparison to displacement and modal methods (D,E). In the second phase (from 0.3s to 0.5s), the displacement methods give significantly higher stress levels.

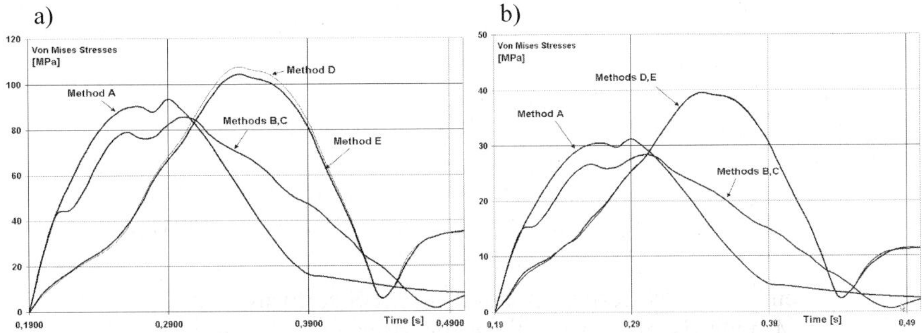

Fig. 7. Stress trajectories in node 1072 (a) and node 915 (b).

However, the maximal values of stresses calculated with force based methods in this time interval are underestimated in comparison with methods D and E.

A similar situation is observed in case of stress evaluation in the node 915 (moderate stress area). However, the relative discrepancy between maximal values is greater than in the previous case and climbs to 30 %. In case of strength evaluation or durability analysis such difference can make several analyses useless.

Significant differences between stresses calculated using different methods are also observed at time 0.49s. The stress levels obtained using methods D and E are significantly greater than in force based methods. These differences are attributable to the fact, that the modal and displacement stress recovery methods include the modal acceleration effects in the inertial terms of the system differential equations, which the force based approach does not.

Finally, it should be pointed out that in method B several variants of statically determinate supports have always been investigated.

4 Multibody Model of the Military Aircraft Main Landing Gear

In this section a third case study - MBS model of the Su-22 military aircraft main landing gear - is presented [14]. The whole landing gear model includes individual models of hydraulic actuators, shock absorber, flexible tire and contacts between some landing gear parts. The absolute coordinates and the DAE formulations and algorithms were used in all calculations [7]. Several types of simulations like landing gear extension and selected ground manoeuvres were performed. On that basis values of the forces which will allow to assess fatigue and durability of landing gear in future experiments were obtained. The received results were compared to the experimental measurements which were carried out on a real military airplane.

4.1 Su-22 landing gear – basic information. Multibody model

The Su-22 aircraft, which is shown in Figure 8a, is equipped with three-strut landing gear. The main landing gear is attached directly to the wings. The nose landing gear is attached to the front part of fuselage. The main landing gear retracts to the recesses which are located in the wings near the fuselage.

The struts of the main landing gear are equipped with oleo-pneumatic shock absorbers. In addition the struts are connected to the two hydraulic actuators, which are in charged of retraction end extension of the main landing gear. The hydraulic actuators use the hydraulic installation which is controlled by the electro-hydraulic valve to retract and to extend the landing gear.

The diagram taken from the technical documentation of the Su-22 main landing gear is shown in Figure 8b [14]. On the basis of technical documentation and some landing gear parts' measurements the model was built using one of the CAD packages [19]. Figure 9 describes the main parts and kinematical scheme of the Su-22 landing gear right strut.

a) b)

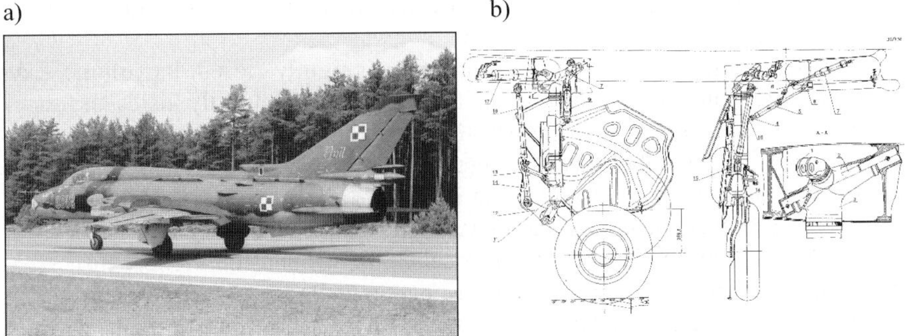

Fig. 8. Military aircraft during take off (a) and technical details of the main landing gear (b).

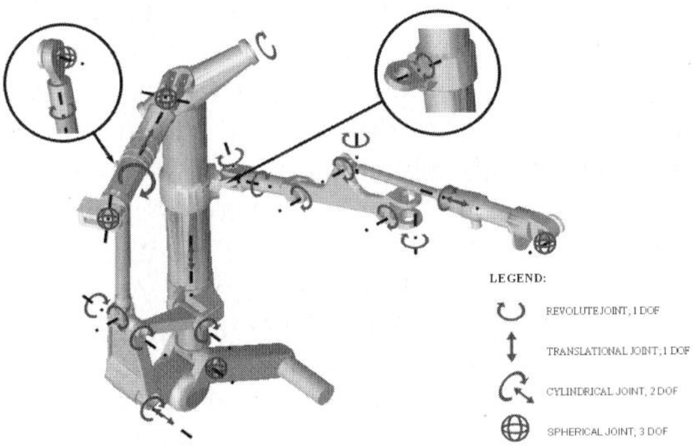

Fig. 9. The kinematics structure of the right strut of the Su-22 main landing gear.

The MBS model of the landing gear was built using rigid MBS formalism. All redundant constraints were eliminated in the modelling process in order to obtain reliable results of joint reactions evaluations and friction forces estimations in dynamical conditions.

Both right and left main strut has six degrees of freedom when the gear is retraced or extended. During landing and taxiing the side brace is blocked and the right and left strut has four degrees of freedom. Due to the fact that some elements of the landing gear adjoin each other in some phases of gear motion the contact of these part was modelled using theory of contact for rigid bodies.

4.2 The shock absorber and tire model. Equation of motion

There are two basic types of shock absorbers in landing gears [5]. The first one uses a spring usually made of steel or rubber. The second type of shock absorber uses a fluid spring of gas or oil, or a mixture of those therefore is called as oleo-pneumatic. Nowadays, most landing gear shock absorber contains a mixture of gas and oil. The shock absorber of the aircraft main landing gear is a kind of oleo-pneumatic element.

The Su-22 main landing gear shock absorber is tightly closed. Its volume is divided into three smaller chambers which are called A, B and C. Inside the shock absorber there is 1700 cm^3 of oil AMG-10 which fulfils a role of damping in system. In upper "A" chamber there is a pressurized nitrogen which acts as a spring. The gas pressure for fully extended strut is 8,33 MPa.

Before landing the landing gear is extended and the strut is fully extended due to pressurized nitrogen. When the tire touches the ground the vertical load starts acting on the piston and the strut stroke increases. In shock absorber the oil is forced from the lower "B" chamber to the upper "A" chamber through the main orifice. The oil flows from "A" chamber to "C" chamber. However this need only be a hole in the main orifice plate, the hole area is often changed by the varying-diameter metering pin. It is important that the oil not only flows through the main orifice but also through numerous orifices in orifice support tube and upper sleeve. Consequently in upper "A" chamber the pressure of nitrogen increases because of decreasing the gas chamber volume. The energy is absorbed by "pushing" a chamber of oil against chamber of dry nitrogen and then the gas and oil is compressed. After an initial impact, the rebound is controlled by the nitrogen pressure forcing the oil to flow back from "C" chamber to "B" chamber through orifices. It is worth knowing that when the oil flows back the upper sleeve is mechanically decreased the number of small orifices from 42 to 2. As a result of that the oil has got smaller area to flows back. Because of that the oil flow speed increases as well as its temperature. In this way the work of external loads is being dissipated.

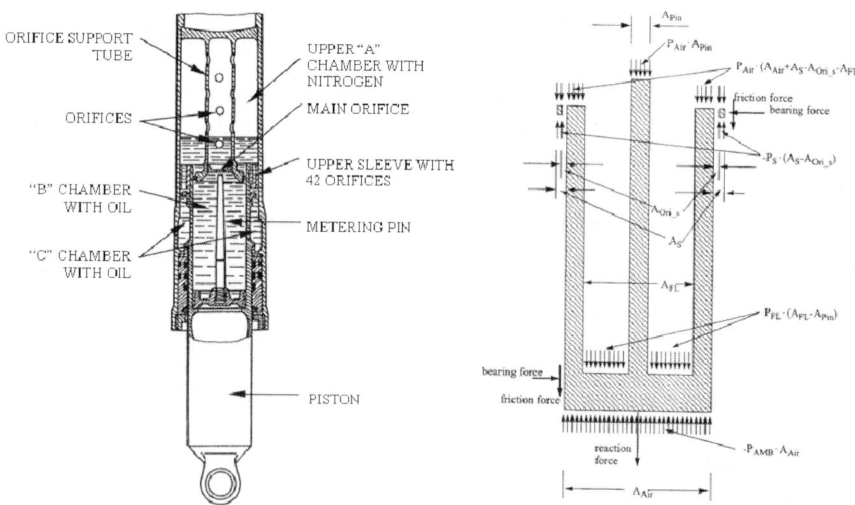

Fig. 10. Main landing gear oleo-pneumatic shock absorber and force model [20].

The oleo-pneumatic element is modelled as single point to point force acting between two points which belong to two different parts *i* and *j*. The parts (*i* and *j* points) move toward or away from each other [20].

The force of oleo-pneumatic element can be calculated from the formula:

$$F_{OLEO} = (P_{Air} - P_{AMB}) \cdot A_{Air} + \Delta P \cdot (A_{FL} - A_{pin}) - \Delta P_s \cdot (A_s - A_{ori\_s}) \tag{11}$$

The P_{AMB} is an ambient pressure and the other pressures like P_{Air}; ΔP; ΔP_s are the pressures acting on the specific areas in shock absorber which are denoted as A_{Air}; A_{FL}; A_{pin}; A_s; A_{ori_s}.

Equation (11) is the result of the summing of the axial pressure forces (pressures action on surfaces normal to the shock strut centreline) on the strut piston shown in Figure 10. It should be pointed out that the strut reaction forces, the bearing force, and the friction force are not included in that type of solution.

The nitrogen pressure can be calculated using equation of the polytropic change however pressures in fluid chambers can be calculated using formulas based on Bernoulli's equations. Detailed calculations are given in [14].

Contact forces between tire and the ground are the major importance for the dynamic behaviour of the aircraft during the landing and taxiing. To model real driving conditions during landing and taxiing it is important to choose the proper tire model [21]. Variety of models which describe the tire forces generated at braking, driving, sliding, etc are applied in communicated research. Some of them are theoretical which means that they describe the physical model of tire. Others are empirical which means that the forces are generated by the functional approximations and experimental data. The theoretical models require usually many parameters which are difficult to obtain.

In this research, to model the tire forces behaviour of the Su-22 main landing gear, the one of the simplest, Fiala Handling Force Model was applied. Detailed calculations for this model can be found in [20] and [21].

Equation of motion of a rigid MBS can be written with the use of absolute coordinates in index-3 DAE formulation (descriptor form) analogous to the formula (5). In the paper Gear (BDF) algorithms were used (index-3, and stabilized index 1 formulations) for numerical integration of the equations of motion. All models (tire, shock absorbers, and MBS model) described in previous sections were included into equations of motion in the form of external forces **Q** or in the form of additional kinematical and driving constraints.

4.3 Simulation analyses. Results

The MBS model of the main landing gear was used in simulations of many different manoeuvres of the airplane on the ground during taxiing, taking off and under landing impact. The results obtained were validated using experimental measurements. In order to obtain stress, strains and forces in the landing gear parts the number of strain gauges were installed on these parts.

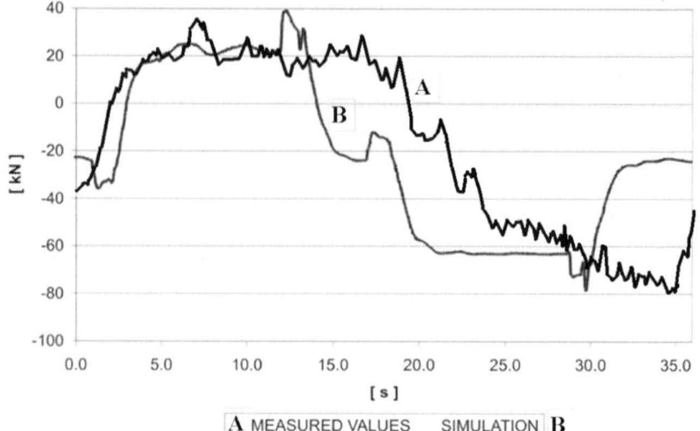

Fig. 11. The axial force of the right lower side brace during taxiing ("eight" manoeuvre).

The strains were measured and recorded in real time by digital data acquisition system mounted on board for specific manoeuvres of the aircraft during taxiing, landing and taking off. The forces and stresses were calculated off line after measurement experiments.

Figure 11 illustrates exemplary comparison of the real and simulation results of axial force in right lower side brace (Figure 9) during the Su-22 aircraft "eight" manoeuvre performed on the ground (the sequence of two turns, right and left similar to printed eight figure).

Despite some small differences which are observed in Figure 11 the results are, in general still acceptable. However it seems the discrepancies are caused by two reasons. The first follows from the fact that due to lack of precise data concerning mass moments of inertia of the aircraft they were obtained using CAD software and very simple geometrical model of the Su-22 airframe.

The second reason of differences is a simplification in simulation of aircraft motion. In the real conditions a pilot steers an aircraft on the ground by engine thrust and using brakes. However the precise values of the engine trust and the forces on the brakes were unknown during ground manoeuvres. Moreover, the real "eight" manoeuvre was performed as precisely as it was possible but the constant velocity and the constant radius during test remain a assumption. To model the constant velocity during the "eight" manoeuvre the motion element attached to the nose landing gear with constant radial velocity was used. Having the constant radial velocity the airplane has got almost (because of the tire friction) constant linear velocity.

5 Conclusions

The following, general conclusions, regarding various MBS formulae in application to dynamical analysis of aircrafts during various manoeuvres can be formulated:

- The presented use of fully Cartesian coordinates (natural) for a aircraft MBS system allows for a simple and efficient formulation. They allow for a simple formulation of constraints, efficient formulation of Jacobian matrix and the equations of motion. The method proposed is reliable and economic in terms of computer resources, as it has been proved with the test example that corresponds to real system.
- For flexible but not slender bodies which are not exposed to very high angular motion, the process of discretizing of a flexible body into small bodies instead of the FEM is less complex in determining the MBS equations of motion and reduces their numbers in comparison to the FEM. This is expected to reduce the computer time and improve the numerical efficiency as well. Therefore, better results can be achieved. Aerodynamics forces can be included using classical vortex methods. Various aerodynamic effects (like ground) effect can be modelled using this method.
- In case of analyses of real subsystems of the airplanes like landing gear fixed to flexible structures of the airplanes FEM and MBS seems the very convenient way of modelling. In case of stress evaluation, durability assessment etc. the consequences of the method of stress calculation must be taken into account. The examples given in the paper (glider analysis) suggest that modal and deformation based stress recovery are recommended methods.
- All formulae presented in the paper lead to DAE of motion. In the first case DAE equation was solved using velocity transformation method. This method leads to ODE. In two other cases it was solved using GEAR algorithm applied directly to index-3 and stabilized index-1 equation. In case of stress estimation significant differences between results obtained with GEAR algorithm applied to index-3 and stabilized index-1 equations were not observed in most cases.
- Natural coordinates with velocity transformation seem to be a less physical formulation than the absolute coordinates formulation but in many cases it proves to be more efficient. It is communicated in many references. However, in case of using fast computer calculations this disadvantage does not seem to be a strong argument against using absolute coordinates for general purpose MBS packages [20] used in aircraft simulations.

Acknowledgement

The work developed in this article has been partially supported by Ministry of Science and Inf. Tech. through project No. 4T07A03329.

References

1. Prashant DK (2003) Simulation of asymmetric landing and typical ground maneouvres for large transport aircraft. Aerospace Science and Technology. 7: 611-619
2. Doyle GA (1986) A review of computer simulations of aircraft-surface dynamics. J. Aircraft. 23 (4)
3. Catt T, Cowling D (1993) Active Landing Gear Control for Improved Ride Quality During Ground Roll, Smart Structure for Aircraft. AGARD CP 531, Stirling Dynamics Ltd. Bristol
4. Hitch HPY (1981) Aircraft ground dynamics. Vehicle System Dynamics. 10: 319-332
5. Kruger WR et al. (1997) Aircraft landing gear dynamics: simulation and control. Vehicle System Dynamics. 28:157-289
6. Pritchard JI (1999) An Overview of Landing Gear Dynamics. Hampton,Virginia, Langley Research Center
7. Garcia de Jalon, Eduardo B (1994) Kinematic and Dynamic Simulation of Multibody Systems. Springer-Verlag
8. Nikravesh P (1988) Computer-Aided Analysis of Mechanical Systems. Prentice-Hall, Englewood Cliffs, New Jersey
9. Schwertassek R, Wallrap O (1999) Dynamics of Flexible Multi Body Systems. Vieweg, Germany
10. Shabana AA (2005) Dynamics of Multibody Systems. Cambridge University Press
11. Bathe KJ (1996) Finite Element Procedures. Prentice Hall, Englewood Cliffs, New Jersey
12. Bertin JJ (2002) Aerodynamics for Engineers. Prentice Hall, Upper Saddle River, NJ 07458
13. Brenan KE, Cambpell SL, Petzold LR (1996) Numerical Solution of Initial Value Problems in DAE. SIAM, Philadelphia 1996
14. Fraczek J, Lazicki P, Leski A (2005) Dynamical Analysis and Experimental Verification of Military Aircraft Landing Gear Using MBS Methods. In Proc. of the ECCOMAS Thematic Conference, 21-24 June, Madrid, Spain
15. Issa SM, Arczewski K (1998) Wings in Steady and Unsteady Ground Effects. Canadian Aeronautics and Space Journal. 44(3): 188-193
16. PW-6. Technical Documentation. DWLKK. Warsaw
17. Arczewski K, Fraczek J (2005) Friction Models and Stress Recovery Methods in Vehicle Dynamics Modelling. Multibody System Dynamics, 14: 205-224
18. Yim HJ, Haug EJ and Dopker B (1989) Computational methods for stress analysis of mechanical components in dynamic systems. In Concurrent Engineering of Mechanical Systems, Vol.1, E.J. Haug (ed.) University of Iowa, Iowa City, 1989, 217-237
19. UNIGRAPHICS, Users Manual, 2003
20. MSC. NASTRAN/ADAMS/AIRCRAFT/TIRE (2005) Users Guide, MSC Software, 2004
21. H.B. Pacejka (Ed.) (1991) Tire Models for Vehicle Dynamics Analysis. In Proc. of 1st International Colloquium on Tyre Models for Vehicle Dynamics Analysis, Swets & Zeitlinger

Modelling of Joint Friction in Robotic Manipulators with Gear Transmissions

J.B. Jonker[1], R.R. Waiboer[2] and R.G.K.M. Aarts[1]

[1]*Laboratory of Mechanical Automation, Department of Engineering Technology, University of Twente, P.O. Box 217, 7500 AE Enschede, The Netherlands; E-mail: {J.B.Jonker, R.G.K.M.Aarts}@utwente.nl*
[2]*Netherlands Institute for Metals Research, P.O. Box 5008, 2600 GA Delft, The Netherlands; E-mail: R.R.Waiboer@alumnus.utwente.nl*

Abstract. This paper analyses the problem of modelling joint friction in robotic manipulators with gear transmissions in the sliding regime, i.e. at joint velocities varying from close to zero until their maximum appearing values. It is shown that commonly used friction models that incorporate Coulomb, (linear) viscous and Stribeck components are inadequate to describe the friction behaviour for the full velocity range. A new friction model is proposed that relies on insights from tribological models. The basic friction model of two lubricated discs in rolling-sliding contact is used to analyse viscous friction and friction caused by asperity contacts inside gears and roller bearings of robot joint transmissions. The analysis shows different viscous friction behaviour for gears and pre-stressed bearings. The sub-models describing the viscous friction and the friction due to the asperity contacts are combined into two friction models; one for gears and one for the pre-stressed roller bearings. In this way, a new friction model is developed that accurately describes the friction behaviour in the sliding regime with a minimal and physically sound parametrisation. The model is linear in the parameters that are temperature dependent, which allows to estimate these parameters during the inertia parameter identification experiments. The model, in which the Coulomb friction effect has disappeared, has the same number of parameters as the commonly used Stribeck model. The model parameters are identified experimentally on a Stäubli Rx90 industrial robot.

Key words: Joint friction modelling, helical gear pair, pre-stressed roller bearing, non-linear viscous friction, asperity friction, Stribeck curve, gear transmission, industrial robot.

1 Introduction

Experimental identification of the inertia parameters of robotic manipulators imposes high demands on the accuracy of the applied friction models as friction contributes significantly to the measured joint torques. The friction models commonly used in

J.C. García Orden et al. (eds), Multibody Dynamics. Computational Methods and Applications, 221–243.

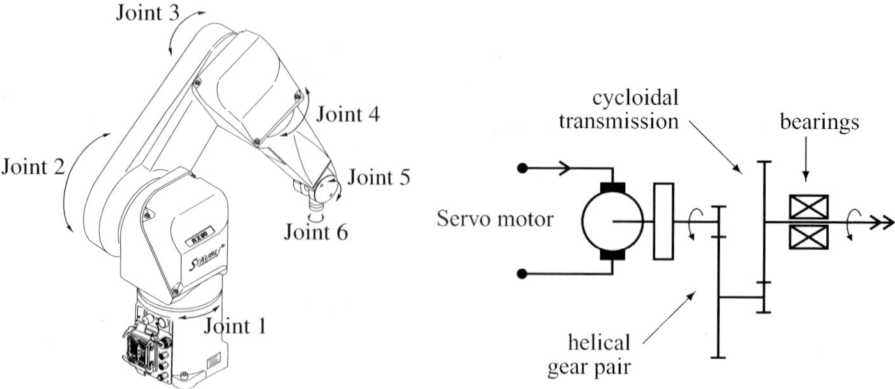

Fig. 1. Stäubli Rx90B six-axes industrial robot. Courtesy of Stäubli, Faverges, France.

Fig. 2. Schematic representation of the gear transmission inside the joint assembly.

robot literature for this purpose, so-called classical friction models, incorporate Coulomb, (linear) viscous and Stribeck components [1]. These models establish that the friction torque is a function of the joint velocity. However, it will be shown in this paper that these models are inadequate to describe the non-linear viscous friction behaviour at high velocities. The models can be improved ad hoc by including extra (non-linear) terms with extra parameters in the friction model, but the physical meaning of such additions is unclear.

This paper analyses the problem of modelling joint friction in industrial robots in the sliding regime, i.e. at velocities varying from close to zero up to the maximum appearing values. In order to get an understanding of friction in the joints of industrial robots, a closer look is taken at the gear transmissions inside the joint assemblies of the Stäubli Rx90 robot, see Figure 1. The first four joints of the robot, see Figure 2, are equipped with a so-called JCS (Stäubli Combined Joint), which is a sophisticated assembly that includes both a cycloidal transmission and the joint bearing support. The cycloidal transmission is driven by a servo motor via a helical gear pair. The gears and bearings in the cycloidal transmission are prestressed in order to eliminate any backlash or play. Both the cycloidal transmission and the helical gear pair are lubricated by means of an oil bath in order to reduce friction losses and to minimise wear. Naturally, all joints are equipped with roller bearings. The remaining two joints in the robot's wrist will not be discussed in this paper as they have a different set-up.

A new friction model is proposed that relies on insights from tribological models. The basic friction model of two lubricated discs in rolling-sliding contact is used to analyse viscous friction and friction caused by asperity contacts inside gears and roller bearings of the robot joints. The sub-models that describe the viscous friction and friction due to the asperities are combined into two friction models: one for gears and one for pre-stressed roller bearings. In this way a new friction model is developed that accurately describes the friction behaviour observed in the Stäubli Rx90 industrial robot.

A brief description of the classical friction models and their applicability in industrial robot identification is given in Section 2. In Section 3, the friction phenomena of a single lubricated contact is discussed on the basis of two lubricated discs in a rolling-sliding contact; expressions for the friction forces due to lubricant viscosity and asperity contacts are presented. These expressions are applied to model friction forces arising in two typical transmission components like the helical gear pair and the pre-stressed roller bearing in Section 4. In Section 5 these models are combined into a friction model that accounts for friction in a single joint. The unknown friction parameters of the friction model are identified by means of experiments in Section 6. It will be shown that the friction model is linear in the parameters that depend on the temperature of the robot joint, which makes it very suitable to use the friction models in inertia parameter identification experiments.

2 Friction Modelling at System Level

2.1 Classical friction models

Most friction models in robot literature are combinations of the classical friction models, see Figure 3. For modelling of friction in robots with revolute joints, friction is usually modelled as a joint torque $\mathcal{T}_j^{(f)}$ which is a function of its angular joint speed $\dot{q}_j$. The subscript j denotes the joint number. The most elementary model is the Coulomb friction model:

$$\mathcal{T}^{(f)} = \operatorname{sign}(\dot{q})\, \mathcal{T}^{(f,C)}, \tag{1}$$

where $\mathcal{T}^{(f,C)}$ is the Coulomb friction torque and $\dot{q}$ is the angular speed. The Coulomb friction model originates to the friction between sliding dry surfaces which generally produce large friction forces. Note that $\operatorname{sign}(\dot{q})$ is not defined for zero velocities. This means that the model is not able to describe the friction torque for a velocity equal to zero. The application of a lubricant between the surfaces results in the presence of a viscous term in the friction model:

$$\mathcal{T}^{(f)} = \operatorname{sign}(\dot{q})\, \mathcal{T}^{(f,C)} + c^{(v)}\dot{q}, \tag{2}$$

where $c^{(v)}$ is the viscous friction parameter, see Figure 3(a). Viscous friction is, in this model, taken as a linear function of the angular joint speed.

It was found by the Swiss scientist Euler (1707–1783) that a higher force was needed to bring the surfaces in a sliding motion than there is needed to keep the surfaces in motion, see Figure 3(b). This so-called static friction effect is taken into account as

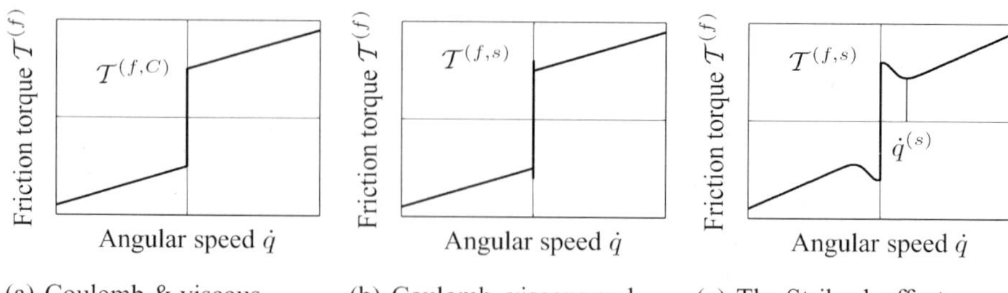

(a) Coulomb & viscous friction, Eq. (2).

(b) Coulomb, viscous and static friction, Eq. (3).

(c) The Stribeck effect, Eq. (5).

Fig. 3. Classic friction models.

$$\mathcal{T}^{(f)} = \begin{cases} |\mathcal{T}^{(f)}| \leq \mathcal{T}^{(f,s)} & \text{if } \dot{q} = 0\,, \\ \text{sign}(\dot{q})\, \mathcal{T}^{(f,C)} + c^{(v)}\dot{q} & \text{if } \dot{q} \neq 0\,, \end{cases} \tag{3}$$

where $\mathcal{T}^{(f,s)}$ is the static friction torque and $\mathcal{T}^{(f,C)} < \mathcal{T}^{(f,s)}$. Note that this model gives a non-unique solution for the coefficient of friction for zero velocities and that it shows discontinuous behaviour in a transition from zero velocity to non-zero velocity.

Stribeck [2] discovered that the drop from static friction to Coulomb friction is not discontinuous for lubricated surfaces but that it is a continuous function of the velocity, see Figure 3(c). Therefore, the graph representing the relation between friction and velocity will hereafter be referred to as the Stribeck curve. A well known model describing the Stribeck effect has been developed by Bo and Pavelescu [3], which has an exponentially decrease from the static friction to the Coulomb friction:

$$\mathcal{T}^{(f)} = \text{sign}(\dot{q})\Big(\mathcal{T}^{(f,C)} + (\mathcal{T}^{(f,s)} - \mathcal{T}^{(f,C)})e^{-|\dot{q}/\dot{q}^{(s)}|^{\delta}}\Big), \tag{4}$$

where $\dot{q}^{(s)}$ is known as the Stribeck velocity, which indicates the velocity range in which the Stribeck effect is effective. According to [3] the empirical exponent δ ranges from 0.5 to 1 for different material combinations.

Armstrong-Hélouvry [4] adopted this Stribeck model and added a viscous term $c^{(v)}\dot{q}$:

$$\mathcal{T}^{(f)} = \text{sign}(\dot{q})\Big(\mathcal{T}^{(f,C)} + (\mathcal{T}^{(f,s)} - \mathcal{T}^{(f,C)})e^{-|\dot{q}/\dot{q}^{(s)}|^{\delta}}\Big) + c^{(v)}\dot{q}\,. \tag{5}$$

This friction model has been applied by many authors, e.g. [1, 5, 6], for the modelling of sliding friction in robotic systems. In the next section, the applicability of this model in the modelling of joint friction in an industrial robot is investigated. The five unknown parameters, $\mathcal{T}^{(f,C)}$, $\mathcal{T}^{(f,s)}$, $\dot{q}^{(s)}$, δ and $c^{(v)}$, are determined experimentally.

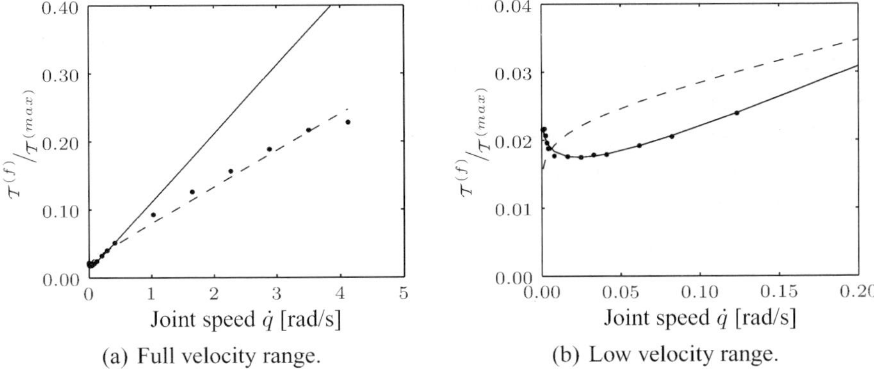

Fig. 4. Stribeck curve for joint 1 of the Stäubli Rx90 industrial robot as function of the joint speed. The dots (•) represent the measurements. Note the non-linear viscous behaviour at high velocities. Model 1 (– –) is estimated in the full velocity range whereas model 2 (—) is estimated in the range from 0 to 0.5 rad/s.

2.2 Stribeck curve measurement

In order to determine the unknown parameters, a Stribeck curve measurement of the first joint of the Stäubli Rx90 is carried out. For this purpose the joint friction torques have to be measured as functions of the joint speed. Experimentally, it appears that the friction torque also depends on e.g. the temperature of the joints and the joint angle [7]. To have an identical temperature during the experiments an initial warmup motion is executed before the actual measurements. The influence of the position and noise are minimised by averaging the measured joint torques during the constant velocity part of a trapezoidal velocity profile. Figure 4 shows the joint torques of one joint that are obtained this way for a number of chosen values of the joint speed $\dot{q}$. The joint torques are normalised with the maximum joint torque. The Stribeck effect is clearly visible in the detailed Figure 4(b). Note that the Stribeck velocity parameter $\dot{q}^{(s)}$ does not necessarily coincide with the joint speed where the friction torque has its minimum.

The friction model is a non-linear function of two of the unknown parameters, namely $\dot{q}^{(s)}$ and δ. In order to estimate all five parameters at once, one has to rely on non-linear optimisation techniques. It is commonly known that non-linear optimisation techniques may lead to local optima in which non-physical parameter values are found. Non-linear optimisation techniques can be applied successfully in cases for which the model is consistent with the observed behaviour combined with a proper first estimate of the parameter values.

To prevent difficulties with non-linear estimation techniques, a linear least squares optimisation technique is used to obtain the values for the parameters $\mathcal{T}^{(f,C)}$,

$\mathcal{T}^{(f,s)}$ and $c^{(v)}$ which are linear in the model. Values for the parameters $\dot{q}^{(s)}$ and δ are selected manually and are assumed to be constant.

In this way the parameters are identified in three steps. In the first step, the parameters for the Stribeck effect $\dot{q}^{(s)}$ and δ have been given a reasonable value. In step two, the remaining parameters $\mathbf{p}^{(f)} = \left[\mathcal{T}^{(f,s)}\ \mathcal{T}^{(f,C)}\ c^{(v)}\right]^T$ are estimated with a linear least square optimisation, which implies minimising the ℓ^2-norm

$$\hat{\mathbf{p}}^{(f)} = \arg\min_{\mathbf{p}^{(f)}} \| \mathcal{T} - \mathbf{A}^{(f)}\mathbf{p}^{(f)} \|_2^2, \tag{6}$$

where $\mathcal{T} = \left[\mathcal{T}_1 \cdots \mathcal{T}_n\right]^T$ is the vector of measured friction torques and matrix $\mathbf{A}^{(f)}$ is defined as

$$\mathbf{A}^{(f)} = \begin{bmatrix} e^{-\left|\dot{q}_1/\dot{q}^{(s)}\right|^{\delta}} & 1 - e^{-\left|\dot{q}_1/\dot{q}^{(s)}\right|^{\delta}} & \dot{q}_1 \\ \vdots & \vdots & \vdots \\ e^{-\left|\dot{q}_n/\dot{q}^{(s)}\right|^{\delta}} & 1 - e^{-\left|\dot{q}_n/\dot{q}^{(s)}\right|^{\delta}} & \dot{q}_n \end{bmatrix}, \tag{7}$$

for n measured velocity values. The least squares estimates $\hat{\mathbf{p}}^{(f)}$ from Equation (6) can be expressed mathematically as

$$\hat{\mathbf{p}}^{(f)} = \left(\mathbf{A}^{(f)^T}\mathbf{A}^{(f)}\right)^{-1}\mathbf{A}^{(f)^T}\mathcal{T}. \tag{8}$$

The last and third step is to fine-tune the manually chosen values for δ and $\dot{q}^{(s)}$. This is an iterative process where the chosen values are changed slightly before the second step is repeated. By inspection of the fit between the modelled Stribeck curve and the measured Stribeck curve the values $\delta = 0.33$ and $\dot{q}^{(s)} = 0.024$ rad/s have been obtained.

In this way, two different parameter sets are estimated; one for the full velocity range from 0 to 4 rad/s and one for a low velocity range from 0 to 0.5 rad/s. As can be observed from Figure 4(a), the models 1 and 2 show different behaviour for the low velocity (below 0.5 rad/s) and high velocity (above 0.5 rad/s) range.

Model 1 is estimated for the full velocity range and shows better performance at higher velocities. The low velocity behaviour is clearly not modelled correctly, as a value for the static friction torque is found which is lower than the value for the Coulomb friction. This is caused by the fact that for higher velocities a lower viscous friction parameter shows a better fit. The mismatch for the viscous friction behaviour at low velocity is compensated by a negative Stribeck effect.

Model 2 on the other hand shows to be quite accurate for the low velocity range, the Stribeck effect is described accurately, but then the extrapolation into the high velocity range is poor. It appears that linear viscous behaviour of the model in Equation (5) does not confirm with the actual viscous behaviour of the robot joint.

From the fact that a poor fit is obtained, it can be concluded that the model is not capable to describe the friction phenomena for the full velocity range in the sliding

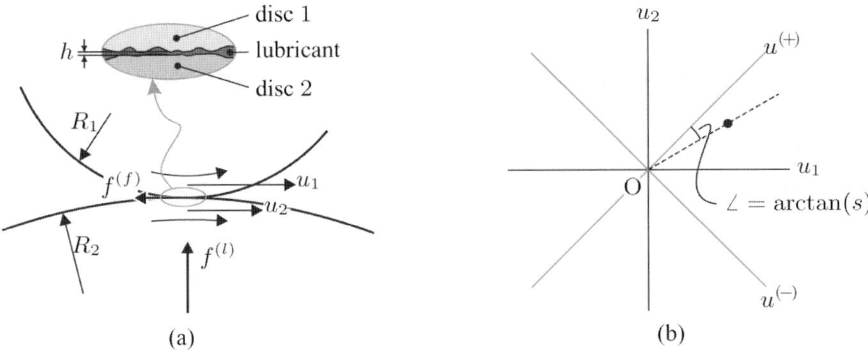

Fig. 5. Two lubricated discs in a rolling–sliding contact (a) and the velocity diagram (b) in which the velocity state (•) of a lubricated contact can be indicated.

regime. The fit can be improved by including extra (non-linear) terms with extra parameters in the friction model [8], but the physical meaning of such additions in unclear.

3 Friction Modelling at Contact Level

In this section the friction phenomena of a single lubricated contact are studied. On system level, the friction is accounted for as a joint torque $\mathcal{T}^{(f)}$. On contact level, it is more convenient to consider friction as a force $f^{(f)}$. Analogously, the surface velocity u is considered in stead of the joint speed $\dot{q}$.

The main components in a robot joint are bearings and gears. In tribology, friction inside gears and bearings is often represented by two lubricated discs in a rolling–sliding contact [9, 10]. The motivation for this representation is that friction in both the roller–raceway contact in roller bearings and the contact between two teeth in a helical or spur gear wheel pair can be represented by the friction behaviour of two lubricated discs in a rolling–sliding contact.

3.1 Two lubricated discs in a rolling–sliding contact

In Figure 5(a), an illustration is given of two lubricated discs in a rolling–sliding contact. The friction force between both discs is defined as $f^{(f)}$. The surface velocities of both discs are defined as u_1 and u_2, respectively. The velocity state of the lubricated contact can be expressed as a function of these surface velocities. It is, however, more convenient to express the velocity state of the contact as a function of the sliding velocity and the sum velocity, see the velocity diagram in Figure 5(b). The sliding velocity is the difference of both velocities

$$u^{(-)} = u_1 - u_2 \tag{9}$$

and the sum velocity $u^{(+)}$ is defined as

$$u^{(+)} = u_1 + u_2. \tag{10}$$

Another frequently used quantity to express the velocity state is the slip ratio s, which is defined as the ratio between the sum and the sliding velocity

$$s = \frac{u^{(-)}}{u^{(+)}}. \tag{11}$$

With these definitions, three typical situations for the velocity state can be distinguished:

i. Perfect rolling. Both velocities, u_1 and u_2, are equal in magnitude and direction. Then the sliding velocity $u^{(-)}$ equals zero and, consequently, there is zero slip. This velocity state is indicated by the $u^{(+)}$-axis.
ii. Full sliding. Both velocities, u_1 and u_2, are equal in magnitude and opposite in direction. Then the sum velocity $u^{(+)}$ equals zero, resulting in infinite slip. This velocity state is indicated by the $u^{(-)}$-axis.
iii. Constant slip. The ratio s between the sum and sliding velocity remains constant. This velocity state is indicated by e.g. the dashed line in Figure 5(b). In fact, it may be any line that crosses the origin O.

In Section 4.1 it will be shown that the frictional behaviour of contacts inside gear transmissions and roller bearings may be characterised by a constant slip ration. The friction behaviour as function of velocity at a constant slip ratio is expressed by the so-called Stribeck curve. As the Stribeck curve is defined for a constant slip ratio, the curve may be plotted as a function of either the sum velocity $u^{(+)}$ or the sliding velocity $u^{(-)}$. Schipper [11] defines a lubrication number

$$\mathcal{L} = \frac{\eta_0 u^{(+)}}{p_{\text{av}} R_{\text{a}}}, \tag{12}$$

where η_0 is the viscosity, p_{av} is the average pressure and R_{a} is the combined surface roughness. Plotting the Stribeck curve as a function of this lubrication number $\mathcal{L}$ yields a so-called generalised Stribeck curve.

The Stribeck curve is characterised by three lubrication regimes: Boundary Lubrication (BL), Mixed Lubrication (ML) and Elasto-Hydrodynamic Lubrication (EHL). In Figure 6, these lubrication regimes are indicated in a typical Stribeck curve for an arbitrary lubricated contact. In the BL regime, at very low velocity, the friction force is mainly caused by the metallic contact between the surface asperities. The height of the lubricant film between the surfaces is in the same order as the height of the surface summits. As the velocity increases, the lubricant film grows

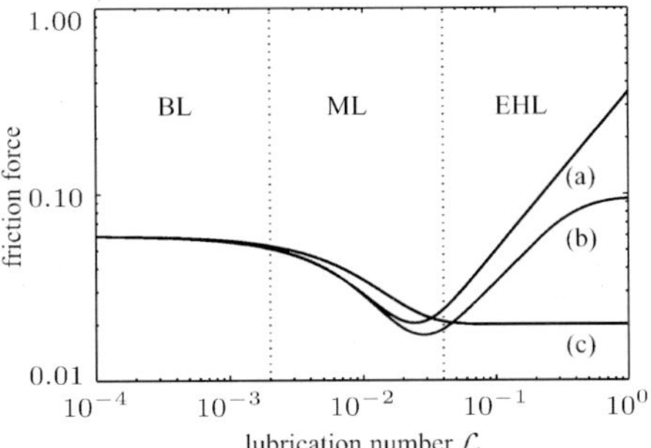

Fig. 6. A typical Stribeck curve for an arbitrary lubricated contact as function of the lubrication number $\mathcal{L}$ and for a constant slip ratio s.

and less asperities are in contact, resulting in a reduction of the friction force caused by the surface summits. On the other hand, viscous friction caused by the lubricant is increasing. This regime is known as the ML regime. Finally, in the EHL regime, the lubricant film has grown such that the surface summits are fully separated. The friction force is the force needed to shear the lubricant film. In Figure 6 three different Stribeck curves are plotted. The curves range from full Newtonian behaviour to full non-Newtonian behaviour of the lubricant. Curve (a) shows a typical Newtonian behaviour of the lubricant at high velocity. In curve (b) the lubricant is mainly Newtonian, but at high velocity the lubricant shows non-Newtonian behaviour. Curve (c) corresponds with full non-Newtonian behaviour. It appears that the viscous properties of the lubricant plays a central role in the friction behaviour in the EHL regime.

3.2 Friction force in the lubrication regimes

In the boundary lubrication regime, the friction force is mainly determined by the friction force due to the asperity contacts, denoted by $f^{(a)}$. On the other hand, in the elasto-hydrodynamic lubrication regime, the friction force $f^{(v)}$ due to the viscosity of the lubricating film is dominant. In the mixed lubrication regime both the asperity contacts and the lubricant viscosity determine the total friction force. As a consequence, the total friction force $f^{(f)}$ is assumed to be the sum of the friction force due to the asperity contacts $f^{(a)}$ and a friction force due to the hydrodynamic component $f^{(v)}$ [12]. This leads to the expression for the total friction force

$$f^{(f)} = f^{(a)} + f^{(v)} = \sum_{i=1}^{n^{(a)}} \iint_{A_i^{(a)}} \tau_i^{(a)} \mathrm{d}A_i^{(a)} + \iint_{A^{(H)}} \tau^{(s)} \mathrm{d}A^{(H)}, \tag{13}$$

where $n^{(a)}$ is the number of asperities in contact, $A_i^{(a)}$ denotes the area of contact of a single asperity i, $\tau_i^{(a)}$ represents the shear stress at the asperity contact i, $\tau^{(s)}$

is the shear stress of the hydrodynamic component and $A^{(H)}$ is the effective area of contact of the hydrodynamic component. In order to model the total friction force, both the friction force due to the asperity contacts $f^{(a)}$ and the friction force due to hydrodynamic component $f^{(v)}$ will be investigated next.

3.2.1 Friction force due to the hydrodynamic component

The force needed to shear a fluid film resembles the sliding friction between two lubricated surfaces. The fact that a force is needed to shear a fluid film was first proposed by Sir Isaac Newton (1642–1727). Newton states that the shear stress $\tau^{(s)}$ is proportional to the shear rate $\dot{\gamma}$ in the film

$$\tau^{(s)} = \eta \dot{\gamma}, \tag{14}$$

where η is known as the viscosity. A lubricant behaviour is called Newtonian when the shear stress–shear rate relation is according to Equation (14) and consequently has a viscosity which is shear rate independent.

Many lubricants, however, show non-Newtonian behaviour at increasing shear rates and show a limiting shear stress for high shear rates. The limiting shear stress implies that the lubricant behaves like a plastic solid for high shear rates. The limiting shear strength is a function of temperature and pressure; it increases at higher pressures and at lower temperature. Furthermore, the value of the shear rate at which the viscous–plastic transition occurs increases with a decrease in pressure and an increase in temperature. At high pressures, such as in roller bearings, most lubricants behave as plastic solids at relative low shear rates.

The model presented in [10] is used to describe the shear stress as a function of the shear rate for a full non-Newtonian fluid,

$$\tau^{(s)} = \tau_l^{(s)}\left(1 - e^{-\frac{\eta_0 \dot{\gamma}}{\tau_l^{(s)}}}\right), \tag{15}$$

where η_0 the viscosity at reference temperature and pressure and $\tau_l^{(s)}$ is the limiting shear stress. Assuming that the sliding velocity $u^{(-)}$ is a continuous linear function of the height h of the lubricating film and that there is no slip at the interface between the fluid film and the solid surfaces, the shear rate $\dot{\gamma}$ in the lubricating film may be approximated by

$$\dot{\gamma} = \frac{u^{(-)}}{h}. \tag{16}$$

The friction force $f^{(v)}$ due to the hydrodynamic component can be be approximated by considering a constant average film height h over a certain hydrodynamic area of contact $A^{(H)}$, yielding

$$f^{(v)} = \iint_{A^{(H)}} \tau^{(s)} \mathrm{d}A^{(H)} \approx \tau^{(s)} A^{(H)}. \tag{17}$$

For the Newtonian case, substitution of Equations (14) and (16) into Equation (17), yields the following expression for the friction force due to the hydrodynamic component:

$$f^{(v)} = \eta A^{(H)} \frac{u^{(-)}}{h}. \tag{18}$$

For non-Newtonian situations, substitution of Equations (15) and (16) into Equation (17) yields

$$f^{(v)} = A^{(H)} \tau_l^{(s)} \left(1 - e^{-\frac{\eta_0 u^{(-)}}{\tau_l^{(s)} h}}\right). \tag{19}$$

Inspection of the above relations shows a dependency of the hydrodynamic friction force on the height h of the lubricating film. It appears that h strongly depends on the sum velocity $u^{(+)}$, as will be outlined in more detail next.

Height of the lubricating film

The calculation of the lubricant film height has been studied intensively in Elasto–Hydrodynamic Lubrication (EHL) research [9, 13, 14, 15, 16]. It was found that the film height depends on six independent variables:

R	the radius of the roller pair,	E	the elastic modulus of a roller pair,
η_0	the viscosity,	w	the load per unit width,
α	the pressure exponent of the lubricant; $\eta = \eta_0 e^{\alpha p}$, with pressure p,	$u^{(+)}$	the sum velocity.

The film height is then expressed as a function:

$$\frac{h}{R} = f\left(\frac{w}{ER}, \frac{u^{(+)}\eta_0}{ER}, \alpha E\right), \tag{20}$$

where the above variables are grouped into four dimensionless parameters. These dimensionless parameters are:

$H = \frac{h}{R}$	the relative film height,	$W = \frac{w}{ER}$	the load parameter,
$U = \frac{u^{(+)}\eta_0}{ER}$	the velocity parameter,	$G = \alpha E$	the material parameter.

It has been found analytically by Dowson and Higginson [9] that the minimum film thickness can fairly accurate be represented by

$$H_{\text{min}} = \frac{1.6G^{0.6}U^{0.7}}{W^{0.13}}. \tag{21}$$

The equation shows that the influence of the material parameter G is quite large. However, G can be considered as constant for a specific combination of materials and lubricant. Furthermore, it can be observed that the load parameter W only weakly influences the film height. The velocity parameter U is clearly the most significant parameter. From expression (21) follows a proportionality between the film height and the sum velocity, expressed as

$$h \propto (u^{(+)})^{0.7}. \tag{22}$$

However, according to experimental results by Crook [14], the film height shows a proportionality to the sum velocity $u^{(+)}$ given by

$$h \propto (u^{(+)})^{0.5}. \tag{23}$$

This indicates that the power in which the film height relates to the sum velocity does not have an unique value, but varies between 0.5 and 0.7 depending on the details of the specific contact. Therefore, it has to be determined for the specific application at hand.

With these observations, it is possible to express the film height h as a function of the sum velocity $u^{(+)}$ as

$$h = h^{(s)} \left(\frac{u^{(+)}}{u^{(s)}} \right)^{\delta}, \tag{24}$$

where the proportionality constant $h^{(s)}$ represents the reference film height, which is a function of the load parameter W, the material parameter G and the radius R. In order to keep proper dimension, a scaling velocity $u^{(s)}$ is introduced, which relates to the lubricant viscosity η, the elastic modulus E and the radius R. From Eqs. (22) and (23) it follows that the power δ can range from 0.5 to 0.7.

With expression (24) and Equation (18) or (19) the viscous friction force $f^{(v)}$ can be described as function of both the sum velocity $u^{(+)}$ and the sliding velocity $u^{(-)}$. The film height of the lubricant also plays a significant role in the friction force due to the asperity contacts.

3.2.2 Friction force due to asperity contacts

In this section, a relation for the friction force due to the asperity contacts in the boundary lubrication regime will be derived. The normal load acting on a lubricated contact is shared between the hydrodynamic component and the interacting asperities of the surfaces. So as the carrying capacity due to the hydrodynamic action increases as function of the film height, the load carried by the asperities decreases. Greenwood and Williamson [17] introduced an approach to the modelling of the friction caused by the asperities which is based on the statistics of the surface roughness of the surfaces in contact. The height distribution of the surface summits can be considered to be Gaussian, but, according to Greenwood and Williamson [17], an exponential distribution shows to be a fair approximation for the uppermost 25% of the asperities of most surfaces. Using the exponential distribution gives the advantage that a fairly simple expression for the number of asperities $n^{(a)}$ in contact can be used. The expression is given by

$$n^{(a)} = d^{(a)} A^{(a)} e^{-\lambda^{(s)}}, \tag{25}$$

where $d^{(a)}$ is the asperity density and $A^{(a)}$ the total area of contact. Exponent $\lambda^{(s)}$ is known as the separation which is the ratio between the film height h and the standard deviation of the height of the surface summits $\sigma^{(s)}$, defined as

$$\lambda^{(s)} = \frac{h}{\sigma^{(s)}} . \tag{26}$$

Using the relation for the film height of Equation (24), the separation can be written as a function of the sum velocity $u^{(+)}$

$$\lambda^{(s)} = \frac{h^{(s)}}{\sigma^{(s)}} \left(\frac{u^{(+)}}{u^{(s)}} \right)^{\delta} . \tag{27}$$

Then the total friction force due to $n^{(a)}$ asperities can be approximated as

$$f^{(a)} = \sum_{i=1}^{n^{(a)}} \iint\limits_{A_i^{(a)}} \tau_i^{(a)} \mathrm{d}A_i^{(a)} \approx n^{(a)} f^{(a,0)}, \tag{28}$$

where $f^{(a,0)}$ is the average force needed to break a single asperity i.

Substitution of Equations (25) and (27) into Equation (28) yields the expression for the friction force $f^{(a)}$ due to the asperity contacts as a function of the sum velocity

$$f^{(a)} = f^{(a,0)} d^{(a)} A^{(a)} \exp\left(-\frac{h^{(s)}}{\sigma^{(s)}} \left(\frac{u^{(+)}}{u^{(s)}} \right)^{\delta} \right) . \tag{29}$$

Note the correspondence of Equation (29) with the exponential part in the model presented by Bo and Pavelescu [3], Equation (4).

4 Friction Models of Elementary Transmission Components

In Sections 3.2.1 and 3.2.2, expressions for the friction force due to lubricant viscosity and due to the asperities for a lubricated contact between two rolling–sliding discs have been presented. In this section, these expressions are applied to model the friction forces arising in the two elementary components; a helical gear-pair and a pre-stressed roller bearing.

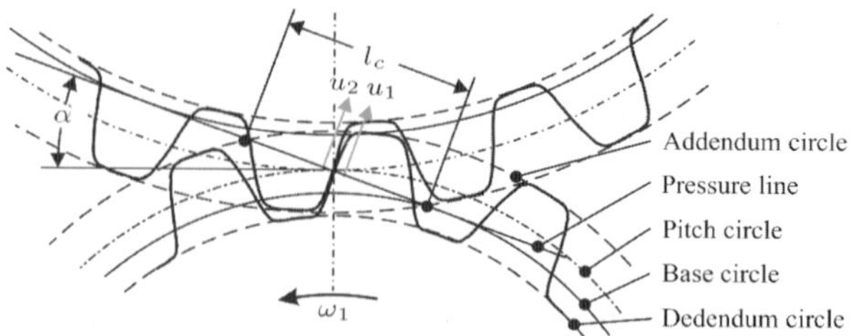

Fig. 7. Schematic representation of a gear pair.

4.1 Friction model of a helical gear pair

In Figure 7 a schematic representation of a helical gear pair is depicted. Here, the lower gear wheel drives the upper gear wheel. The figure shows that several teeth are in contact at the same instance in a gear pair. Furthermore, since the gear pair is of the helical type, a single teeth is in contact over the full width of the gear wheel along a large part of the contact line l_c.

The two surface velocities u_1 and u_2 are depicted in the teeth contact as illustrated in Figure 7. At the point where the contact line l_c crosses the pitch circle the velocities u_1 and u_2 are equal in magnitude which results in a pure rolling motion ($u^{(-)} = 0$). However, at all other points along the contact line, the velocities are not equal in magnitude, leading to a non-zero sliding velocity. At the moment of interconnection of the teeth, the surface velocity u_1 associated with the driving gear is lower in magnitude compared with u_2. Towards the pitch circle they become gradually equal in magnitude. After the moment of pure rolling, u_1 gets larger in magnitude compared with u_2. Consequently, there is a large part during the interconnecting phase where there is a non-zero sliding velocity.

Since the contact pressures in a helical gear pair are sufficiently low, it is assumed that the lubricant behaviour will be Newtonian. Consequently, the expression for the viscous friction, Equation (18), derived in Section 3.2.1, will be applied. The film height is highly dependent on the sum velocity, as was shown in Section 3.2.1. So substitution of the expression for the film height, Equation (24), into Equation (18) yields the viscous friction force for Newtonian behaviour of the lubricant

$$f^{(v)} = \frac{\eta A^{(H)} u^{(-)}}{h^{(s)}} \left(\frac{u^{(s)}}{u^{(+)}} \right)^{\delta}. \tag{30}$$

According to Equation (13), summation of the viscous friction force, Equation (30) and the friction force due to the asperities, Equation (29), leads to the combined friction force in a single teeth contact

$$f^{(f)} = f^{(a)} + f^{(v)}$$
$$= f^{(a,0)} d^{(a)} A^{(a)} \exp\left(-\frac{h^{(s)}}{\sigma^{(s)}}\left(\frac{u^{(+)}}{u^{(s)}}\right)^{\delta}\right) + \frac{\eta A^{(H)} u^{(-)}}{h^{(s)}}\left(\frac{u^{(s)}}{u^{(+)}}\right)^{\delta}. \tag{31}$$

The next step is to derive an expression for the total joint friction torque that is caused by the friction in a gear pair as function of the joint speed $\dot{q}$. The angular velocity ω_1 of the driving gear can be related to the joint speed $\dot{q}$ as

$$\omega_1 = n\dot{q}, \tag{32}$$

where n is the gear ratio of the transmission. With this relation the sum velocity $u^{(+)}$ in a single teeth contact is approximated according to Waiboer [7] by

$$u^{(+)} = r\dot{q}, \tag{33}$$

where

$$r = 2nR_1^{(p)} \sin\alpha. \tag{34}$$

Here $R_1^{(p)}$ is the radius of the pitch circle of the driving gear and α is the pressure angle of the gear pair. The sliding velocity $u^{(-)}$ varies during the messing phase. In the interconnecting phase $u^{(-)}$ is negative, it increases to zero at the point where the pressure line crosses the pitch circle and in the separating phase $u^{(-)}$ becomes positive. According to Equation (30) the viscous friction force depends linearly on $u^{(-)}$. Therefore, since the friction force is dissipative, an averaged positive sliding velocity $\overline{u}^{(-)}$, defined by [7]

$$\overline{u}^{(-)} = s_0 r\dot{q}, \tag{35}$$

may be used to describe the viscous friction force for a single teeth contact. Here s_0 is an averaged constant slip factor. This implies that the viscous friction inside gear transmissions may be approximated by a Stribeck curve.

The total friction torque $\mathcal{T}^{(f)}$ can be derived by multiplying the averaged friction force $\overline{f}^{(f)}$ with constant r and the number of teeth k in contact. Accordingly, substitution of the Equations (33) and (35) for the sum and averaged sliding velocity in Equation (31), respectively, yields the expression for the friction torque of a helical gear pair

$$\mathcal{T}^{(f)}_{gear} = \mathcal{T}^{(a)} + \mathcal{T}^{(v)} = \underbrace{\mathcal{T}^{(a,0)} e^{-\left(\dot{q}/\dot{q}^{(s)}\right)^{\delta}}}_{\text{asperities}} + \underbrace{c^{(v)}\dot{q}^{(1-\delta)}}_{\text{viscosity}}. \tag{36}$$

with the parameters

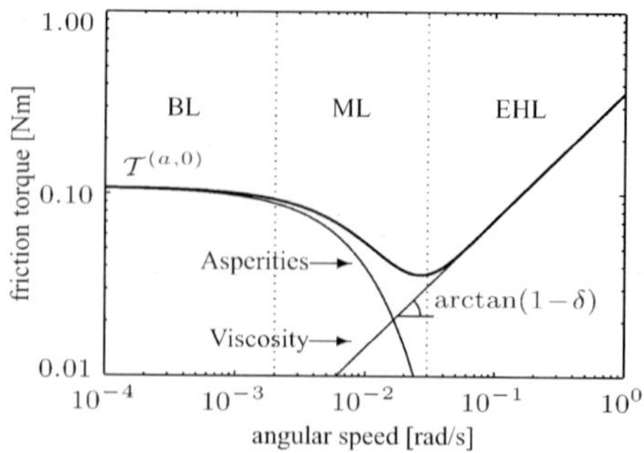

Fig. 8. A typical Stribeck curve for a gear pair.

$$\mathcal{T}^{(a,0)} = rkf^{(a,0)}d^{(a)}A^{(a)}, \tag{37a}$$

$$\dot{q}^{(s)} = \frac{u^{(s)}}{r}\left(\frac{\sigma^{(s)}}{h^{(s)}}\right)^{1/\delta}, \tag{37b}$$

$$c^{(v)} = \frac{r^2 k s_0 \eta A^{(H)}}{h^{(s)}}\left(\frac{u^{(s)}}{r}\right)^{\delta}. \tag{37c}$$

A typical Stribeck curve for a helical gear pair, computed by the model in Equation (36) with an arbitrary parameter set, is illustrated in Figure 8. The contributions to the friction torque from both the asperity contacts and the hydrodynamic component are also depicted. It shows clearly that the asperity contacts are responsible for the friction force in the BL regime and that the hydrodynamic component dominates the friction force in the EHL regime.

4.2 Friction model of a pre-stressed roller bearing

In Figure 9, a single lubricated roller bearing is illustrated. The bearing is a assembly of two concentric circular raceways, the inner ring and the outer ring. The difference in velocity between the inner ring and the outer ring is covered by the rolling of the rollers in between these raceways. Small differences between the velocities (u_1, u_2), and also between the velocities (u_1', u_2') cause friction forces in the contact surfaces between a roller and both the inner and outer rings. These friction forces form a torque that brings the roller in motion. Since the bearing is highly pre-stressed the lubricant behaviour will be non-Newtonian, even for small slip ratios [18].

The friction force can be computed as the sum of the viscous friction and the friction force caused by the asperities. Due to the non-Newtonian behaviour of the lubricant in the bearing, Equation (19) is used to compute the viscous friction force.

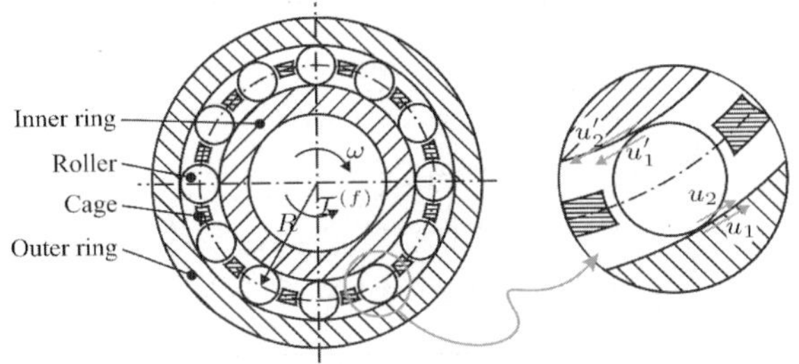

Fig. 9. Schematic representation of a roller bearing.

Equation (29) will account for the friction force due to the asperity contact. Combining these equations yields an expression for the friction force in a single contact between a roller and a raceway inside a bearing

$$\begin{aligned} f^{(f)} &= f^{(a)} + f^{(v)} \\ &= f^{(a,0)} d^{(a)} A^{(a)} e^{-\frac{h^{(s)}}{\sigma^{(s)}}\left(\frac{u^{(+)}}{u^{(s)}}\right)^{\delta}} + A^{(H)} \tau_l^{(s)} \left(1 - e^{-\frac{\eta_0 u^{(-)}}{\tau_l^{(s)} h}}\right). \end{aligned} \tag{38}$$

The sum velocity $u^{(+)}$ depends on the average raceway radius R and the angular velocity ω of the inner ring, see Figure 9. This angular velocity ω in turn depends linearly on the joint speed $\dot{q}$. According to Waiboer [7] these relations can be combined to express the velocities $u^{(+)}$ and $u^{(-)}$ as functions of the angular joint velocity $\dot{q}$

$$u^{(+)} \approx r\dot{q}, \qquad u^{(-)} \approx r s_0 \dot{q}, \tag{39}$$

where a constant slip ratio s_0 is assumed and r is a constant.

The total friction torque $\mathcal{T}^{(f)}$ generated by a prestressed roller bearing is considered as the sum of all torques generated by the friction forces at all roller–raceway contacts, which is achieved by multiplying the friction force in Equation (38) with the constant r and the number of rolling elements k. Substitution of the velocity expressions (39) into Equation (38) yields the expression for the friction torque of a pre-stressed roller bearing

$$\mathcal{T}^{(f)}_{\text{bearing}} = \mathcal{T}^{(a)} + \mathcal{T}^{(v)} = \underbrace{\mathcal{T}^{(a,0)} e^{-(\dot{q}/\dot{q}^{(s)})^{\delta}}}_{\text{asperities}} + \underbrace{\mathcal{T}^{(l,\infty)} \left(1 - e^{-\dot{q}/\dot{q}^{(l)}}\right)}_{\text{viscosity}}. \tag{40}$$

with the parameters

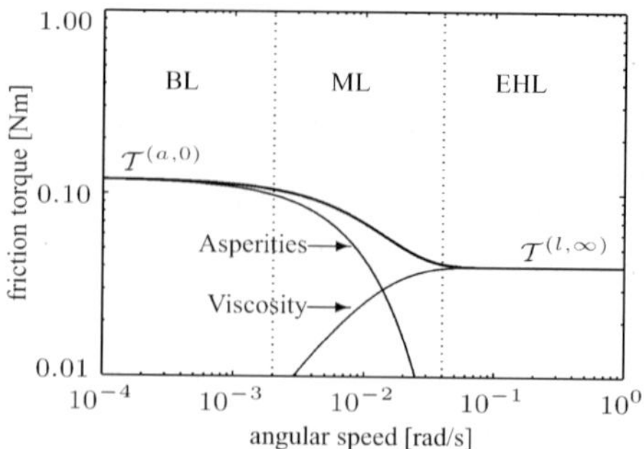

Fig. 10. A typical Stribeck curve for a pre-stressed roller bearing.

$$\mathcal{T}^{(a,0)} = rkf^{(a,0)}d^{(a)}A^{(a)}, \qquad \dot{q}^{(s)} = \frac{u^{(s)}}{r}\left(\frac{\sigma^{(s)}}{h^{(s)}}\right)^{1/\delta}, \tag{41a}$$

$$\mathcal{T}^{(l,\infty)} = rk\tau_l^{(s)}A^{(H)}, \qquad \dot{q}^{(l)} = \frac{\tau_l^{(s)}h_l}{rs_0\eta_0}. \tag{41b}$$

A typical Stribeck curve for an arbitrary prestressed roller bearing represented by the model in Equation (40) is illustrated in Figure 10. The curves for the friction torque generated by the asperity contacts and the friction torque due to lubricant viscosity are also plotted separately. Note that the friction torque decreases from the static friction torque $\mathcal{T}^{(a,0)}$ towards the (limiting shear stress) viscous friction torque $\mathcal{T}^{(l,\infty)}$.

The film height h is not included as a function of the rolling velocity in the expression for the viscous friction force in Equation (40). This simplification is introduced since the dependency of the film height on the rolling velocity only influences the shape of the exponential function associated with the non-Newtonian behaviour. Furthermore, in the velocity region where the exponential function has an effect, the friction torque is dominated by the friction torque due to the asperity contacts. Therefore, the film height h for the viscous part can in this case be approximated by a constant film height h_l.

5 The Joint Friction Model

In Sections 4.1 and 4.2, friction models for both a gear pair and a prestressed roller bearing have been derived. It is shown that the frictional behaviour may be characterised by a constant slip ratio and may be represented by a Stribeck curve. The models are based on physical models from tribology literature in which elementary variables such as lubricant viscosity, contact topology and material properties have been taken into account. These elementary variables have been combined into a new set of parameters, some of which are constant for a specific contact, while others may change during operation, e.g. due to temperature variations. The number of model parameters is the minimal number that is required to describe the Stribeck behaviour associated with the modelled transmission component. With the derived models it has been shown that it is possible to calculate Stribeck curves for both the gear pair and the pre-stressed roller bearing.

The next step is to combine these sub-models into friction models that account for the friction that arises in a single robot joint. The joint friction model can be considered to be a combination of the friction models associated with the joint transmission components. However, summation of all sub-models will lead to a large friction model which includes many parameters. Instead, only the friction characteristics of the components will be evaluated and only the most significant effects will be taken into account.

The first four robot joints are constructed according to the schematic representation given in Figure 2. The assembly contains three main components: a helical gear pair, a cycloidal transmission and the joint bearings. The joint bearings are highly pre-stressed and therefore it is expected that the bearings are responsible for the main part of the asperity friction torque.

The viscous friction torque of the bearings, however, is much lower than its asperity friction torque, as can be observed from the Stribeck curve of the roller bearing, see Figure 10. Taking into account that the helical gear pair in the joint is operating at a high angular velocity due to the high transmission ratio, it can be expected that its viscous friction torque will be dominant with respect to the viscous friction torque of the bearing.

The cycloidal gears are operating at a low angular velocity and are prestressed as well. This will results in a small viscous friction torque in comparison with the helical gear pair. Furthermore, the friction behaviour at low velocity will be similar to the asperity friction behaviour of a roller bearing.

The final joint friction model will be a combination of the asperity part of the model of a roller bearing, Equation (40), and the viscous part of the model model of a helical gear pair, Equation (36). This yields then the combined friction model for joint j:

$$\mathcal{T}_j^{(f)} = \mathcal{T}_j^{(a,0)} e^{-(\dot{q}_j/\dot{q}_j^{(s)})^{\delta_j^{(a)}}} + c_j^{(v)} \dot{q}_j^{(1-\delta_j^{(v)})}. \tag{42}$$

Note the different values for $\delta_j^{(a)}$ and $\delta_j^{(v)}$ as the friction torque from the asperities and the viscous friction torque are generated at different elements and may therefore show a different film height–velocity behaviour.

For each joint j, there are five unknown parameters; the static asperity friction torque $\mathcal{T}_j^{(a,0)}$, the Stribeck velocity $\dot{q}_j^{(s)}$, the Stribeck velocity power $\delta_j^{(a)}$, the viscous friction coefficient $c_j^{(v)}$ and $\delta_j^{(v)}$ denotes the viscous friction power. The parameters

$\delta_j^{(a)}$, $\delta_j^{(v)}$ and the Stribeck velocity $\dot{q}_j^{(s)}$ depend on the configuration of the friction contacts. As the configuration of these contacts is assumed to be time-invariant, these parameters are assumed to be constant. The viscous friction coefficient $c_j^{(v)}$ depends on the lubricant viscosity η and as a result it depends on temperature. The values for all unknown parameters will be obtained by means of experimental identification. It appears that the static asperity friction torque $\mathcal{T}_j^{(a,0)}$ only slightly depends on temperature [7].

Comparing the friction model in Equation (42) with the "classical" friction model of [4] as presented in Equation (5), two main differences can be noticed. The first difference is shown in the viscous friction part, where the new model shows a non-linear velocity–viscous friction relation in terms of $c^{(v)}\dot{q}^{(1-\delta)}$, as opposed to a linear velocity relation expressed by $c^{(v)}\dot{q}$. The second item in which the new model differs from the standard model is that the Coulomb friction term has disappeared. This is due to the fact that the new friction model is based on lubricated surfaces and that Coulomb friction generally is associated with dry contacts.

6 Friction Parameter Estimation

The values of the parameters are estimated based on the measured values of the Stribeck curve. The measured values are the mean friction torques at constant joint speed. Three measurements series are used in which the measurements at each robot joint are carried out after an initial warmup motion of the robot.

All friction models are non-linear functions of the parameters $\delta^{(a)}$, $\dot{q}^{(s)}$ and $\delta^{(v)}$, and linear functions of the temperature dependent parameters $\mathcal{T}^{(a,0)}$ and $c^{(v)}$. Estimating these parameters in a single optimisation requires a non-linear optimisation technique. However, non-linear optimisation techniques may lead to local optima in which non-physical parameter values are found, as was already concluded in Section 2.2. To prevent difficulties with non-linear estimation techniques, the values are obtained in four steps by means of linear least squares techniques. These four steps are concisely described below. For a more detailed description of the identification steps, the reader is referred to [7].

The first step of the identification process is to determine the power $1 - \delta^{(v)}$ and the magnitude $c^{(v)}$ of the viscous part. Taking the natural logarithms of the joint torques and the joint speeds allows for the application of a linear least squares estimation technique to find the viscous friction parameters. Since only the high velocity region, from 0.5 to 4.5 rad/s, is considered, the influence of the asperity friction torques may be neglected.

The second step involves the selection of a proper value for the power $\delta^{(a)}$. As a logical fist estimate, it is set to the same value as $\delta^{(v)}$. Additionally, a value for the Stribeck velocity $\dot{q}^{(s)}$ needs to be chosen. A good first estimate is a value close to the joint speed where the friction torque is at its minimum.

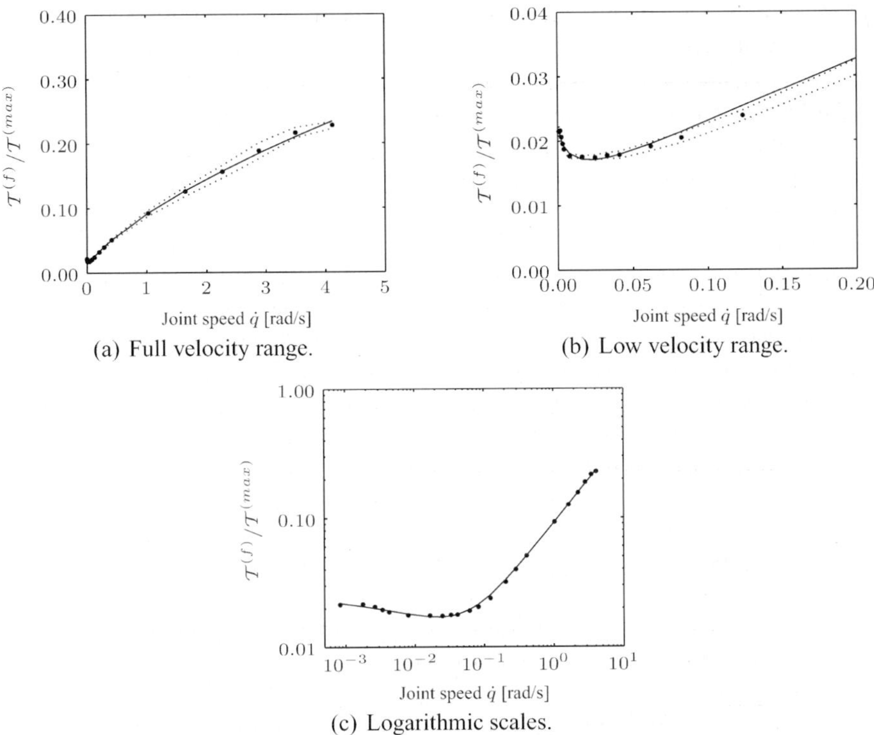

(a) Full velocity range.

(b) Low velocity range.

(c) Logarithmic scales.

Fig. 11. Measured (•) and modelled (—) Stribeck curve for joint 1. The sample standard deviation s of the measured values is indicated by $(\cdots)$.

During the third step, the magnitude of the static asperity friction torque $\mathcal{T}^{(a,0)}$ is determined by means of a linear least squares estimation analogue to the estimation technique described in Section 2.2. For best model fits, the magnitude of the viscous friction coefficient $c^{(v)}$ will again be included in the estimation, which can be done because $c^{(v)}$ is linear in the model.

During the fourth and final step, the manually chosen values for $\delta^{(a)}$ and $\dot{q}^{(s)}$ are manually fine-tuned in an iterative process. The resemblance between the modelled Stribeck curve and the measured Stribeck curve is inspected visually and by modifying the values for $\delta^{(a)}$ and $\dot{q}^{(s)}$, a set of appropriate vales is obtained.

Figure 11 shows the measured Stribeck curve as well as the estimated Stribeck curve. Note that the measured Stribeck is obtained by averaging the joint friction torques $\mathcal{T}^{(f)}$ from $N = 3$ measurements at each joint speed. In Figure 11(a) the full velocity range is shown and in Figure 11(b) a detail of the low (Stribeck) velocity range is shown. In both figures also the sampled standard deviation s of the measurements is also plotted. In Figure 11(c), both the measured and modelled Stribeck curves for joint 1 are plotted on logarithmic scales. It shows that the relative errors between the model and the measurements are equally small across the full velocity range. Consequently, the model accurately describes joint friction for both the low and the high velocity range.

7 Conclusions

In this paper it is shown that classical friction models, commonly used in robot literature, are inadequate to model the viscous friction behaviour for the full velocity range with sufficient accuracy. Therefore, a new joint friction model is developed that relies on insights from sophisticated tribological models. The basic friction model of two lubricated discs in a rolling–sliding contact is used to analyse the different contacts inside the gears and bearings of the robot joint transmissions.

It is shown that the film height of the lubricant is a function of the sum velocity, which causes a non-linear relation between the joint angular speed and the viscous friction torques. The analysis shows different behaviour for gears and pre-stressed bearings. In agreement with Greenwood and Williamson [17], it is shown that friction torques caused by the asperity contacts depend on the ratio between lubricant film height and the height distribution of the surface summits. Increasing the joint speed leads to a decrease of the asperity friction torque.

Sub-models for the viscous friction and the friction due to the asperity contacts are combined into two friction models; one for gears and one for prestressed roller bearings. The sub-models describing the asperity part of the roller bearings and the viscous friction part of the helical gear pair are combined into a joint friction model. In this way, a new friction model is developed that accurately describes the friction behaviour in the sliding regime with a minimal and physically sound parametrisation.

The model is linear in the parameters that are temperature dependent, which allows to estimate these parameters during the inertia parameter identification. The model, in which the Coulomb friction effect has disappeared, has exactly the same number of unknown parameters as the commonly used Stribeck model [5].

Acknowledgements

This work was carried out under the project number MC8.00073 in the framework of the Strategic Research programme of the Netherlands Institute for Metals Research in the Netherlands (http://www.nimr.nl/).

The authors wish to thank Stäubli Faverges, France for the support of the research.

References

1. Olsson H, Åström KJ, Canudas de Wit C, Gäfvert M, and Lischinsky P (1998) Eur J Control 4(3):176–195.
2. Stribeck R (1902) Zeitschrift des Vereines deutcher Ingenieure 46(38,39):1342–1348, 1432–1437.
3. Bo LC and Pavelescu D (1982) Wear 82(3):277–289.
4. Armstrong-Hélouvry B (1991) Control of Machines with Friction. Kluwer Academic Publishers, Boston Dordrecht.
5. Canudas de Wit C, Olsson P, Åström K, and Lischinsky P (1995) IEEE Trans Automat Contr 40(3):419–425.
6. Hensen RHA, van de Molengracht MJG, and Steinbuch M (2002) IEEE T Contr Syst T 10(2):191–195.
7. Waiboer RR (2005) Dynamic modelling, identification and simulation of industrial robots for off-line programming of robotised laser welding. PhD thesis (to be published), University of Twente, Enschede, The Netherlands.
8. Grotjahn M (2003) Kompensation nichtlinearer dynamischer Effecte bei seriellen und parallelen Robotern zur Erhöhnung der Bahngenauigkeit. PhD thesis, Universität Hannover, Germany.
9. Dowson D and Higginson GR (1977) Elasto-Hydrodynamic Lubrication. Pergamon Press Ltd., Oxford.
10. Bhushan B (1999) Principles and Applications of Tribology. John Wiley & Sons, New York.
11. Schipper DJ (1988) Transitions in the lubrication of concentrated contacts. PhD thesis, University of Twente, Enschede, The Netherlands.
12. Gelinck ERM and Schipper DJ (2000) Tribol Int 33:175–181.
13. Grubin AN and Vinogradova IE (1949) Central scientific research institute for technology & mechanical engineering. Book No. 30 (D.S.I.R Translation No.337).
14. Crook AW (1961) Philos Tr R Soc S-A 254(1040):223–236.
15. Dowson D (1995) Wear 190(2):125–138.
16. Spikes HA (1999) P I Mech Eng J-J Eng 213(5):335–352.
17. Greenwood JA and Williamson JBP (1966) Proc R Soc Lon Ser-A 295(1442):300–319.
18. Bair S and Winer WO (1979) J Lubric Tech-T ASME 101:251–257.

Multibody Dynamics of Biomechanical Models for Human Motion via Optimization

Jorge A.C. Ambrósio[1] and Andrés Kecskeméthy[2]

[1]*IDMEC, Instituto Superior Técnico, Technical University of Lisbon, Av. Rovisco Pais 1, 1049-001 Lisboa, Portugal; E-mail: jorge@dem.ist.utl.pt*
[2]*Institute of Mechatronics and System Dynamics, University of Duisburg-Essen, Lotharstrasse 1, 47057 Duisburg, Germany; E-mail: andres.kecskemethy@uni-due.de*

Abstract. The human motion analysis, for gait or for most of other activities, relies mostly on the use of multibody formulations applied as kinematic or dynamic tools. In many biomechanical applications to gait analysis the choice between using direct or inverse dynamics to obtain the solution of the problem, even in pure kinematics, only depends on the personal preference of the user and not in any particular form of the data available or structure of the equations to be solved. In this work the structure of the equations of a multibody system are reviewed for direct and inverse dynamic analysis. It is shown that if the time dependencies of all degrees-of-freedom of the system are known the inverse dynamics is equivalent to a direct dynamics problem. This equivalence is particularly useful when the problem of the biomechanical analysis consists in finding the muscle forces in an over-actuated biomechanical model that leads to a prescribed motion, which is obtained by using video data acquisition or simply by designing such motion. The problem can then be solved by using optimization procedures in which the objective functions are physiological criteria and, eventually, a measure of matching the prescribed motion. If not used as part of the objective function the prescribed motion is introduced in the optimization problem as nonlinear constraints. The variables of the optimization problem are, for all type of analysis, the muscle forces, directly, or their corresponding muscle activations. It is shown that the natural choice for design variables of the optimal problem is the muscle activations. Two representations of the time history of the muscle actuation are tested in this work: the input sampling where the activations are found in a finite number of time instants and then linearly interpolated in between; the smooth exponential function approach where the actuation is described by a sum of exponential functions being the width and the size of the bumps of each of the functions the unknown quantities. Then the muscle forces are simply obtained by using a Hill type muscle model where the state of force-velocity and the force-length relations are obtained directly from the kinematics of the biomechanical model. The methods presented in this work are demonstrated and discussed in the framework of two problems associated to the human locomotion apparatus.

1 Introduction

To support traditional diagnostics and therapy planning, surgeons are seeking computer tools that allow for the prediction of therapy consequences prior to their

J.C. García Orden et al. (eds), Multibody Dynamics. Computational Methods and Applications, 245–272.

application to the specific patient. Trainers, sports specialists, designers of exercise machines and of prosthesis equipment also look into numerical tools that allow optimizing the athletes' performance or the equipment. The growing number applications of the human motion analysis, new demands computer planned surgical interventions and rising quality standards require a better understanding of the methodologies used and the improvement of their reliability and efficiency.

The analysis of the pathological gait associated to cerebral palsy is an example where the need for computer tools to support the medical intervention is of major importance. Typical effects of the cerebral palsy are pathologic motion performances due to range restrictions caused by muscle shortening and/or constant muscle contraction and/or joint acampsia. A surgical intervention, such as (1) tendotomy (lengthening of tendons), myotomy (notching of muscle tissue) and tendon transfer (change of muscle origin and/or insertion points) in order to release a dynamical or fixed contraction of muscle tissue, (2) neurotomy (cutting specific nerves) to completely relax a spastic palsy irreversibly and (3) ostetomy (manual deformation of bones) if the grade of relaxation of muscle tissue achieved by tendotomy and/or myotomy is insufficient, is chosen if classical therapies failed or are assumed to be insufficient. Most of the surgical approaches are irreversible which creates the need for methods that help surgeons to assess a priori the consequences of changes in the biological system either leading to alternative therapy approaches avoiding a surgical intervention or, if intervention is unavoidable, reducing stress for the patient by an optimized operation planning [1]Current diagnosis methods have such a large tolerance interval (approx. 30%-40%), that computer simulations can provide a major contribution if dynamics can be predicted with tolerances of 10%-20%. In this setting, methods for providing a first rough estimate of muscle activation profiles that can be refined in further optimization stages would render a tool with which typical medical decisions could be already supported better than by conventional methods. Certainly, this situation is illustrative of what aims to be achieved with more flexible, reliable and efficient numerical tools for biomechanical analysis.

The construction of the biomechanical models suitable to the human motion analysis requires the use of a formulation to describe their dynamic equilibrium equations. Multibody formulations such as those based on Cartesian coordinates [2], natural coordinates [3], joint coordinates [4], Kane's formalisms [5] are all suitable for applications to biomechanical modeling, being most of the codes used in biomechanical analysis based on these approaches [6–9]. However, the type of dynamic analysis that are required for different application cases to human motion may not be available in the computer implementations provided by normal commercial codes. Most of these codes only allow for direct dynamic analysis to be used and if, for instance, an inverse dynamics analysis needs to be carried the only solution is to drive kinematically all degrees of freedom of the model with a prescribed motion. An important limitation of the application of the standard commercial computer codes to biomechanics studies that involve redundant muscle actions concerns the impossibility of their use in the framework of optimization.

The biomechanical models applied on the study of the human locomotion require that the major anatomical segments of the lower part of the human body are represented. The multibody description of each segment requires that one or more rigid bodies are associated to it. The anatomical joints are represented either by kinematic joints or by contact joints in the multibody model, depending on the objectives of the analysis. The ligaments and other passive tissues required to provide stability or stiffness to the anatomical joints are typically represented as spring-damper elements with linear or nonlinear characteristics. The muscles of the locomotion apparatus need also to be represented in the models. In particular, the use of a detailed description of the muscles is required for the calculation of the redundant forces produced by the muscle apparatus. Finally, the use of multibody biomechanical models for gait analysis also requires a comprehensive description of the contact between the different segments of the models and external objects, such as the ground.

A purpose of this work to present multibody based methodologies that, together with the use of optimization procedures, allow for the calculation of the redundant muscle forces, generated in a particular muscle apparatus of the human body. A whole body biomechanical model is constructed using rigid bodies interconnected by revolute and universal joints. The biofidelity of the model is improved using the subject's anthropometric link lengths together with biomechanical information regarding the physical characteristics of the anatomical segments. This information is collected from a general database and scaled for the subject dimensions and total body mass [10, 11].

The biomechanical model is driven through an acquired motion by two different types of kinematic constraints: joint actuators, that drive the degrees-of-freedom of the biomechanical model associated with joints, and muscle actuators that drive the degrees-of-freedom of the joints crossed by the muscles. When the aim of the analysis is to calculate net moments-of-force of particular joints only the joint actuators need to be used and the solution to the dynamics problem is unique and non-redundant [12]. When the aim is to evaluate the muscle forces it is required the use of muscle actuators in the biomechanical models and the system becomes redundant, i.e., it contains more unknown forces than the equations available to solve them. Note that the redundant or determined nature of the solution does not depend on the use of direct or of inverse dynamic analysis.

The problem of finding the internal forces of the biomechanical system that lead to a prescribed motion can be defined as an optimal problem where the objective is to find joint net-moments-of force or muscle forces that lead to a target motion. In this case a direct dynamic analysis is carried and the objective is to minimize the distance between the obtained and the prescribed motion of the biomechanical model, and eventually to minimize some physiological criteria. Alternatively, the same problem can be solved by an inverse dynamic analysis where the objective is simply to minimize the physiological criteria, being the motion matching implicitly ensured by the type of analysis used. However, if the objective is to find the best motion for a given task the problem is defined as an optimization of a direct dynamics case and the criteria used may be physiological or other, but not that of motion matching.

The muscle actuators are associated to a muscle model that simulates their activation-contraction dynamics [13–15]. Here a Hill type muscle model is applied, being the force produced by the muscle contractile element calculated as a function of the muscle activation, maximum isometric peak force, muscle length and muscle rate of shortening. The equations of motion of the biomechanical system and the performance criteria used in de optimization procedure are expressed in terms of muscle activations instead of muscle forces. Different types of activation functions are used here to describe the muscle activation profiles. In particular, an approach for the parameterization of muscle activation profiles based on smooth C^{∞} base functions that can be combined to mimic typical activation profiles of leg muscles during gait is explored in this work [1].

The approaches presented here are demonstrated through applications to cases of gait analysis. In both application examples described the motion is prescribed.

2 Multibody Formulations for Biomechanical Modeling

A multibody system is a collection of bodies that is acted upon by forces and moments. The rigid bodies are interconnected to each other by kinematic joints that constrain their relative motion in different forms. The human body is a multibody system where the anatomical segments are the rigid bodies, the muscles, tendons, ligaments are responsible for some of the internal forces, and the anatomical joints are modeled either as kinematic joints or as force transmitting elements with compliance. The biomechanical model defined as a multibody system is acted upon by external forces such as the ground reaction forces during walking. To describe this multibody system different formulations may be used, with relative advantages and disadvantages among them [2-5,16]. However, because the objective here is to emphasize the needs for biomechanical modeling and analysis, the use of Cartesian coordinates is implied in the formulation that follows.

For a constrained multibody system the kinematic joints can be described by a set of algebraic equations in the form

$$\mathbf{\Phi}(\mathbf{q},t) = \mathbf{0} \tag{1}$$

where $\mathbf{q}$ is the generalized coordinates vector and t is the time variable. The anatomical joints are typically modeled as time independent constraints while the prescribed motion of the joints or the length variation of the muscles are typical examples of time dependent constraints in the multibody models presented here.

Differentiating Equation (1) with respect to time yields the velocity constraint equation. After a second differentiation with respect to time the acceleration constraint equation is obtained,

$$\mathbf{\Phi}_{\mathbf{q}}\ddot{\mathbf{q}} = \gamma \tag{2}$$

in which $\mathbf{\Phi_q}$ is the Jacobian matrix of the constraint equations, $\ddot{\mathbf{q}}$ is the acceleration vector and $\ddot{\mathbf{q}}$ is the right hand side of acceleration equations, which contains the terms that are exclusively function of velocity, position and time.

The translational and rotational equations of motion for an unconstrained multibody system of rigid bodies are written as,

$$\mathbf{M}\ddot{\mathbf{q}} = \mathbf{g} \tag{3}$$

where $\mathbf{M}$ is the global system mass matrix, containing the mass and moments of inertia of all bodies, and $\mathbf{g}$ is the generalized force vector that contains all external forces and moments applied on the system and all internal forces that are not due to kinematic constraints. Ground reaction forces exemplify external forces while ligament and contact between anatomical segments exemplify internal forces.

Using the Lagrange multipliers technique the constraint equations (1) are added to the equations of motion (3). The equations of motion are written together with the second time derivative of constraint equations (2) yielding a system of equations written as,

$$\begin{bmatrix} \mathbf{M} & \mathbf{\Phi}_\mathbf{q}^T \\ \mathbf{\Phi_q} & \mathbf{0} \end{bmatrix} \begin{Bmatrix} \ddot{\mathbf{q}} \\ \boldsymbol{\lambda} \end{Bmatrix} = \begin{Bmatrix} \mathbf{g} \\ \boldsymbol{\gamma} \end{Bmatrix} \tag{4}$$

where $\boldsymbol{\lambda}$ is the vector of Lagrange multipliers, which physically are related to the joint reaction forces. The reaction forces, owing to the kinematic joints are expressed as [2],

$$\mathbf{g}^{(c)} = -\mathbf{\Phi}_\mathbf{q}^T \boldsymbol{\lambda} \tag{5}$$

If a direct dynamics solution is used to solve Equation (4) the set of differential-algebraic equations has to be solved and the resulting accelerations integrated in time. However, these do not explicitly use the position and velocity constraint equations allowing for drifts in the system constraints to develop. The Baumgarte stabilization technique, the Augmented Lagrangian formulation or the coordinate partition method are some of the procedures that can be employed to keep the constraint violations under control during the numerical integration [17]. By using the Baumgarte stabilization Equation (4) is modified as,

$$\begin{bmatrix} \mathbf{M} & \mathbf{\Phi}_\mathbf{q}^T \\ \mathbf{\Phi_q} & \mathbf{0} \end{bmatrix} \begin{Bmatrix} \ddot{\mathbf{q}} \\ \boldsymbol{\lambda} \end{Bmatrix} = \begin{Bmatrix} \mathbf{g} \\ \boldsymbol{\gamma} - 2\alpha\dot{\mathbf{\Phi}} - \beta^2\mathbf{\Phi} \end{Bmatrix} \tag{6}$$

where α and β are prescribed positive constants that represent the feedback control parameters for the velocities and positions constraint violations [18]. The dynamic response of multibody systems involves the evaluation of the vectors $\mathbf{g}$ and $\boldsymbol{\gamma}$, for each time step. Then, Equation (6) is solved for the system accelerations $\ddot{\mathbf{q}}$. These accelerations together with the velocities $\dot{\mathbf{q}}$ are integrated in order to obtain the new velocities $\dot{\mathbf{q}}$ and positions $\mathbf{q}$ for the time step. This process is repeated until the complete description of system motion is obtained.

2.1 Kinematic constraint to model joint moments-of-force

A joint moment-of-force is a simplified representation of the lumped moment caused by all muscles that cross a particular anatomical joint. Its mechanical model is a joint actuator, such as that represented for the knee in Figure 1, where the angle between two adjacent bodies about the axis of the joint is a function of time. The kinematic constraint associated to the joint moment-of-force is

$$\mathbf{\Phi}^{(joint,1)}(\mathbf{q},t) \equiv \mathbf{s}_{P_i}^{\prime T}\mathbf{s}_{P_j}' - |\mathbf{s}_{P_i}'||\mathbf{s}_{P_j}'|\cos(\theta(t)) = \mathbf{0} \tag{7}$$

where vectors $\mathbf{s}_{P_i}'$ and $\mathbf{s}_{P_j}'$ are fixed to the adjacent bodies of the joint and $\theta(t)$ may be a prescribed function of time, such as for the case of its application in inverse dynamics problems, or a unknown function of time that is calculated during the solution of a direct dynamics analysis.

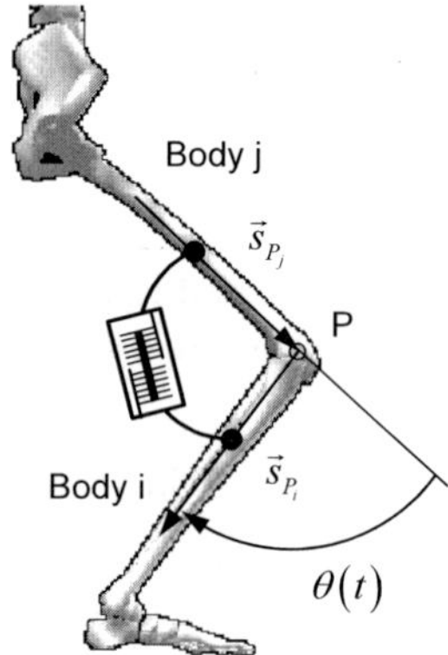

Fig. 1. Joint actuator associated with the knee joint.

The moments-of-force are obtained from the definition of the kinematic constraint through their relation with the Lagrange multipliers expressed by Equation (5). Due to this relation between joint angles and joint moments any biomechanical analysis that requires that torques between adjacent bodies are applied can be described by using equivalent joint actuators instead.

2.2 Kinematic constraint to model muscle actions

Muscles are introduced in the equations of motion of the multibody system as point-to-point kinematic driver actuators, also designated by myoactuators. Two of the muscles of the lower extremity muscle apparatus are shown Figure 2 to illustrate different complexities in their path. The *semimembranosus* is a two-point muscle, the origin and insertion points, and the *tensor fasciae latea* is a muscle defined by multiple points for an accurate characterization of its curvature.

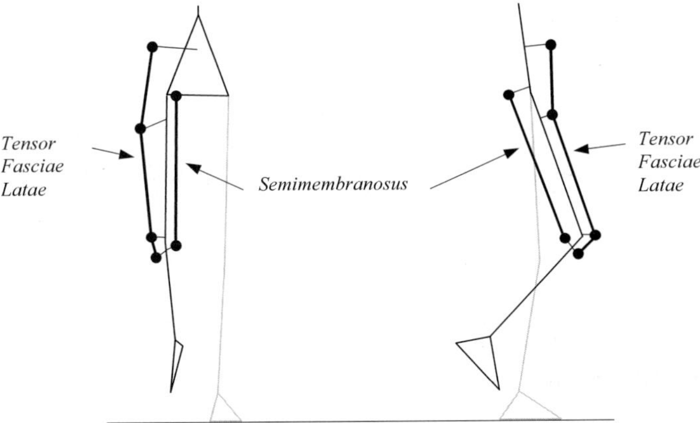

Fig. 2. Muscle actuators defined with two or more points.

A constraint equation that forces the distance between two generic points of different rigid bodies to change according to a length-time function is associated to each muscle. Considering a two-point muscle actuator, with an origin located in point n of rigid body i, and an insertion located in point m of rigid body j, as depicted by Figure 3, the kinematic constraint is

$$\mathbf{\Phi}^{(MA,1)}(\mathbf{q},t) = \left(\mathbf{r}_m - \mathbf{r}_n\right)^T \left(\mathbf{r}_m - \mathbf{r}_n\right) - L_{nm}^2(t) = 0 \tag{8}$$

where $\mathbf{r}_m$ and $\mathbf{r}_n$ are the global position vectors of the origin and insertion points of the muscle, respectively, and $L_{nm}(t)$ is the muscle total length, calculated for each time step of the analysis.

A Lagrange multiplier is associated to each muscle actuator of the locomotion apparatus through Equation (5). The physical dimension of this multiplier, used in the context of these actuators, is a force per unit of length. Muscle actuators defined with more than two points are introduced in the Jacobian matrix of the constraints as a sum of several two-point muscle actuators. Consider, for example, the muscle *tensor fasciae latea* presented in Figure 4. This muscle is described using three two-point muscle actuators, labeled respectively m_1, m_2 and m_3 in Figure 7. The Lagrange multipliers λ_{m1}, λ_{m2} and λ_{m3}, calculated by using equation (4) and appearing in the reaction force equation (5), are associated to muscle actuators m_1, m_2 and m_3, respectively.

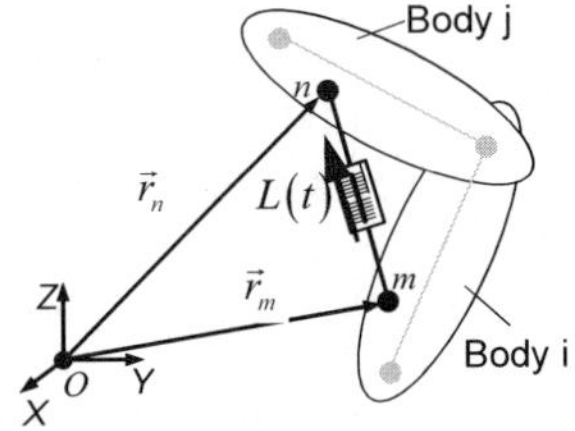

Fig. 3. Muscle actuator defined between points n and m of rigid bodies i and j.

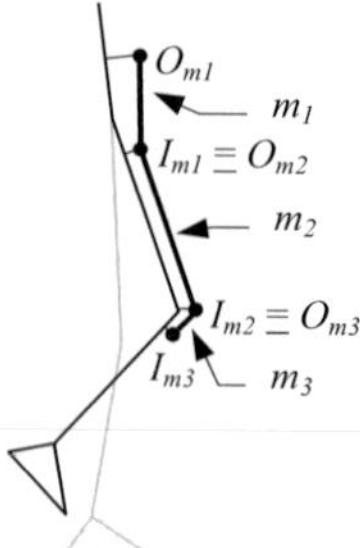

Fig. 4. Describing more complex muscle actuators.

With the information presented before, the term $\mathbf{\Phi}_{\mathbf{q}}{}^{T}\boldsymbol{\lambda}$ of Equation (4) is assembled for muscle actuators m_1, m_2 and m_3, and written as:

$$\mathbf{g}^{(\mathbf{c})(muscle)} = -\mathbf{\Phi}_{\mathbf{q}}^{(muscle)T}\boldsymbol{\lambda} \equiv \begin{pmatrix} \mathbf{q}_3 \rightarrow \\ \mathbf{q}_4 \rightarrow \\ \mathbf{q}_5 \rightarrow \\ \mathbf{q}_6 \rightarrow \\ \mathbf{q}_7 \rightarrow \end{pmatrix} \overset{\begin{matrix} m_1 & \quad\quad m_2 & \quad\quad m_3 \\ \downarrow & \quad\quad \downarrow & \quad\quad \downarrow \end{matrix}}{\begin{bmatrix} \partial\Phi^{m1}/\partial\mathbf{q}_3 & \mathbf{0} & \mathbf{0} \\ \partial\Phi^{m1}/\partial\mathbf{q}_4 & \partial\Phi^{m2}/\partial\mathbf{q}_4 & \mathbf{0} \\ \mathbf{0} & \partial\Phi^{m2}/\partial\mathbf{q}_5 & \partial\Phi^{m3}/\partial\mathbf{q}_5 \\ \mathbf{0} & \mathbf{0} & \mathbf{0} \\ \mathbf{0} & \mathbf{0} & \partial\Phi^{m3}/\partial\mathbf{q}_5 \end{bmatrix}} \begin{Bmatrix} \lambda_{m1} \\ \lambda_{m2} \\ \lambda_{m3} \end{Bmatrix} \tag{9}$$

where $\mathbf{q}_3$ to $\mathbf{q}_7$ mean the rows of the Jacobian matrix and represent the set of coordinates defining the rigid bodies interconnected by the muscle, and $\partial\mathbf{\Phi}^{mi}/\partial\mathbf{q}_j$ are the partial derivatives of muscle actuator equation m_i with respect to coordinates $\mathbf{q}_j$.

Because a muscle must have a constant force per unit of length from its origin to its insertion the Lagrange multipliers associated with each segment of the muscle must be equal. In the *tensor fasciae latea* the Lagrange multipliers must be

$$\lambda_{m1} = \lambda_{m2} = \lambda_{m3} = \lambda_{TFL} \tag{10}$$

Substituting Equation (10) in Equation (9) leads to the expression for a single myoactuator of a curved muscle, written as

$$\mathbf{g}^{(\mathbf{c})(muscle)} = -\mathbf{\Phi}_{\mathbf{q}}^{(muscle)T}\boldsymbol{\lambda} \equiv \begin{pmatrix} \mathbf{q}_3 \rightarrow \\ \mathbf{q}_4 \rightarrow \\ \mathbf{q}_5 \rightarrow \\ \mathbf{q}_6 \rightarrow \\ \mathbf{q}_7 \rightarrow \end{pmatrix} \overset{\begin{matrix} (m_1 + m_2 + m_3) \\ \downarrow \end{matrix}}{\begin{bmatrix} \partial\Phi^{m1}/\partial\mathbf{q}_3 + \mathbf{0} + \mathbf{0} \\ \partial\Phi^{m1}/\partial\mathbf{q}_4 + \partial\Phi^{m2}/\partial\mathbf{q}_4 + \mathbf{0} \\ \mathbf{0} + \partial\Phi^{m2}/\partial\mathbf{q}_5 + \partial\Phi^{m3}/\partial\mathbf{q}_5 \\ \mathbf{0} + \mathbf{0} + \mathbf{0} \\ \mathbf{0} + \mathbf{0} + \partial\Phi^{m3}/\partial\mathbf{q}_5 \end{bmatrix}} \{\lambda_{TFL}\} \tag{11}$$

In a biomechanical model the muscle actions in some joints may be represented with the joint actuator while the myoactuators may be used in other joints.

3 Solution Methods: Forward versus Inverse Dynamics

Many of the biomechanical problems that involve the human motion require the use of optimization methods that are set differently for various types of dynamic analysis. The solution of a biomechanical problem by optimization that requires the use of inverse dynamics is designated by static optimization, while if direct dynamics analysis is used in the optimal problem the procedure is designated by dynamic optimization. Problems for which the target is to find the best motion for a particular task exemplify typical applications for dynamic optimization procedures. Problems for which the target is to find the internal forces of the biomechanical system when developing a prescribed motion provide typical applications for static optimization approaches.

3.1 Direct dynamics

The direct dynamic analysis requires the solution of Equation (4) or (5) and the integration of the resulting accelerations and of the velocities of the system forward in time to obtain the new velocities and positions for the next time step. The process continues until the motion of the system during the complete analysis period is calculated. The direct dynamic analysis flowchart is depicted by Figure 5.

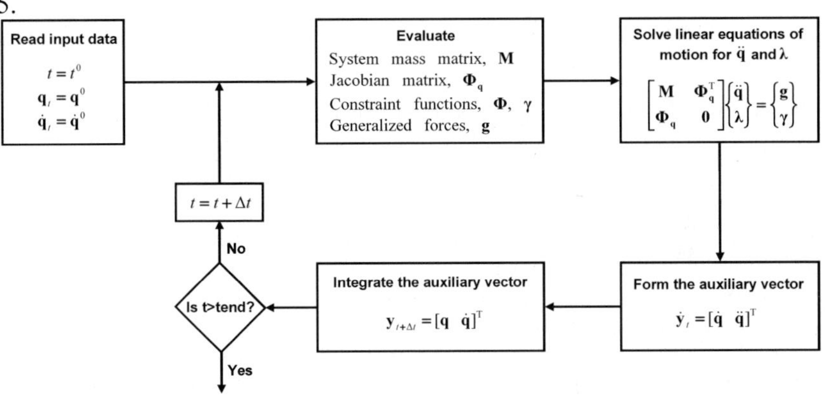

Fig. 5. Flowchart for the direct dynamic analysis.

Most of the biomechanical models for which a direct dynamic analysis is used have a number of independent kinematic constraints lower than the number of coordinates. Not all degrees-of-freedom of the biomechanical models are guided and the motion obtained is the one that results from the known forces applied to the system No particular change on the solution of the equations of motion is required for this type of analysis when redundant muscle forces are involved. The work by Pandy and Anderson [19] exemplifies the use of direct dynamics procedures in biomechanical studies of the human motion.

3.2 Inverse dynamics

Inverse dynamics procedures are used in the solution of biomechanical problems when the motion is completely known and only the forces that cause such motion have to be found. The motion of the biomechanical model is generally acquired using experimental techniques based on video imaging [20] or is defined by some other criteria. In this work it is assumed that the reader is familiar with the motion acquisition techniques, which are described in some detail in [21].

To set up the inverse dynamics problem let it be assumed that the prescribed motion of the model is fully known and consistent with the kinematic constraints of the biomechanical model. Furthermore, let it be assumed that the force vector is partitioned into a vector of know forces $\mathbf{g}_{known}$ and unkown forces $\mathbf{g}_{unknown}$. Let the vector of unknown forces be represented by

$$\mathbf{g}_{unknown} = \mathbf{C}^T \mathbf{f}_{unknown} \tag{12}$$

where matrix **C** is used to map the space of the forces into the space of coordinates that describes the system and, consequently, its structure is dependent on the particular type of force applied. In Equation (4) the only unknowns are the Lagrange multipliers □ and the unknown applied forces $\mathbf{f}_{unknown}$. Therefore, the first line of Equation (4) is now re-written in the form

$$\begin{bmatrix} \mathbf{\Phi}_{\mathbf{q}}^T & -\mathbf{C}^T \end{bmatrix} \begin{Bmatrix} \boldsymbol{\lambda} \\ \mathbf{f}_{unknown} \end{Bmatrix} = \left\{ -\mathbf{M}\ddot{\mathbf{q}} + \mathbf{g}_{known} \right\} \tag{13}$$

The solution of Equation (13) is obtained for a finite number of time instants which depends on the sampling required for the solution and, eventually, on the sampling of the system kinematics. The solution obtained for a particular time instant is fully independent from the solution obtained for any other time instant. Furthermore, this form of the inverse dynamic analysis requires that any unknown moment is applied about known axis. Therefore, the modeling of anatomical joints through spherical joints is not possible when using this formulation. The use of this type of approaches is exemplified by the work of Silva [15].

The solution of the linear system of Equations (13) is unique if the number of independent kinematic constraints and unknown forces is equal to the number of coordinates of the biomechanical system. Otherwise, the solution is not unique due to the redundant set of forces and/or constraints used and the solution of the problem has to be obtained by defining suitable criteria and using optimization methodologies. It must be noted that the joint moments-of-force and the muscle actions can be introduced in Equation (13) either as kinematic constraints, as defined by Equations (7) and (8), or by using directly the vector of unknown forces. In what follows it is always assumed that the muscle actions and joint moments-of-force are represented by kinematic constraints.

3.3 Inverse dynamics through direct dynamics

The solution of the inverse dynamics problem may be obtained by using a direct dynamics approach provided that all degrees-of-freedom of the system are driven by suitable joint or muscle actuators. The procedure in this case starts by acquiring the motion of the biomechanical model, as in the case of inverse dynamics. Then the time history of all joint or muscle actuators is defined. The procedure used for the direct dynamics analysis can now be applied as illustrated in Figure 5. The use of this approach is exemplified in the work Strobach et al. [1].

It can be shown that in the case of a biomechanical model fully driven by joint and myoactuators the results of the analysis by direct and inverse dynamics are generally the same [22]. However, some caution must be used when applying this procedure for cases in which the biomechanical system is inherently unstable, such as in gait analysis, because the resulting motion of the model may differ visibly from the target motion [23].

4 Biomechanical Modeling of the Locomotion Apparatus

A biomechanical model suitable for human motion analysis requires that the different anatomical segments and their relative mobility are described, the muscle activation and corresponding forces are represented and that the skeletal-muscle apparatus is included in the model. In what follows a description of each part of the biomechanical model is provided.

Table 1. Physical characteristics of anatomical segments and rigid bodies for the 50th-percentile human male. The dimensions and positions of the center of mass locations, with respect to the proximal joint with reference to Figure 6.

Description	Body	Length	CM Location		Mass (Kg)	Moments of Inertia (10^{-2} Kg.m^2)
	i	L_i (m)	d_i (m)	$\underline{d}_i$ (m)	m_i	$(I_{xx}/I_{yy}/I_{zz})_i$
Lower Torso	1	0.275	0.064	0.094	14.200	26.220/13.450/26.220
Upper Torso	2	0.294	0.101	0.161	24.950	24.640/37.190/19.210
Head	3	0.128	0.020	0.051	4.241	2.453/2.249/2.034
R Upper Arm	4	0.295	0.153	-	1.992	1.492/1.356/0.248
R Lower Arm	5	0.250	0.123	-	1.402	1.240/0.964/0.298
Hand	13	0.185	0.093	0.045	0.489	0.067/0.146/0.148
L Upper Arm	6	0.295	0.153	-	1.992	1.492/1.356/0.248
L Lower Arm	7	0.376	0.180	-	1.892	1.240/0.964/0.298
Hand	14	0.185	0.093	0.045	0.489	0.067/0.146/0.148
R Upper Leg	8	0.434	0.215	-	9.843	1.435/15.940/9.867
R Lower Leg	9	0.439	0.151	-	3.626	1.086/3.830/3.140
Foot	15	0.069	0.271	0.035	1.182	0.129/0.128/2.569
L Upper Leg	10	0.434	0.215	-	9.843	1.435/15.940/9.867
L Lower Leg	11	0.439	0.151	-	3.626	1.086/3.830/3.140
Foot	16	0.069	0.271	0.035	1.182	0.129/0.128/2.569
Neck	12	0.122	0.061	-	1.061	0.268/0.215/0.215

4.1 Skeleton Model

The biomechanical model of the human body is defined using 16 anatomical segments and their corresponding rigid bodies is presented in Table 1 and illustrated in Figure 6. Considering this kinematic structure, an open loop topology can be identified, with a base body described by rigid body number 1, and 5 kinematic branches defined by the 4 limbs and the head/neck. The model has 44 degrees-of-freedom that correspond to 38 rotations about 26 revolute joints and 6 universal joints, plus 6 degrees-of-freedom that are associated with the free body rotations and translations of the base body.

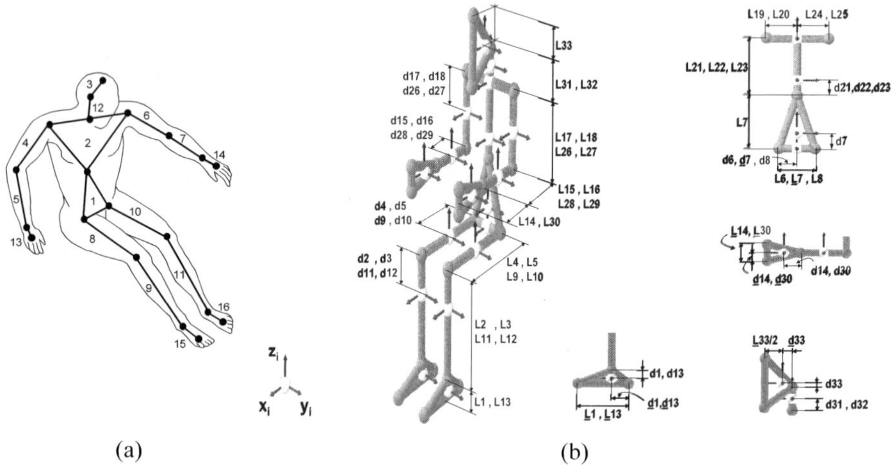

Fig. 6. Biomechanical model with 16 anatomical segments: (a) Topology of the model; (b) Reference for the length and center of mass of each anatomical segment.

This is a general-purpose biomechanical model that can be applied to any type of kinematic or dynamic analysis. Due to the kinematic structure, where no spherical joints is used it can be applied in inverse dynamic analysis. The model presented is only one of many that can be used to the biomechanical analysis of different human motion tasks. However, its application to a particular individual requires that its anatomical segments are properly scaled. In the present work, the scaling procedure used calculates for each anatomical segment, non-dimensional scaling factors, based on measured data from the subject and equivalent data from the 50th percentile human male. These scaling factors are defined as [10]:

$$\chi_{L_i} = \frac{L_i^{n^{th}}}{L_i^{50^{th}}}; \quad \chi_{m_i} = \frac{m_i^{n^{th}}}{m_i^{50^{th}}}; \quad \chi_{I_i} = \chi_{m_i} \cdot \chi_{L_i}^2 \tag{14}$$

where χ_{L_i}, χ_{m_i} and χ_{I_i} are respectively the scaling factors of the length, mass and moments of inertia calculated for segment i. It should be noted that if the length and mass of each segment of the subject are not available, the calculation of χ_{L_i} and χ_{m_i} can be performed using the ratio between heights and the ratio between total body weights, respectively. The length scaling factor is used to scale all the dimensions, including the location of each center-of-mass. This procedure should only be used to scale subjects of the same gender and with anthropometric characteristics not far from the reference model.

4.2 Muscle Model

The dynamics of muscle tissue can be divided into activation dynamics and muscle contraction dynamics [13], as schematically indicated in Figure 7. The activation dynamics generates a muscle tissue state that transforms the neural excitation produced by the central nervous system, into activation of the contractile apparatus. The activation dynamics, although not implemented in this work, describes the time lag between neural signal and the corresponding muscle activation [15].

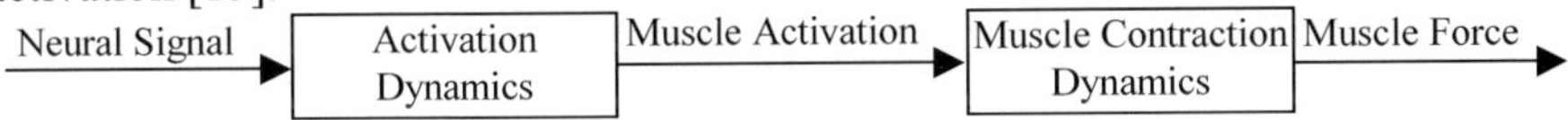

Fig. 7. Dynamics of muscle tissue.

The muscle contraction dynamics requires that a mathematical model of the muscle is introduced. In the present work the Hill muscle model is applied to the simulation of the muscle contraction dynamics. The model, depicted in Figure 8, is composed of an active Hill contractive element (CE) and a passive element (PE). Both elements contribute to the total muscle force $F^m(t)$. In the present work, the series elastic element (SEE), usually associated with cross-bridge stiffness, is not included in the model since it can be neglected in coordination studies not involving short-tendon actuators [13].

In the Hill muscle model, the contractile properties of the muscle tissue are controlled by its current length $l^m(t)$, rate of length change $\dot{l}^m(t)$ and activation $a^m(t)$. The force produced by the active Hill contractile element, for muscle m, is

$$F_{CE}^m(a^m(t), l^m(t), \dot{l}^m(t)) = \frac{F_l^m(l^m(t)) F_{\dot{l}}^m(\dot{l}^m(t))}{F_0^m} a^m(t) \tag{15}$$

where F_0^m is the maximum isometric force and $F_l^m(l^m(t))$ and $F_{\dot{l}}^m(\dot{l}^m(t))$ are two functions that represent the muscle force-length and force-velocity dependency, respectively [13,15]. These two functions are approximated analytically by [15]

$$F_l^m(l^m(t)) = F_0^m e^{-\left[\left[-\frac{9}{4}\left(\frac{l^m(t)}{l_0^m} - \frac{19}{20}\right)\right]^4 - \frac{1}{4}\left[-\frac{9}{4}\left(\frac{l^m(t)}{l_0^m} - \frac{19}{20}\right)\right]^2\right]} \tag{16}$$

and

$$F_l^m(\dot{l}^m(t)) = \begin{cases} 0 & -\dot{l}_0^m > \dot{l}^m(t) \\ -\dfrac{F_0^m}{\arctan(5)}\arctan\left(-5\dfrac{\dot{l}^m(t)}{\dot{l}_0^m}\right) + F_0^m & 0.2\dot{l}_0^m \geq \dot{l}^m(t) \geq -\dot{l}_0^m \\ \dfrac{\pi F_0^m}{4\arctan(5)} + F_0^m & \dot{l}^m(t) \geq 0.2\dot{l}_0^m \end{cases} \quad (17)$$

where l_0^m is the muscle resting length and $\dot{l}_0^m$ is the maximum contractile velocity above which the muscle cannot produce force. The passive element is independent of the activation and it only starts to produce force when stretched beyond its resting length l_0^m. The force produced by the passive element is approximated by [15]:

$$F_{PE}^m(l^m(t)) = \begin{cases} 0 & l_0^m > l^m(t) \\ 8\dfrac{F_0^m}{l_0^{m3}}\left(l^m - l_0^m\right)^3 & 1.63 l_0^m \geq l^m(t) \geq l_0^m \\ 2F_0^m & l^m(t) \geq 1.63 l_0^m \end{cases} \quad (18)$$

Equation (18) shows that the force produced by the passive element is only a function of the muscle length. The force produced by the passive element is not unknown being treated as an external force directly applied to the rigid bodies interconnected by the muscle. The forces produced by the contractile element are the only unknown forces. A muscle actuator equation associated to each contractile element is accomplished multiplying each actuator equation by a proper scalar factor, so that the Lagrange multiplier associated to the actuator, represents muscle force or muscle activation. The factors for the muscle force and activation are

$$\mathsf{C}_\lambda^m = 1/2l^m \quad ; \quad \mathsf{C}_\lambda^m = F_l^m F_l^m / \left(2F_0^m l^m\right) \quad (19)$$

As the Lagrange multiplier represents muscle activation the associated muscle force is calculated using Equation (11).

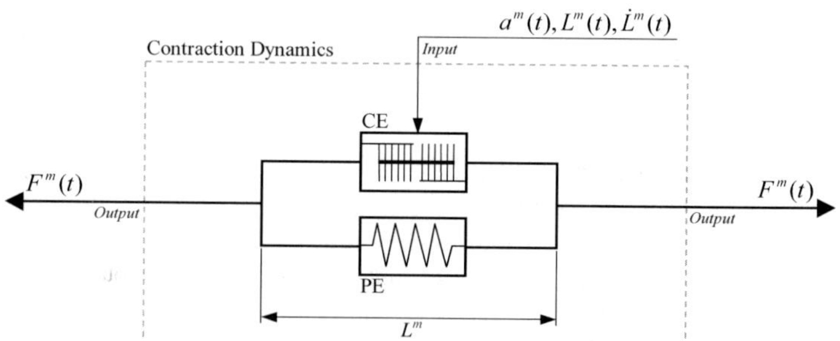

Fig. 8. Contraction dynamics using a Hill-type muscle model.

4.3 Skeleton-muscle model of locomotion apparatus

A muscle locomotion apparatus, with thirty-five muscle actuators can be used to simulate the right lower extremity intermuscular coordination. The muscle apparatus and a brief description of each muscle action [24] are presented in Table 2. The physiological information regarding the muscle definition is obtained from the literature [25,26] and compiled in a muscle database. This information consists in the maximum isometric force, resting length, attachment points, wrap-around bodies and the local coordinates of the origin, insertion and via points. The whole muscle apparatus is presented in Figure 9.

Table 2. List and description of the lower extremity muscle apparatus [27].

No	Muscle Name	Muscle Action
1	Adductor Brevis	Adducts, flexes and helps to laterally rotate the thigh.
2	Adductor Longus	Adducts and flexes the thigh; helps to laterally rotate the hip.
3	Adductor Magnus	Thigh adductor; superior horizontal fibers also help to flex the thigh, while vertical fibers help extend the thigh.
4	Biceps Femoris (long head)	Flexes the knee, and rotates the tibia laterally; long head extends the hip joint.
5	Biceps Femoris (short head)	Flexes the knee, and rotates the tibia laterally; long head extends the hip joint.
6	Extensor Digitorum Longus	Extend toes 2 – 5 and dorsiflexes ankle.
7	Extensor Hallucis Longus	Extends great toe and dorsiflexes ankle.
8	Flexor Digitorum Longus	Flexes toes 2 – 5; also helps in plantar flexion of ankle.
9	Flexor Hallucis Longus	Flexes great toe, helps to supinate ankle; weak plantar flexor of ankle.
10	Gastrocnemius (lateral head)	Plantar flexor of ankle.
11	Gastrocnemius (med. head)	Plantar flexor of ankle.
12	Gemellus (inf. and superior)	Rotates the thigh laterally and helps to abduct the flexed thigh.
13	Gluteus Maximus	Major extensor of hip joint; rotates laterally the hip; superior fibers abduct the hip; inferior fibers tighten the iliotibial band.
14	Gluteus Medius	Abductor of thigh; anterior fibers help to rotate hip medially; posterior fibers help to rotate hip laterally
15	Gluteus Minimus	Abducts and medially rotates the hip joint.
16	Gracilis	Flexes the knee, adducts the thigh, helps to medially rotate the tibia on femur.

Table 2. List and description of the lower extremity muscle apparatus [27] (continued).

No	Muscle Name	Muscle Action
17	Iliacus	Flexes the torso and thigh with respect to each other.
18	Pectineus	Adducts the thigh and flexes the hip joint.
19	Peroneus Brevis	Everts foot and plantar flexes ankle.
20	Peroneus Longus	Everts foot and plantar flexes ankle; helps to support the transverse arch of the foot.
21	Peroneus Tertius	Dorsiflexes, everts and abducts foot.
22	Piriformis	Lateral rotator of the hip joint; helps abduct the hip if it is flexed.
23	Psoas	Flex the torso and thigh with respect to each other.
24	Quadratus Femoris	Rotates the hip laterally; also helps adduct the hip.
25	Rectus Femoris	Extends the knee.
26	Sartorius	Flexes and laterally rotates the hip joint and flexes the knee.
27	Semimembranosus	Extends the thigh, flexes the knee, and also rotates the tibia medially, especially when the knee is flexed.
28	Semitendinosus	Extends the thigh and flexes the knee, and also rotates the tibia medially, especially when the knee is flexed.
29	Soleus	Powerful plantar flexor of ankle.
30	Tensor Fasciae Lata	Helps stabilize and steady the hip and knee joints by putting tension on the iliotibial band of fascia.
31	Tibialis Anterior	Dorsiflexor of ankle and invertor of foot.
32	Tibialis Posterior	Principal invertor of foot; adducts foot, plantar flexes ankle, helps to supinate foot.
33	Vastus Intermedius	Extends the knee.
34	Vastus Lateralis	Extends the knee.
35	Vastus Medialis	Extends the knee.

A set of data for the muscles described in Table 2, including the coordinates of their insertion points, the physiological cross section area and the coordinates for the via points of curved muscles, are found on the work by Carhart [26].

5 Solution of the Force Redundant Problem by Optimization

The human muscle system is highly redundant, being possible to identify several muscles that guide the same degree-of-freedom of the same anatomical joint. Indeterminate systems have an infinite set of possible solutions, being the aim of optimization techniques to find, from all the possible solutions, the one that minimizes a prescribed objective function, subjected to a certain number of restrictions or constraints. Mathematically, the optimization problem is stated as:

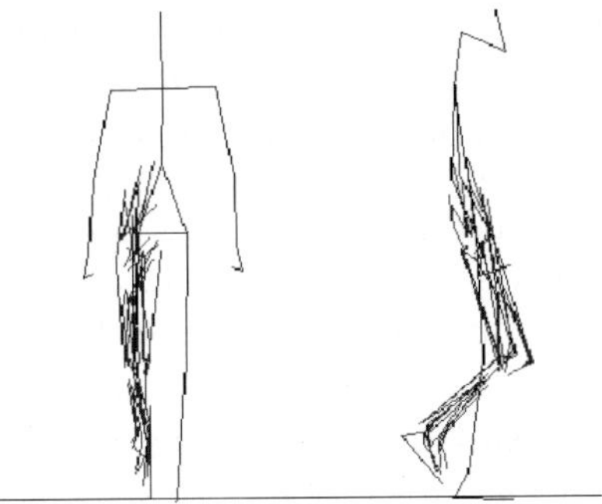

Fig. 9. Lower extremity muscle apparatus.

$$\text{minimize} \quad F_0(\mathbf{b})$$

$$\text{subject to:} \begin{cases} f_j(\mathbf{b}) = 0 & j = 1,\ldots,n_{ec} \\ f_j(\mathbf{b}) \geq 0 & j = (n_{ec}+1),\ldots,n_{tc} \\ b_i^{lower} \leq b_i \leq b_i^{upper} & i = 1,\ldots,n_{sv} \end{cases} \tag{20}$$

where the vector of the unknown parameters, or design variables, is **b** with the components b_i bounded respectively by b_{lower} and b_{upper}, $F_0(\mathbf{b})$ is the objective or cost function to minimize and $f_j(\mathbf{b})$ are constraint equations that restrain the state variables. In Equation (20), n_{sv} represents the total number of design variables and n_{tc} the total number of constraint equations in which n_{ec} are of the equality type.

The minimization of cost functions simulates the criteria adopted when deciding which muscles to recruit and defines the level of activation that produce the adequate motion or posture for a specific task. The selection of the most appropriate criterion to use in the optimization process resides upon several aspects such as the type of motion under analysis, the objectives to achieve or the presence of any type of pathology.

5.1 Objective functions associated to prescribed motion

Assume that the objective of the analysis of a given human movement task is to find the muscle forces that are developed in order to achieve a given motion that is experimentally acquired. The problem can be set as the minimization of an objective function that measures the distance between the motion obtained and the acquired target motion, i.e., the minimization of the sum of the squares of the differences between the motions evaluated at a given number of time instants

$$F_0(\mathbf{b}) = \sum_{i=1}^{m} w_i \left(\varphi^t(t_i) - \varphi^c(t_i, \mathbf{b})\right)^2 \tag{21}$$

where the target angle $\varphi^{t}(t_i)$ is obtained by sampling the measured angle $\varphi(t)$ at m discrete instants in time. Likewise, the computed angle $\varphi^{c}(t_i,\mathbf{b})$ is obtained by performing a forward dynamics simulation of the complete model and sampling this function at the prescribed output points in time t_i. Appropriate penalty functions can be added to the objective function represented in Equation (21) on either the design variables directly, to facilitate the optimization process, or on the behavior of the biomechanical system, such as a physiological criteria.

5.2 Objective functions associated to physiological criteria

A cost function must reflect the inherent physical activity or pathology and to include relevant physiological characteristics and functional properties, such as the maximum isometric force or the electromyographic activity [28]. From the computational point of view, a cost function must be numerically stable and fast to evaluate. Some of the most commonly used cost functions are:

$$F_0(\mathbf{b}) = \sum_{m=1}^{n_{ma}} \left(F_{CE}^{m}\right)^2 \tag{22}$$

Designated by sum of the square of the individual muscle forces, when applied to the study of human locomotion this cost function is considered to fulfill the objective of energy minimization. This cost-function does not include any physiological or functional capabilities [28];

$$F_0(\mathbf{b}) = \sum_{m=1}^{n_{ma}} \left(\sigma_{CE}^{m}\right)^3 \tag{23}$$

Known as sum of the cube of the average individual muscle stress this cost function, introduced by Crowninshield and Brand [29], is based on a quantitative force-endurance relationship and on experimental results. It includes physiological information, namely the value of the physiological cross sectional area of each muscle and it is reported to predict co-activation of muscle groups in a more physiologically realistic manner [28]. The interested reader may find a list of other suitable objective functions and their description in [27, 28].

5.3 Optimization methods

Different optimization packages can be used in the solution of the optimization of the redundant muscle forces: DOT 5.0 [30], DNCONG from IMSL Library [31], NAG library [32] and MMA – Method of Moving Asymptotes [33] are examples of some of these packages. The first three methods use successive quadratic programming algorithms, while the fourth one uses the globally convergent method of moving asymptotes with inner and outer iterations. In all cases the optimization problem is subject to linear and/or nonlinear constraints.

5.4 Parameterization of the activation profile

In the optimization problems described here the muscle activations are the design variables used to find the optimal muscle forces. Measurement data suggests that an activation curve typically involves a limited number of maxima during full gait cycle, typically 1-2 bumps [29,34]. Because the signal strength in surface electromyography is related to several factors such as subdermal muscle depth, skin preparation and interference of crossing muscles, its use to draw conclusions on exact muscle activation profiles is strongly limited. However, this observation can be used to reduce the complexity of the function search space. A procedure to discretize activation profiles is to sample the functions at given points and interpolating the values between sampling points by piecewise linear functions, leading to

$$a(t) = A_i + (t - t_i)\frac{A_{i+1} - A_i}{t_{i+1} - t_i}, \quad i = 0,\ldots n; \quad t \in [t_i, t_{i+1}) \tag{24}$$

where t_i are the sampling times, A_i are the sampling values and $a(t)$ is the resulting activation function. By using the sampled values A_i as design parameters, the optimization routine can find the best fit such that the resulting motion matches prescribed values. This approach, shown in Figure 10(a), is called input sampling. Equation (24) provides a large flexibility to the problem solution due to an arbitrary number of design variables A_i. However, the same large number n of design parameters can prevent the optimizer of converging.

Another approach, proposed by Strobach et al. [1], employs a limited number of exponential functions that render smooth bump behavior. The goal of the optimization is to determine location, amplitude and width of these exponential bumps, and therefore using them as design parameters. For a n-bump activation profile the activation function is written as

$$a(t) = A_1\, e^{-C_1(t-T_1)^2} + A_2\, e^{-C_2(t-T_1)^2} + \ldots + A_n\, e^{-C_n(t-T_1)^2} \tag{25}$$

This approach is denoted as C^∞ in what follows. The principle of this parameterization is depicted in Figure 10(b).

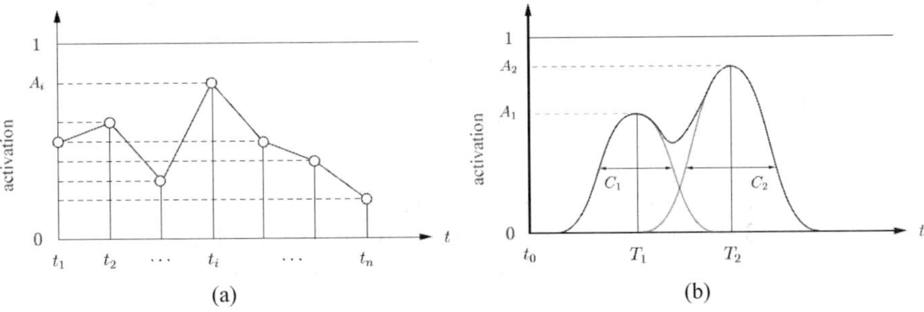

Fig. 10. Parameterization of muscle activation by (a) input sampling, i.e., discretization and linear interpolation and (b) smooth exponential function approach C^∞.

6 Application to Cases of Human Locomotion

To appraise the different types of dynamic analysis of biomechanical systems presented in this work and their use in the framework of the human locomotion two application cases are presented. The use of inverse dynamics analysis in human motion to study an exercise of jumping is presented and discussed. This application is based on [35]. To illustrate the use of the direct dynamics analysis in gait analysis a case of the extension and flexion of a leg, based on [1], is presented and discussed. In both cases the muscle redundancy is taken into account in the models and the muscle activation levels are used as the unknown parameters that have to be calculated in order to find the individual muscle forces.

6.1 Inverse dynamics analysis of a jumping exercise

The methodology proposed here is applied to solve the muscle force distribution problem in the supported leg during the take-off phase to aerial trajectories in a ballistic motion. A 1.68 m male with a body mass of 68 kg performed a jump, represented in Figure 11 by two frames that coincide with the beginning and end of the take-off phase of the motion. The ground reaction forces are measured using a Kistler 9281B force platform with a sampling frequency of 1000 Hz, while the body motion is videotaped at 50 Hz by 4 synchronized cameras [35].

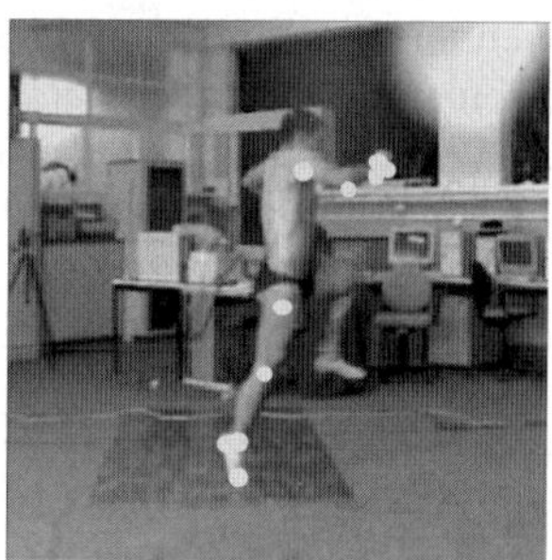

Fig. 11. Beginning and ending of the take-off phase of a jumping exercise.

The time characteristics of the ground reaction forces measured in the trial are shown in Figure 12. The data differ from typical gait data both in the shapes and magnitudes of the force curves. The maximal vertical reaction is almost four times larger than for its gait counterpart, and the peak value of the medial-lateral component is seven times larger than for its corresponding in a gait trial [15].

The inverse dynamic analysis of the biomechanical model is carried out firstly to find the net moments-of-force in the anatomical joints of the lower extremities. The results obtained in the inverse dynamic analysis of the jump, presented in

Figure 13, show a considerable loading of the ankle joint of the supporting leg. As expected, a strong activity of the plantar-flexors muscles spanning this joint is also observed. A large peak of the net torque at the hip joint occurring at the beginning of the contact phase of the foot with the ground, i.e., at t = 0.08 s is observed. An oscillating time-force characteristic for some of the muscles that span the knee and hip joints can be anticipated as a consequence of this behavior.

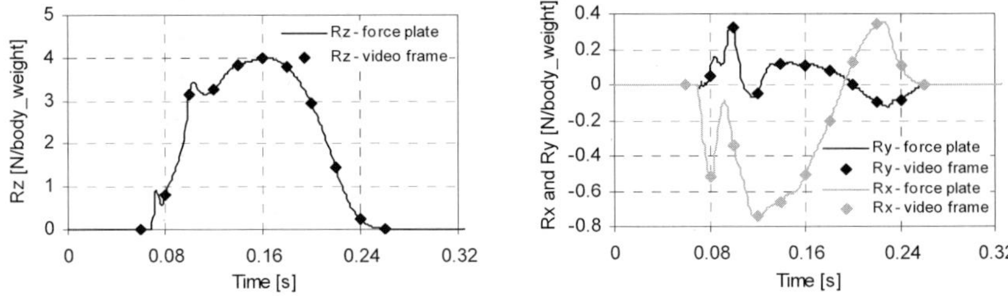

Fig. 12. Vertical (Rz), anterior-posterior (Rx) and medial-lateral (Ry) components of the ground reaction force measured in the force platform. The markers denote the time instants for which the video frames are available.

The results of the optimization procedure are the activations of the muscles of the locomotion apparatus. The forces observed in each muscle are calculated by using the individual muscle characteristics and the respective muscle force-length and force-velocity curves. The force versus time characteristics of the selected muscle forces obtained are presented in Figure 14. Some of the force time histories refer to groups of muscles obtained by lumping muscles with similar functionalities. Muscle groups that result from the lumping process are the iliopsoas, which results from the iliacus and the psoas, the hip adductors, which include the adductor longus, adductor brevis and adductor magnus, the hamstrings that account for the semimembranosus, semitendinosus and biceps femoris long head, the vasti muscles, which include the vastus medialis, vastus intermedius and vastus lateralis, and the triceps surae that is the lumping of the gastrocnemius lateral, gastrocnemius medial and soleus.

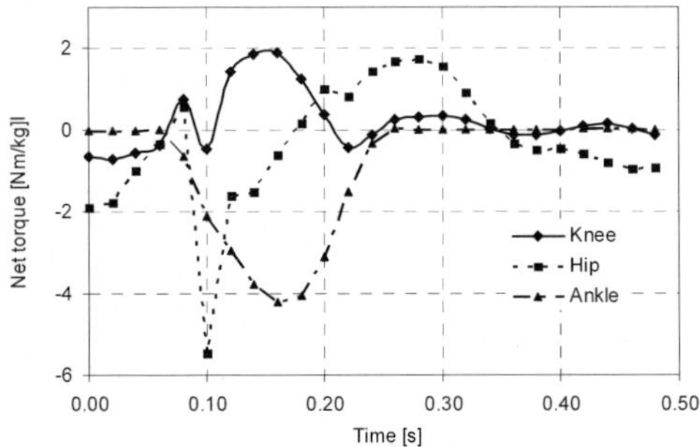

Fig. 13. Resultant net torques at the anatomical joints of the lower extremity.

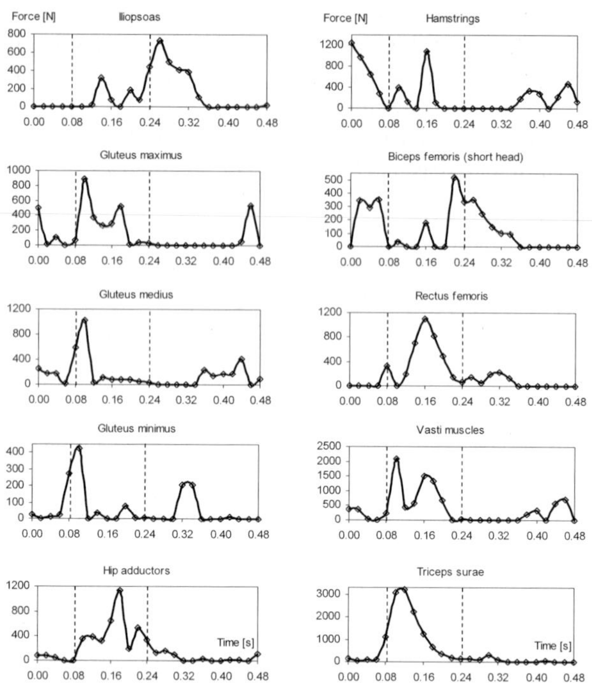

Fig. 14. Forces of selected muscles of the supporting leg of the jumper during the trial. The vertical dashed lines indicate the supporting phase of the jump. The markers represent the results obtained at the time frames.

The analysis of the trial starts with the jumper in the air, when his trunk is in a forward-lean position. The hamstrings, the short head of the biceps femoris (with a small time delay) and the gluteus maximus are therefore working at a maximum force to raise the trunk to an upright position, which is achieved when a foot touches the ground. At this moment, very strong contractions of glutei and vasti muscles occur to neutralize impact ground reaction forces. The gluteus maximus works as a hip extensor and the vasti group extends the knee joint. The rectus femoris, which spans two joints, is not involved in this action. A considerable activity of the hamstrings, which stabilize the movement with their antagonistic action to the knee extensors, can also be noticed. Simultaneously, the triceps surae generates a powerful ankle plantar flexion to push the body forward. The common action of all these muscles stretches out the lower leg, which is very useful when carrying the maximal ground force reactions in the middle of the support phase. A remarkable excitation of hip adductors at this time stabilizes the movement of the jumper and helps to put the legs together in the air later. The final activity of the glutei, biceps femoris (short head), iliopsoas and hamstrings causes knee bending and thigh uprising that allow the body to achieve an appropriate airborne position.

The activation levels of selected knee flexors and extensors are depicted in Figure 15. Not shown in this figure, the characteristics of the sartorius and gracilis, which are the remaining knee flexors, are similar in shape to the ones presented.

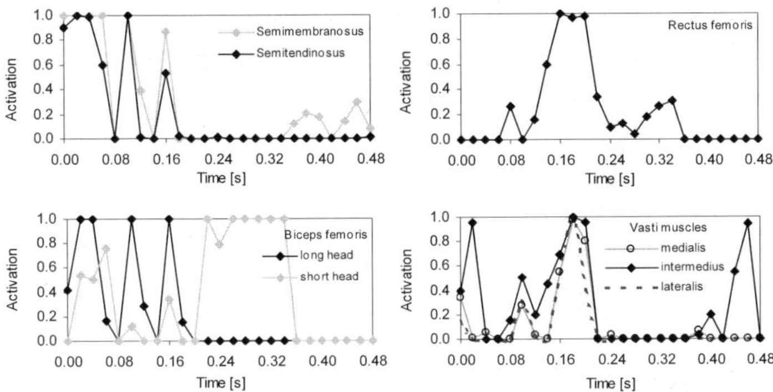

Fig. 15. Activation level of the knee flexors, semimembranosus, semitendinosus and biceps femoris and of the knee extensors, rectus femoris and vasti muscles.

During the take-off phase there are considerable changes in activation levels of muscles spanning the knee joint. Two peaks of force occur in the beginning and in the middle of the support phase, which match those already observed in Figure 14. The first peak corresponds to the steep increase of the resultant ground reaction force, whereas the other one is associated to the maximal vertical ground reaction. The period of time used to increase the activation of a muscle is estimated to be on the order of 10 ms, while that for deactivation is on the order of 50 ms [13]. It is rather unlikely that semitendinosus has been fully activated/deactivated within 40 ms, which corresponds to the first peak in Figure 15.

Fast deactivation of some muscles is a consequence of the static optimization procedure used where the calculation of the activations that lead to the minimization of the cost function at a single instant of time is unrelated to what happens in other instants of time. Several models of the muscle apparatus of the lower limb have been used in the solution of the redundant problem, starting from a 35 muscles structure. It has been observed that models using a lower number of muscles lead to larger oscillations for the force-time response of the muscles. The general trend observed is that a larger number of muscles involved in the solution of the redundant problem leads to smoother results for the muscle force time behavior.

6.2 Direct dynamics in gait analysis

The case presented here to illustrate the use of the procedure for a case where direct dynamics is used in the framework of a gait analysis, consists of a two-joint subsystem of the right leg, comprising pelvis (fixed in space), femur, tibia and fossa pedis [1]. The model is driven by two pairs of antagonistic muscles, adductor longus, gluteus maximus, biceps femoris caput brevis, vastus intermedius. The hip and knee joint are assumed to supply one degree of freedom for flexion-extension each, as presented in Figure 16. The time interval for the integration was set to T=3 s.. The biometric data of the leg concerns a 1.78 m tall male specimen of 77.9 kg. The mass and inertia properties are scaled from that provided in Table 1 and the muscle origins, insertions and geometry are reported in [25].

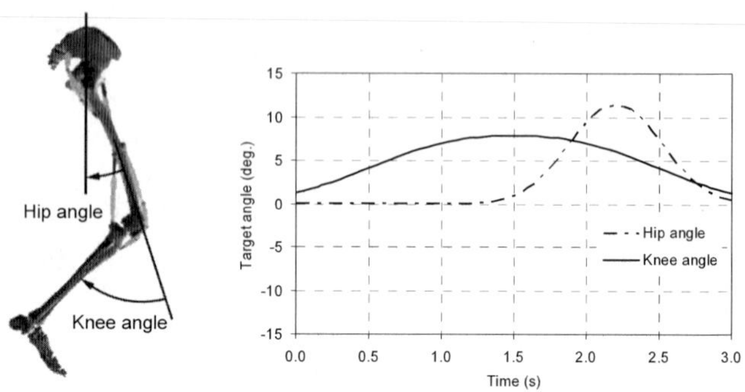

Fig. 16. Two-joint subsystem of hip and knee and prescribed target motion for the hip and knee angles.

The fulfillment of a prescribed target extension and flexion motion of the leg is the objective of the analysis. The objective function is given by

$$F_0(\mathbf{b}) = \tfrac{1}{2}\sum_{i=1}^{m}\left(\varphi_h(t_i) - \varphi_h^c(t_i,\mathbf{b})\right)^2 + \tfrac{1}{2}\sum_{i=1}^{m}\left(\varphi_k(t_i) - \varphi_k^c(t_i,\mathbf{b})\right)^2 \tag{26}$$

where $\varphi_h(t_i)$, $\varphi_k(t_i)$, $\varphi_h^c(t_i,\mathbf{b})$ and $\varphi_k^c(t_i,\mathbf{b})$ denote the sampled target angles for hip and knee and the simulated joint angles for hip and knee for a design parameter set **b** for the m time instants during the optimization process, respectively. Note that in the objective function described in Equation (26) no physiological criteria is used for the purpose of this example. A more general objective function may use a measure of the precision of the motion obtained with respect to the target motion and a physiological criterion, represented by Equations (22) or (23). Although the influence of the output sampling step size has been included in the analysis all results reported here concern a sampling step of □t= 0.1 s. The number of muscles included in the system lead to 16 design parameters in case of the smooth exponential function C^∞ approach, with two bumps, where the muscle activation is represented by Equation (25). When the input sampling approach is used, being the muscle activations described by Equation (24), a total of 40 design parameters are required when the number of amplitudes per muscle is set to 10.

To assess quality of matching between target and simulated functions, the objective function is divided by the number of output sampling points, leading to a normalized objective function defined by $\overline{f}(\mathbf{b}^*) = f(\mathbf{b}^*)/m$, where $\mathbf{b}^*$ is the set of design parameters at the optimum point and m=T/(□t+1). The performance of the methodology is measured as the CPU time used in the analysis. Figure 18 shows the computation time and normalized final cost function for the cases in which the input sampling and the smooth exponential function C^{∞} are used.

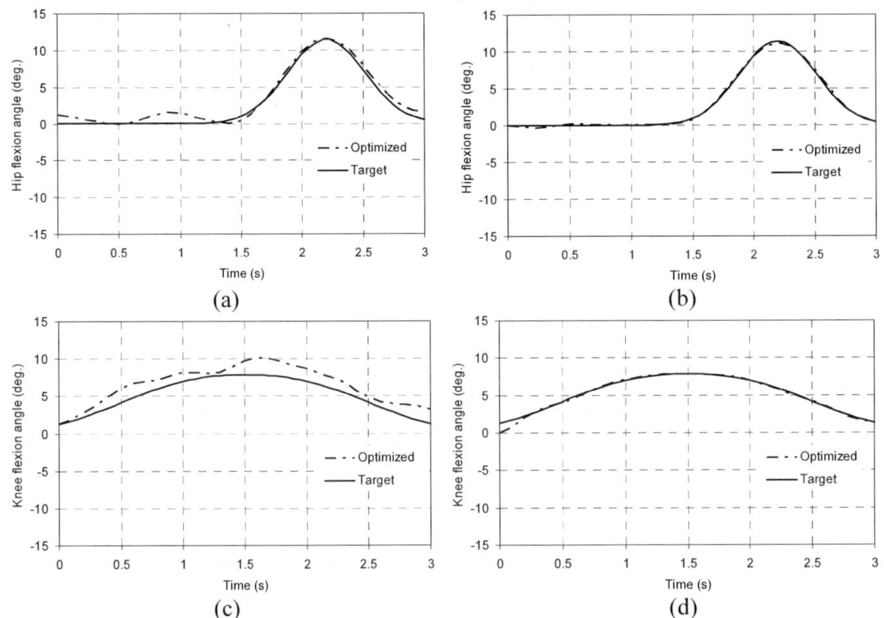

Fig. 17. Optimization results, measured in terms of the hip and knee angles, for the biomechanical system shown in Fig. 16 for: (a) hip flexion using the input sampling; (b) hip flexion using the C^{∞} function; (c) knee flexion using the input sampling; (d) knee flexion using the C^{∞} function

Both in terms of the CPU time and value of the optimum the smooth exponential function C^{∞} renders better results as observed in Figure 18. These conclusions are consistent with the observation of Figure 17 where the target angles for both anatomic joints are better matched in the procedure using the smooth function.

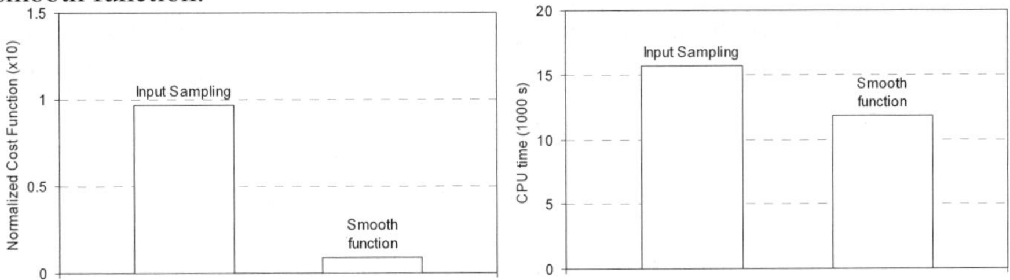

Fig. 18. Comparison of computational performance in terms of quality and necessary CPU-time on an Intel Pentium IV 2.4 GHz PC running Linux Redhat 8.0.

7 Conclusions

Multibody dynamics methodologies for developing biomechanical models and for analyzing them have been presented and discussed here in the framework of human motion activities, with emphasis in human gait analysis. It has been shown that although both direct and inverse dynamic approaches can be used to analyze biomechanical problems with prescribed motion they pose different requirements for the development of biomechanical models. The definition of the biomechanical analysis as an optimal problem is the only form to solve the muscle force redundancy problem in general. It was shown that to solve the problem, regardless being defined as a static or a dynamic optimization, the muscle actuations are better design variables than the muscle forces directly. Furthermore, it was shown that for cases of prescribed motion the minimization of objective functions that reflect the distance between the actual motion of the biomechanical model and such target motion allows for the solution of a problem where inverse dynamics methods are typically applied by direct dynamics methods. Furthermore, when the analyst adds to the distance objective function the physiological criteria not only the biomechanical analysis leads to a motion equal or similar to the prescribed one but also the muscle forces are shared in a optimal form from the physiological point of view. All the optimization procedures are more efficient and more robust when proper descriptions of the activation profiles are used. It has been shown that the use of smooth exponential functions to describe the muscle activation profiles has advantages over the traditional piecewise linear interpolation both in terms of precision and computational effort. All procedures described have been exemplified by human motion activities with the locomotion apparatus.

Acknowledgements

The work reported here represents the part of the research carried at the Institute of Mechanical Engineering in Lisbon and in the Institute for Mechatronics and System Dynamics in Duisburg. The authors are grateful to Miguel Silva, Adam Czaplicki, Matthew Kaplan, Daniel Strobach, Gerhart Steinwender and Bernhard Zwick for their contribution to this research. Finally, the support by Fundação para a Ciência e Tecnologia through the project POCTI/EME/39976/2001 is gratefully acknowledged.

References

1. Strobach D, Kecskeméthy A, Steinwender G, Zwick B (2005) A Simplified Approach for Rough Identification of Muscle Activation Profiles Via Optimization and Smooth Profile Patches. In Proc of Multibody Dynamics 2005, ECCOMAS Thematic Conference (Goicolea J, Cuadrado J, Garcia Orden J, eds.), June 21-24, Madrid, Spain, 1-17
2. Nikravesh P (1988) Computer-Aided Analysis of Mechanical Systems. Prentice Hall, Englewood-Cliffs, New Jersey
3. Jalon J G, Bayo E (1994) Kinematic and Dynamic Simulation of Mechanical Systems – The Real-Time Challenge. Springer-Verlag, Berlin, Germany
4. Nikravesh P, Gim G (1993) Systematic Construction of the Equations of Motion for Multibody Systems Containing Closed Kinematic Loops, Journal of Mechanical Design 115(1): 143-149
5. Kane T, Levinson D (1985) Dynamics: Theory and Applications. McGraw-Hill, New York
6. MDI (1998) ADAMS User's Manual. Mechanical Dynamics, Ann Arbor, Michigan
7. Jimenez J M., Avello A, Garcia-Alonso A, García de Jalón J (1990) COMPAMM: A simple and Efficient Code for Kinematic and Dynamic Simulation of 3D Systems with Realistic Graphics. In Multibody Systems Handbook (Schiehlen W, ed.) Springer-Verlag, Berlin, Germany, 285-304
8. TNO Automotive (1999) MADYMO Theory manual vs. 5.4,TNO Automotive, Delft, The Netherlands
9. Kecskeméthy A. (2002) M□BILE1.3 User's Guide. Lehrtuhl Mechanik, University Duisburg-Essen, Germany
10. Laananen D, Bolokbasi A, Coltman J (1983) Computer simulation of an aircraft seat and occupant in a crash environment – Volume I: technical report, US Dept of Transp., Federal Aviation Administration, Report n DOT/FAA/CT-82/33-I
11. Ambrósio J, Silva M, Abrantes J (1999) Inverse Dynamic Analysis of Human Gait Using Consistent Data. In Proc of the IV Int. Symp. on Computer Methods in Biomechanics and Biomedical Engng, October13-16, Lisbon, Portugal
12. Silva M, Ambrósio J (2004) Sensitivity of the Results Produced by the Inverse Dynamic Analysis of a Human Stride to Perturbed Input Data. Gait and Posture 19(1): 35-49
13. Zajac F (1989) Muscle and tendon: properties, models, scaling, and application to biomechanics and motor control. Critical Reviews in Biomedical Engineering 17(4): 359-411
14. Hatze H (1984) Quantitative Analysis, Synthesis and Optimization of Human Motion. Human Movement Science 3: 5-25
15. Silva M (2003) Human Motion Analysis Using Multibody Dynamics and Optimization Tools. Ph.D. Dissertation, Instituto Superior Técnico, Technical University of Lisbon, Lisbon, Portugal
16. Hiller M, Kecskeméthy A (1989) Equations of Motion of Complex Multibody Systems Using Kinematical Differentials. In: Proc. Of 9th Symposium on Engineering Applications of Mechanics, 13, London, Canada, 113-121
17. Neto M, Ambrósio J (2003) Stabilization Methods for the Integration of DAE in the Presence of Redundant Constraints. Multibody System Dynamics 10: 81-105

18. Baumgarte J (1972) Stabilization of Constraints and Integrals of Motion in Dynamical Systems. Computer Methods in Applied Mechanics and Engineering, 1: 1-16
19. Pandy M (2001) Computer modeling and simulation of human movement. Annual Review Biomedical Engineering 3: 245-273
20. Ambrósio J, Lopes G, Costa J, Abrantes J (2001) Spatial reconstruction of the human motion based on images from a single stationary camera, J. Biomech. 34:1217–1221
21. Allard P, Stokes I, Blanchi J (1995) Three-Dimensional Analysis of Human Movement. Human Kinetics, Champaign, Illinois
22. Anderson F, Pandy M (2001) Static and Dynamic Optimization Solutions for Gait are Practically Equivalent. J. Biomech. 34: 153-161
23. Barbosa I, Ambrósio J, Silva M (2003) Inverse Versus Forward Dynamic Analysis of the Human Locomotion Apparatus. In: Proc.of the VII Congresso de Mecânica Aplicada e Computacional (Barbosa J, ed.), April 14-16, Évora, Portugal: 525-534
24. Richardson M (2001) Lower Extremity Muscle Atlas, in internet address http://www.rad.washington.edu/atlas2/, University of Washington - Department of Radiology, Washington
25. Yamaguchi G (2001) Dynamic Modeling of Musculoskeletal Motion. Kluwer Academic Publishers, Boston, Massachussetts
26. Carhart M (2000) Biomechanical Analysis of Compensatory Steping: Implications for paraplegics Standing Via FNS., Ph.D. Dissertation, Department of Bioengineering, Arizona State University, Tempe, Arizona
27. Ambrósio J, Silva M (2005) A Biomechanical Multibody Model with a Detailed Locomotion Muscle Apparatus. In: Advances in Computational Multibody Systems (Ambrósio J, ed.), Springer, Dordrecht, The Netherlands: 155-184
28. Tsirakos D, Baltzopoulos V, Bartlett R (1997) Inverse Optimization: Functional and Physiological Considerations Related to the Force-Sharing Problem. Critical Reviews in Biomedical Engineering 25(4-5): 371-407
29. Crowninshield R, Brand R (1981) Physiologically Based Criterion of Muscle Force Prediction in Locomotion. J. Biomech. 14(11): 793-801
30. Vanderplaats R&D (1999) DOT – Design Optimization Tools – USERS MANUAL – Version 5.0, Colorado Springs, Colorado
31. V. Numerics (1995) IMSL FORTRAN Numerical Libraries – Version 5.0, Microsoft Corp.
32. Numerical Analysis Group (2003) The NAG Fortran Library Manual Mark 20. Wilkinson House, Oxford, United Kingdom
33. Svanberg K (1999) The MMA for Modeling and Solving Optimization Problems. In Proceedings of the 3rd World Congress of Structural and Multidisciplinary Optimization, May 17-21, New York
34. Perry J (1992) Gait Analysis: Normal and Pathological Function. McGraw-Hill, New York, New York
35. Czaplicki A, Silva M, Ambrósio J (2004) Biomechanical Modeling for Whole Body Motion Using Natural Coordinates, Journal of Theoretical and Applied Mechanics, 42(4): 927-944

From Multibody Dynamics to Multidisciplinary Applications

Martin Arnold[1] and Andreas Heckmann[2]

[1] *Institute of Mathematics, Martin-Luther University Halle-Wittenberg, D-06099 Halle (Saale), Germany; E-mail: martin.arnold@mathematik.uni-halle.de*
[2] *Institute of Robotics and Mechatronics, DLR German Aerospace Center, P.O. Box 1116, D-82230 Wessling, Germany; E-mail: andreas.heckmann@dlr.de*

Abstract. With the increasing integration of mechanical, electrical and hydraulical components in advanced engineering systems, the integrated analysis of coupled physical phenomena and coupled technical systems gets more and more important. The methods and software tools of multibody dynamics are used successfully as integration platform for these multidisciplinary investigations. The present paper summarizes some multidisciplinary applications in the context of multibody dynamics and considers common problems and solution strategies. A novel modal multifield approach for coupled field effects like thermoelasticity is discussed in more detail.

1 Introduction

In the early days of technical simulation, the analysis of coupled physical phenomena was restricted to high-end applications like aeroelastic problems that require the investigation of fluid-structure interaction. Sophisticated models for the coupled problems were studied using highly developed specialized simulation tools on the most powerful computer hardware being available at that time.

Today, the state-of-the-art is characterized by substantially improved and standardized methods for model setup and numerical solution and by the dramatically improved power of modern computer hardware. Multidisciplinary applications involving coupled technical systems and coupled physical problems are handled by adapting and extending standard simulation tools on standard PC or workstation hardware. Nevertheless, there are still challenging open problems and a large potential to study multidisciplinary problems as robustly and efficiently as classical monodisciplinary ones.

The trend from monodisciplinary to multidisciplinary applications in technical simulation was strongly pushed by the increasing integration of mechanical, elec-

J.C. García Orden et al. (eds), Multibody Dynamics. Computational Methods and Applications, 273–294.

trical and hydraulical components in advanced engineering systems and by the increasing success of linear and nonlinear model-based control devices.

Block-oriented approaches like Matlab/Simulink[1] or Modelica[2] [1] play an important part in network based modelling strategies for coupled problems. On the other hand, the standard tools for coupled field effects that are distributed in space are based on finite element discretizations [2].

Therefore, it might be surprising that also methods and software tools for classical monodisciplinary applications like multibody dynamics are very successfully used in the analysis of multidisciplinary problems. Furthermore, in several disciplines a common network approach is used for the modelling of complex systems. Monodisciplinary network models of system components may be combined to one large model of a multidisciplinary problem that is analysed by block-oriented tools or following a co-simulation approach. These topics will be discussed in more detail in Section 2.

The numerical methods of multibody dynamics require much less computational effort than methods that are based on finite elements. The discussion of a modal multifield approach for thermoelastic problems in Section 3 illustrates that well approved numerical methods from multibody dynamics can be carried over to more complex multidisciplinary applications including coupled field effects. In Section 4 the new approach is applied to a high-performance machine tool with thermal loads caused by linear induction drives.

The material of Sections 3 and 4 is the condensed and revised version of a paper that both authors contributed to the ECCOMAS Thematic conference "Multibody Dynamics 2005" in June 2005 in Madrid [3].

2 Methods and Software Tools of Multibody Dynamics in Multidisciplinary Applications

The term "multidisciplinary" is used to characterize applications that are beyond the bounds of classical monodisciplinary methods and tools for modelling and for system analysis. Because of the continuous development of monodisciplinary methods and tools, these bounds are not fixed a priori.

2.1 Classical multibody dynamics

Multibody dynamics may be considered as a typical example for continuous development and extension of monodisciplinary methods and tools. In its most simple

[1] Matlab and Simulink are trademarks of The MathWorks, Inc.

[2] Modelica is a trademark of the Modelica Association.

form, the multibody system consists of N rigid bodies that are connected by (massless) joints and by force elements. The equations of motion form a second order differential-algebraic equation (DAE) of the general form [4]

$$\boldsymbol{M}(\boldsymbol{p},t)\,\ddot{\boldsymbol{p}}(t) = \boldsymbol{h}(\boldsymbol{p},\dot{\boldsymbol{p}},t) - \boldsymbol{G}^{\top}(\boldsymbol{p},t)\boldsymbol{\lambda}, \tag{1a}$$

$$\boldsymbol{0} = \boldsymbol{g}(\boldsymbol{p},t), \tag{1b}$$

with (unknown) position coordinates $\boldsymbol{p}(t)$. The matrix $\boldsymbol{M}(\boldsymbol{p},t)$ represents the symmetric inertia matrix of the multibody system. The generalized Coriolis forces and the applied forces are summarized in $\boldsymbol{h}(\boldsymbol{p},\dot{\boldsymbol{p}},t)$.

In general, the position coordinates $\boldsymbol{p}(t) \in \mathbb{R}^{n_{\boldsymbol{p}}}$ have to satisfy $n_{\boldsymbol{\lambda}} < n_{\boldsymbol{p}}$ holonomic constraints (1b) that are coupled to the dynamical equations (1a) by constraint forces $-\boldsymbol{G}^{\top}(\boldsymbol{p},t)\boldsymbol{\lambda}$ with Lagrangian multipliers $\boldsymbol{\lambda}(t) \in \mathbb{R}^{n_{\boldsymbol{\lambda}}}$ and the constraint matrix $\boldsymbol{G}(\boldsymbol{p},t) := (\partial\boldsymbol{g}/\partial\boldsymbol{p})(\boldsymbol{p},t)$. For a minimum set of generalized coordinates $\boldsymbol{p}(t)$, there are no constraints (1b) and therefore (1) simplifies to the second order ordinary differential equation (ODE)

$$\boldsymbol{M}(\boldsymbol{p},t)\,\ddot{\boldsymbol{p}}(t) = \boldsymbol{h}(\boldsymbol{p},\dot{\boldsymbol{p}},t) \tag{2}$$

with $n_{\boldsymbol{\lambda}} = 0$.

2.2 Force elements with inner state variables

The compact form of the equations of motion (1) is favourable for textbook presentations and is used, e.g., throughout the textbook literature on time integration methods for constrained mechanical systems, see [5].

However, none of the multibody system simulation tools that are used in real-life applications has ever been restricted to the pure mechanical behaviour of conservative rigid N-body systems [6], see also [7]. Standard extensions are dissipative forces $\boldsymbol{h}(\boldsymbol{p},\dot{\boldsymbol{p}},\boldsymbol{\lambda},t)$ that may depend linearly or nonlinearly on the constraint forces $-\boldsymbol{G}^{\top}\boldsymbol{\lambda}$ (joint friction) and force elements with *inner* state variables $z(t)$ describing, e.g., the state of electrical or hydraulical system components [8]. The numerical solvers of state-of-the-art industrial multibody system tools are therefore tailored to model equations that combine constrained second order differential equations (1) with first order differential equations for $z = z(t)$, see [8, 9]:

$$\boldsymbol{M}(\boldsymbol{p},t)\ddot{\boldsymbol{p}}(t) = \boldsymbol{h}(\boldsymbol{p},\dot{\boldsymbol{p}},z,\boldsymbol{\lambda},t) - \boldsymbol{G}^{\top}(\boldsymbol{p},t)\boldsymbol{\lambda}, \tag{3a}$$

$$\dot{z}(t) = \boldsymbol{c}(\boldsymbol{p},\dot{\boldsymbol{p}},z,\boldsymbol{\lambda},t), \tag{3b}$$

$$\boldsymbol{0} = \boldsymbol{g}(\boldsymbol{p},t). \tag{3c}$$

Equations (3) are not restricted to rigid bodies. For flexible multibody systems, the vector $\boldsymbol{p}$ of position coordinates has to be extended by the coordinates that describe the deformation of the elastic bodies [10], see also Section 3 below.

2.3 Analysis of mechatronic systems by multibody system tools

The problem class that results in model equations (3) covers not only rigid and flexible multibody systems but also many *mechatronic* systems combining mechanical, electrical and hydraulical components [11]. Formally, the methods and software tools of multibody dynamics are *mono*disciplinary ones but in practice they provide also a powerful integration platform for the solution of these *multi*disciplinary problems.

The number of library elements for non-mechanical system components is, however, much smaller than for the classical mechanical ones. Often, tailormade force elements (*user elements*) have to be developed and implemented. Another common problem are interfaces for data import (model setup) and data export (postprocessing of simulation data, visualization) for non-mechanical components.

In principle, the above comments on model setup and simulation of "non-mechanical" components apply as well to any (mechanical or non-mechanical) system component that is beyond the classical fields of application of industrial multibody system tools (vehicle dynamics, dynamics of machines and mechanisms, robotics). For a typical and challenging high-end application in biomechanics the interested reader is referred to the PCM (polygonal contact model) website at `http://pcm.hippmann.org`, see also [12].

2.4 Network models and block-oriented tools

In more complex multidisciplinary applications, the extension of *one* existing monodisciplinary tool by additional user elements that handle the system components from all other disciplines is a very time-consuming approach that is furthermore prone to errors and software bugs. It is much more favourable to bundle the methods, software tools and model libraries of *several* disciplines in a unified framework.

The combination of model components from various disciplines is less complicated than one might expect at first sight. Most simulation tools in technical simulation are strongly influenced by the methods of linear and nonlinear system dynamics and follow a common network approach to set up models for complex technical systems [13].

The models are composed of fairly simple basic elements that are made available in model libraries. The basic elements are connected to each other by their inputs and outputs. There are physical laws and phenomenological and empirical rules that describe the behaviour of the individual elements. The behaviour of the full system is furthermore determined by conservation laws like conservation of energy, mass and momentum.

In multidisciplinary applications, network models from several disciplines are combined to one large model for the full system. In a first step, the multidisciplinary problem is decomposed into several monodisciplinary subproblems. Then, each of these subproblems is modelled separately following the network approach of this

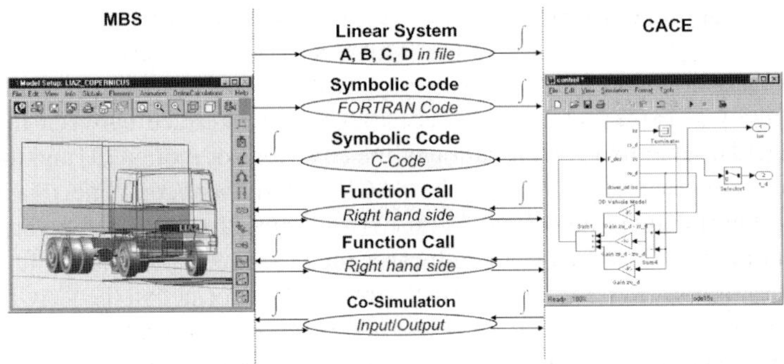

Fig. 1. Typical interfaces between multibody system tools and CACE tools [14].

specific discipline. Connecting inputs and outputs of the subsystems, the model for the full coupled system is obtained in a final third step.

In this *block-oriented* approach the multidisciplinary model is composed of monodisciplinary submodels ("blocks") and their connections.

In classical block-oriented tools like Matlab/Simulink the definition of inputs and outputs is part of the block definition itself (causal modelling). More advanced methodologies like the modelling language Modelica [1] are non-causal. They are simply based on blocks with interfaces. Input-output relations between blocks may be defined later and individually for each simulation task. The separate definition of blocks and input-output relations gives greater flexibility and offers a large potential for the re-use of well approved model components in model libraries.

2.5 Example: Controller design in vehicle system dynamics

From the view point of system dynamics, the modelling of monodisciplinary subsystems as blocks that are implemented as S-functions in Simulink and the analysis of the coupled system by Matlab/Simulink solvers is a very natural approach. The network approach is however not restricted to this implementation.

As a typical example, Figure 1 shows a simulation environment for the design and optimization of control structures in vehicle system dynamics [14], see also the detailed discussion of the controller design for a heavy duty truck with semi-active axle and cabin suspensions in [15].

The truck on the left hand side symbolizes a vehicle or vehicle component that is modelled as multibody system in an industrial multibody system tool that provides powerful model libraries and a CAD like 3D user interface for model setup and visualization of simulation data.

Modelling elements for standard controllers are available in multibody system tools. The optimization of more complex controller structures and the optimal choice of controller parameters would, however, be a very time consuming task in the working environment of a multibody system tool. Therefore it is performed in a block-oriented Computer aided control engineering (CACE) tool like Matlab/Simulink.

The first line in Figure 1 shows as most simple interface between multibody system and CACE tool the linearization of the equations of motion (3) and the export of the system matrices $\boldsymbol{A}, \boldsymbol{B}, \boldsymbol{C}, \boldsymbol{D}$ to the CACE tool. In the second line, this interface is extended to nonlinear model equations (3) exporting the FORTRAN source code of the multibody system model. In both cases the time integration for the full system including multibody system model and controller is performed in the CACE tool.

Since the solvers of CACE tools are not tailored to the differential-algebraic model equations (3) it is often more attractive to perform the time integration of the full system including the controller in the multibody system tool. State-of-the-art CACE tools support this approach by a code export interface. A typical example is the Real-Time Workshop of The MathWorks, Inc., that generates and executes stand-alone C code for developing and testing algorithms modelled in Simulink.

As an alternative to code export the tools of multibody system analysis and CACE may be coupled as well by a function call interface that evaluates during time integration repeatedly the right hand sides of the model equations for various given input data, see [14] for a more detailed discussion. All these approaches have in common that the full set of model equations is solved numerically in *one* simulation tool by *one* solver.

The last line in Figure 1 indicates, that the modular structure of the coupled problem may also be exploited by coupling the multibody system tool and the CACE tool for model setup *and* for time integration (*co-simulation* or *simulator coupling*).

Co-simulation

Co-simulation is a rather general approach for the analysis of multidisciplinary problems [16]. As before, the coupled problem is decomposed into a set of monodisciplinary subproblems. Well established monodisciplinary tools are used for the model setup of the subproblems. Additionally, these monodisciplinary tools are coupled by a co-simulation interface to handle each subproblem during time integration by its own solver in its own monodisciplinary tool. Therefore the subproblems are solved by *different* solvers and each solver may be tailored to its specific subproblem.

The communication between subsystems is restricted to discrete synchronization points T_j. For each subsystem all necessary information from other subsystems has to be provided by interpolation or – if data for interpolation have not yet been computed – by extrapolation from $t \leq T_j$ to the current *macro step* $T_j \to T_{j+1}$. Typical macro stepsizes $H = T_{j+1} - T_j$ are in the range of 1.0 ms.

From a practical viewpoint, co-simulation is very attractive for teams of specialists who are experts in their individual monodisciplinary tools because each expert may remain in his usual working environment. From the numerical viewpoint, the separate time integration of subsystems in a co-simulation framework needs special care.

The additional errors that are introduced by the interpolation and extrapolation steps for data exchange between subsystems have to be kept small, e.g., by higher order inter- and extrapolation schemes and sufficiently small macro stepsizes H. Furthermore, the extrapolation steps make the time integration method in part explicit. Special stabilization techniques have been proposed to avoid numerical instability [16, 17].

3 A Modal Multifield Approach for Thermoelastic Problems

Coupled field effects are a classical application of the finite element (FE) method [2, 18]. A typical example are thermoelastic effects like the elastic deformation of a body due to thermal expansion.

State-of-the-art FE tools provide all necessary elements and tools for model setup, numerical solution and pre- and postprocessing of thermoelastic problems. The computational effort is, however, in the typical range of a FE analysis. Furthermore, the modelling of joints for connecting bodies is substantially more complicated than in multibody dynamics [19] and the coupling to force elements with inner state variables $z(t)$ resulting in additional first order differential equations (3b) is less straightforward than in network models, see also [20] for ongoing research in this field.

The key to the efficient simulation of flexible multibody systems are low dimensional modal representations that describe the elastic deformation of the bodies [10]. The positive experience with this approach motivates the development of a modal multifield representation for the distributed phenomenon thermoelasticity [21], see also [3].

3.1 Thermoelastic problems in multibody systems

From the thermodynamic point of view, the deformations of a flexible body in multibody simulation are usually assumed to proceed isothermally and adiabatically. Even though this concept is thermodynamic contradictory, it proved to be an adequate description for most problems in multibody dynamics.

However, if a mechanical process is associated with a remarkable heat generation or load, the validity of these premises has to be reviewed. In a wide range of applications such as friction brakes, thermal buckling phenomena, machine tools with thermal loads, micro-mechanical devices with resistive heating, the heat energy flow and the thermoelastic coupling cannot be ignored or are even of major concern. Elaborate simulation environments are required for these applications.

The design of high performance machine tools is an appropriate example that will be considered in more detail in Section 4 below. Working tasks in this field combine high speed motion with high demands on the accuracy. But the unavoidable losses in power transmission and the heat generation due to the working task necessarily lead to thermal loads. Industrial experience shows that beyond a specific level additional quality improvements require a combined elastic and thermal description of the system.

It has been state-of-the-art to investigate the coupled thermal and thermoelastic behaviour in elaborate finite element studies that provide high resolution results and give essential information on the design of machine components. However, the large computational effort prevents the application of the finite element method for a system dynamical analysis of the complete system. And looking one step further, a control setup which accounts for thermally induced tool center point displacements can only be built up on an efficient multifield description.

3.2 Material constitution and weak field equations

Thermoelasticity deals with two physical fields, each one specified by a pair of field variable terms. The mechanical state of a material particle is quantified by its stress tensor $\boldsymbol{\sigma}$ and its strain tensor ε and the thermal state by its temperature Θ and entropy density η.

In order to describe the properties and the influence of the material, it is presumed that the thermodynamic state of the material only depends on the current values of the field variables but not on their histories. The constitutive relation between the four field terms is supposed to define the thermodynamic state of a material point uniquely, no matter which process, which change of state variables has led to the current configuration.

Consequently it makes sense to base the material constitution on a thermodynamic potential. If strain ε and temperature Θ are chosen as independent variables, the free energy F arises as associate function

$$\mathrm{d}F = \boldsymbol{\sigma}^\top \mathrm{d}\varepsilon - \eta\, \mathrm{d}\Theta,$$

see [22]. In practice, the introduction of a new variable ϑ, replacing the absolute temperature Θ by the increment w.r.t. the linearisation temperature Θ_0 proved to be advantageous:

$$\vartheta = \Theta - \Theta_0.$$

The free energy, approximated by its second order Taylor expansion at a natural state, in which ϑ and ε vanish, enables the formulation of a linear constitutive equation in matrix form:

$$\begin{pmatrix} \sigma \\ \eta \end{pmatrix} = \begin{pmatrix} \boldsymbol{H}_c & -\boldsymbol{H}_\lambda^\top \\ \boldsymbol{H}_\lambda & \boldsymbol{H}_a \end{pmatrix} \begin{pmatrix} \varepsilon \\ \vartheta \end{pmatrix} = \boldsymbol{H} \begin{pmatrix} \varepsilon \\ \vartheta \end{pmatrix}. \tag{4}$$

The main diagonal blocks of $\boldsymbol{H}$ specify the material properties of the monodisciplinary effects. $\boldsymbol{H}_c$ can be identified as the classical 6 by 6 elasticity tensor relating stress to strain. $\boldsymbol{H}_a = \varrho c/\Theta_0$ involves the specific heat capacity c, the density ϱ and Θ_0 to relate temperature and entropy density.

In most of the usual engineering applications, the influence of the mechanical on the thermal state, the so-called Gough–Joule effect [23], may be neglected. Then the first row of (4) may be rewritten to extract the isotropic thermal strain ε_ϑ, see Equation (4.26) in [2, Vol. 1]:

$$\boldsymbol{\sigma} = \boldsymbol{H}_c(\varepsilon - \varepsilon_\vartheta) \qquad \text{with} \qquad \varepsilon_\vartheta = \boldsymbol{H}_c^{-1}\boldsymbol{H}_\lambda^\top \vartheta = (\alpha\ \alpha\ \alpha\ 0\ 0\ 0)^\top \vartheta. \tag{5}$$

Here, α denotes the thermal expansion coefficient.

The weak equation for the absolute position of a particle $\boldsymbol{r}(\boldsymbol{c}, t)$ as function of the Lagrange coordinate $\boldsymbol{c}$ and time t may be deduced from d'Alembert's principle, see, e.g., Equation (1.6) in [2, Vol. 2]:

$$\int_V [-\varrho\delta\boldsymbol{r}^\top\ddot{\boldsymbol{r}} - \boldsymbol{\sigma}^\top\delta\varepsilon + \boldsymbol{f}_V^\top\delta\boldsymbol{r}]\,\mathrm{d}V + \oint_B \boldsymbol{f}_B^\top\delta\boldsymbol{r}\,\mathrm{d}B = 0. \tag{6a}$$

$\boldsymbol{f}_V$ and $\boldsymbol{f}_B$ denote external forces acting on the volume element $\mathrm{d}V$ or boundary element $\mathrm{d}B$, respectively.

The weak equation of the temperature field results from the principle of virtual temperature [24]

$$\int_V [-(\nabla\delta\Theta)^\top\boldsymbol{q} + (\Theta\dot{\eta} - S)\delta\Theta]\,\mathrm{d}V + \oint_B \boldsymbol{q}_B^\top\boldsymbol{n}_B\delta\Theta\,\mathrm{d}B = 0, \tag{6b}$$

where $\boldsymbol{q}$ denotes the heat flow, S symbolises the heat source density and $\boldsymbol{q}_B$ represents the heat flow at the boundary element $\mathrm{d}B$ with the outer unit normal vector $\boldsymbol{n}_B$. BIOT referred to (6b) as the complementary variational principle in heat transfer [25].

On first sight the equations (6a) and (6b) look like two uncoupled field descriptions from monodisciplinary engineering textbooks. But the coupling becomes obvious by eliminating the dependent field variables using (4).

3.3 A modal multifield approach for thermoelastic problems

The kinematical description is based on a floating frame of reference formulation [10] and gets the form

$$\boldsymbol{r} = \boldsymbol{r}_R + \boldsymbol{c} + \boldsymbol{u}. \tag{7}$$

The position vector $\boldsymbol{r}$ is decomposed into the absolute position vector of the floating frame of reference $\boldsymbol{r}_R$, the Lagrange coordinate of a particle $\boldsymbol{c}$ and its displacement $\boldsymbol{u}$. All vectors in (7) are resolved w.r.t. the body's frame of reference.

The displacement $\boldsymbol{u} = \boldsymbol{u}(\boldsymbol{c}, t)$ is described with separated variables as product of time independent modal functions $\boldsymbol{\Phi}_u(\boldsymbol{c})$ and coefficients $z_u(t)$. Within this approximation the evaluation of the strain field is feasible by means of the differential operator ∇_u, see Equation (6.9) in [2, Vol. 1]:

$$\boldsymbol{u} = \boldsymbol{\Phi}_u z_u, \qquad \varepsilon = \nabla_u \boldsymbol{u} = (\nabla_u \boldsymbol{\Phi}_u) z_u = \boldsymbol{B}_u z_u. \tag{8}$$

The analogous approach is chosen for the scalar temperature field. According to Fourier's law of heat conduction, a ∇-operation multiplied by the thermal conductivity matrix $\boldsymbol{\Lambda}$ leads to the heat flux vector $\boldsymbol{q}$, see [26]:

$$\vartheta = \boldsymbol{\Phi}_\vartheta z_\vartheta, \qquad \nabla\vartheta = (\nabla\boldsymbol{\Phi}_\vartheta) z_\vartheta = \boldsymbol{B}_\vartheta z_\vartheta \quad \Rightarrow \quad \boldsymbol{q} = -\boldsymbol{\Lambda}\boldsymbol{B}_\vartheta z_\vartheta. \tag{9}$$

Now matrices $\boldsymbol{K}_{uu}$ and $\boldsymbol{K}_{u\vartheta}$ are introduced for volume dependent integrals which can be preprocessed and accessed during time integration:

$$\boldsymbol{K}_{uu} := \int_V \boldsymbol{B}_u^\top \boldsymbol{H}_c \boldsymbol{B}_u \, \mathrm{d}V, \qquad \boldsymbol{K}_{u\vartheta} := \int_V \boldsymbol{B}_u^\top \boldsymbol{H}_\lambda^\top \boldsymbol{\Phi}_\vartheta \, \mathrm{d}V. \tag{10}$$

From the mechanical point of view the thermal field generates internal, distributed mechanical loads. Obviously there is no direct influence on the inertia properties of the body. That is why the mass, gyroscopic and centripetal terms within the equations of motion can be adopted from literature.

3.4 Model equations for a thermoelastic body

The generalised Newton–Euler equations for the unconstrained motion of a deformable body that undergoes large reference displacements are given in [10] and [27]. A comparison of (6a) with these references yields the extended equations of motion

$$\begin{pmatrix} \boldsymbol{M}_{aa} & \boldsymbol{M}_{a\alpha} & \boldsymbol{M}_{au} \\ & \boldsymbol{M}_{\alpha\alpha} & \boldsymbol{M}_{\alpha u} \\ \text{sym.} & & \boldsymbol{M}_{uu} \end{pmatrix} \begin{pmatrix} \boldsymbol{a}_R \\ \alpha_R \\ \ddot{z}_u \end{pmatrix} = \begin{pmatrix} \boldsymbol{h}_a \\ \boldsymbol{h}_\alpha \\ \boldsymbol{h}_u \end{pmatrix} + \begin{pmatrix} 0 \\ 0 \\ -\boldsymbol{K}_{uu} z_u + \boldsymbol{K}_{u\vartheta} z_\vartheta \end{pmatrix}. \tag{11}$$

The mass matrix on the left hand side of (11) is formulated as 3×3 block matrix with submatrices that specify the inertia coupling between the translational acceleration $\boldsymbol{a}_R$ of the body's reference frame, the angular acceleration α_R of the reference frame and the second time derivative of the elastic coordinates $\ddot{z}_u$. The right hand side terms $\boldsymbol{h}_a$, $\boldsymbol{h}_\alpha$ and $\boldsymbol{h}_u$ summarise all inertia, damping and external forces.

The added term $\boldsymbol{K}_{u\vartheta} z_\vartheta$ represents the influence of the thermal field on the equations of motion. It may be interpreted as modal force acting on the elastic body.

Although the thermal loads do not influence the inertia quantities in (11), the displacements caused by these loads do, since the mass matrix and the vectors $\boldsymbol{h}_a$ and $\boldsymbol{h}_\alpha$ depend on the deformation state of the body.

The model equations (11) of the elastic body are extended by the thermal equation that describes the changes of the thermal state variables $z_\vartheta(t)$. In (6b), the natural boundary conditions are represented by the heat flux through the boundary surface. It depends on the physical circumstances how this term has to be introduced into the thermal equation.

For Neumann conditions the boundary heat flux q_B is given explicitly. If convection occurs on the boundary surface, a Robin or mixed boundary condition is imposed, specified by the film coefficient h_f and the bulk temperature ϑ_∞ of the fluid [26]. Although this list is not complete, we confine ourselves to these two cases:

$$\boldsymbol{q}_B^\top \boldsymbol{n}_B = -q_B - h_f(\vartheta_B - \vartheta_\infty). \tag{12}$$

In addition to the thermal-mechanical coupling matrix $\boldsymbol{C}_{\vartheta u} = \Theta_0 \boldsymbol{K}_{u\vartheta}^\top$, the following notations are used for geometric integrals:

$$\boldsymbol{C}_{\vartheta\vartheta} := \int_V \Theta_0 \boldsymbol{\Phi}_\vartheta^\top \boldsymbol{H}_a \boldsymbol{\Phi}_\vartheta \, \mathrm{d}V, \quad \boldsymbol{K}_{\vartheta R} := \oint_B h_f \boldsymbol{\Phi}_\vartheta^\top \boldsymbol{\Phi}_\vartheta \, \mathrm{d}B, \quad \boldsymbol{Q}_{\vartheta R} := \oint_B \boldsymbol{\Phi}_\vartheta^\top h_f \, \mathrm{d}B,$$

$$\boldsymbol{K}_{\vartheta\vartheta} := \int_V \boldsymbol{B}_\vartheta^\top \boldsymbol{\Lambda} \boldsymbol{B}_\vartheta \, \mathrm{d}V, \qquad \boldsymbol{Q}_{\vartheta S} := \int_V \boldsymbol{\Phi}_\vartheta^\top \, \mathrm{d}V, \qquad \boldsymbol{Q}_{\vartheta N} := \oint_B \boldsymbol{\Phi}_\vartheta^\top \, \mathrm{d}B.$$

With these notations, the coupled linearised thermal equation is given by

$$\boldsymbol{C}_{\vartheta\vartheta}\dot{\boldsymbol{z}}_\vartheta + \boldsymbol{C}_{\vartheta u}\dot{\boldsymbol{z}}_u + (\boldsymbol{K}_{\vartheta\vartheta} + \boldsymbol{K}_{\vartheta R})\boldsymbol{z}_\vartheta = \boldsymbol{Q}_{\vartheta S}\, S_u + \boldsymbol{Q}_{\vartheta N}\, q_B + \boldsymbol{Q}_{\vartheta R}\, \vartheta_\infty. \tag{13}$$

The generalised velocities $\dot{z}_u$ in (13) indicate that the temperature field depends on the displacements and the strains. Whereas the thermal effect on the displacements is well-established and widely accounted for in finite element analysis, the feedback from displacements on temperatures (Gough–Joule effect) is frequently neglected, see the above comment on (5) and the more detailed discussion in [3].

Equations (11) and (13) are the model equations for a general elastic and heat conducting body $(\ )^{(j)}$. Compared to a pure mechanical description, the model setup of a thermoelastic body requires the definition of one additional, uniquely assigned element. The thermal element reflects (13) and evaluates the thermal state of the body presuming that the thermal field of body $(\ )^{(j)}$ does not interfere with those of other bodies. If the thermal fields of two bodies interact, the mutual influence has to be modelled explicitly defining appropriate boundary conditions (12).

Based on (11) and (13), the model equations (3) of a multibody system with thermoelastic bodies may be assembled by the classical algorithms of multibody dynamics, see, e.g., [28]. In (3), the vector $\boldsymbol{p}$ of position coordinates contains the position coordinates of all rigid bodies and the position coordinates of the reference frames and the elastic coordinates z_u for all (thermo-)elastic bodies. The thermal state variables z_ϑ of all thermoelastic bodies are components of the vector z of first order state variables in (3b), see (13).

3.5 Thermal response modes

The proper selection of modal functions $\boldsymbol{\Phi}_u(\boldsymbol{c})$ and $\boldsymbol{\Phi}_\vartheta(\boldsymbol{c})$ in (8) and (9) is essential for accuracy and efficiency of the modal multifield approach. Classical mode functions $\boldsymbol{\Phi}_u$ of flexible multibody dynamics like eigenmodes, static modes and frequency response modes [29] have to be accompanied with modal functions that consider the coupling of displacement and temperature field in a thermoelastic body.

A detailed investigation of the thermoelastic coupling motivates substantial simplifications, see [21] and Section 3 of [3]. The first simplification is the separate consideration of the thermal eigenvalue problem and the mechanical eigenvalue problem. Furthermore, all inertia terms that correspond to deflections due to thermal loads may be neglected.

These considerations lead to a specific modal reduction scheme that assigns to each temperature mode $\vartheta_i(\boldsymbol{c})$, i.e., to each column of $\boldsymbol{\Phi}_\vartheta = [\ldots\ \vartheta_i(\boldsymbol{c})\ \ldots]$, a displacement mode $\boldsymbol{u}_i(\boldsymbol{c})$ in $\boldsymbol{\Phi}_u = [\ldots\ \boldsymbol{u}_i(\boldsymbol{c})\ \ldots]$. These *thermal response modes* $\boldsymbol{u}_i(\boldsymbol{c})$ have to satisfy

$$\nabla_u \boldsymbol{u}_i(\boldsymbol{c}) = \boldsymbol{H}_c^{-1}\boldsymbol{H}_\lambda^\top \vartheta_i(\boldsymbol{c}) = (\alpha\ \alpha\ \alpha\ 0\ 0\ 0)^\top \vartheta_i(\boldsymbol{c}), \tag{14}$$

see (5) and (8). Each thermal response mode corresponds to exactly one temperature mode and represents the related particular thermoelastic displacement solution. The strain field of this mechanical mode reflects the thermal strain field that corresponds to the related thermal mode.

For bodies with "simple" geometry, the evaluation of $\boldsymbol{\Phi}_u$ with columns $\boldsymbol{u}_i$ satisfying (14) could in principle be performed analytically. For bodies with complex geometry, the thermal mode functions $\vartheta_i(\boldsymbol{c})$ and the corresponding thermal response modes $\boldsymbol{u}_i(\boldsymbol{c})$ are computed by a finite element analysis in a preprocessing step, see [21, Section 3.2.2].

In this approach, the only remaining crucial point is the definition of the thermal mode functions $\vartheta_i(\boldsymbol{c})$ that has to be tailored to the specific modelling task. The evaluation of the corresponding thermal response modes $\boldsymbol{u}_i(\boldsymbol{c})$ by a finite element analysis and the computation of the system matrices $\boldsymbol{K}_{uu}, \ldots$ in (10) are straightforward and may be organized as an automated process.

The accuracy and the convergence properties of the modal multifield representation rely on an appropriate thermal field approximation. On the other hand, a substantial reduction of the number of degrees of freedom may be achieved that way. We refer again to [21] for more details on the definition of thermal response modes and some verification examples that demonstrate their application.

In the present paper the definition of thermal response modes is illustrated by the finite element model of a circular disc shown in Figure 2 that has been selected as sample structure. Figure 3 shows the thermal response modes by the deformed mesh compared to the undeformed outer circle contour. For this sample structure, the thermal response modes turned out to be orthogonal and a mechanical frequency $\bar{\omega}_i$ could be assigned to each thermal response mode.

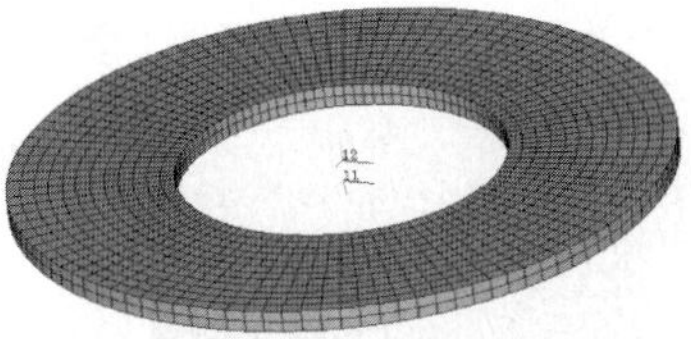

Fig. 2. Finite element model illustrating the definition of thermal response modes.

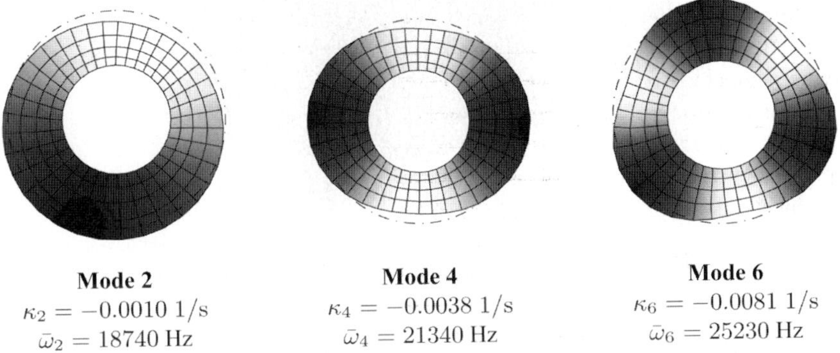

Fig. 3. Thermal modes and thermal eigenvalues κ_i, thermal response modes and associated mechanical frequencies $\bar{\omega}_i$ of the sample structure in Fig. 2.

4 Case Study: A Machine Tool with Thermoelastic Deformations

Modern machine tool drives show excellent dynamical properties and allow high accelerations of slides and tool heads. However, for point-to-point working tasks the accumulation of high power inputs near frequently used start and stop positions cannot be avoided for physical reasons. Due to performance losses localised thermal loads may be generated and result in an inhomogeneous temperature field of the machine base or other machine components.

The corresponding inhomogeneous thermal expansion causes tool centre point displacements that are difficult to be measured. In industrial applications these thermally induced displacements are either accepted to be unavoidable or costly cooling devices are designed to ensure a homogeneous thermal state of the machine. However, with increasing demands on the economic efficiency and the accuracy there is a necessity for smart, mechatronic concepts to handle this problem for future generations of high accuracy machine tools.

As a first step towards such a mechatronic concept, an industrial multibody simulation environment has been extended to deal with the thermoelastic deformation of machine tools. Therefore, the methodological base is provided to develop new measurement and control strategies that account for thermally induced displacements.

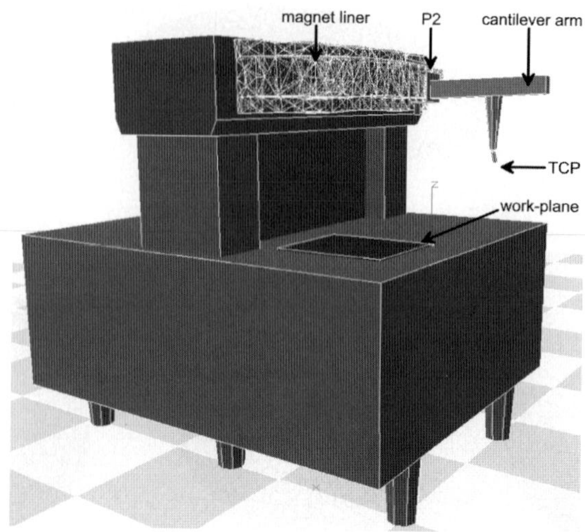

Fig. 4. 3D visualization of the machine tool in SIMPACK.

4.1 Simulation scenario

The feasibility study was defined in cooperation with an industrial partner who provided a finite element model of the machine base. The welded construction of the machine is sketched in the SIMPACK model in Figure 4.

The machine is symmetric w.r.t. a vertical plane and is assembled with two cantilever arms, one at each side. Each cantilever arm is driven by a linear induction device and moves along a magnet liner which is parallel to the y-axis of the machine. The cantilever carries x- and z-drives and the tool head with the tool centre point (TCP) at its tip. The flat workpiece on the machine table in Figure 4 demonstrates the position of the work plane.

This study is based on the assumption that the working task of the machine is repeated very often and varies periodically. The objective of the study is to reproduce a constant thermal operating state of the machine that is reached after a sufficiently large time span. This is a frequently observed operating condition in the industrial use of machine tools.

Figure 5 shows the positioning loop of the cantilever arm along the y-axis that was predefined and taken as the starting point of the feasibility study.

In order to model the thermal behaviour, a heat source of intensity $\bar{q}$ at the position $\bar{y} = \bar{y}(t)$ is considered to move along an isotropic one-dimensional continuum, described by the coordinate y. For mathematical representation, the formulation of a point source by means of the Dirac function $\eth(y - \bar{y})$, see [22], is extended by a term that accounts for the geometrical dimensions of the heat source, i.e., the drive head on the cantilever arm. The heat flux is assumed to be distributed as a Gaussian

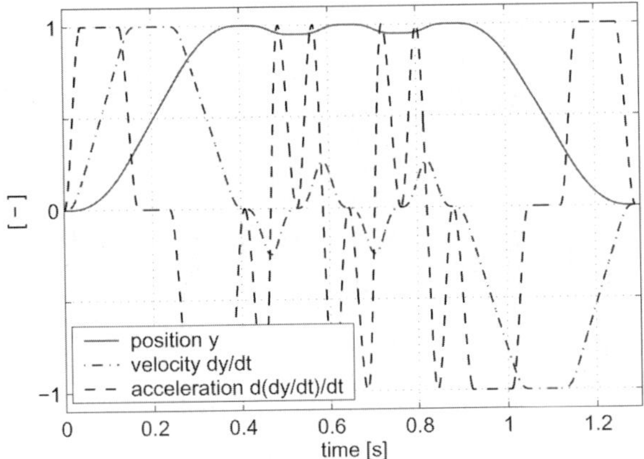

Fig. 5. Predefined positioning loop $\bar{y} = \bar{y}(t)$ and its time derivatives (normalised).

bell shaped curve with a parameter a that reflects the length of the drive head:

$$-\Lambda\vartheta_{,yy} + \varrho c\dot{\vartheta} = Q(t, y, \bar{y}) = \begin{cases} \bar{q}\ \eth(y - \bar{y}) & \text{point source,} \\ \bar{q}\ \sqrt{\frac{a}{\pi}}\ \exp(-a(y - \bar{y})^2) & \text{distributed source.} \end{cases} \tag{15}$$

For stationary hot running conditions the time dependent terms in (15) have to vanish. The localised heat supply $q = q(y)$ is obtained as time average over one positioning loop with period T:

$$\left.\begin{aligned} \bar{q}(t) &= \bar{q}(t + nT) \\ \bar{y}(t) &= \bar{y}(t + nT) \\ n &\to \infty \end{aligned}\right\} \Rightarrow\ -\Lambda\vartheta_{,yy} \approx \frac{1}{T}\int_0^T Q(t, y, \bar{y})\ \mathrm{d}t = q(y). \tag{16}$$

Since the specific design of the cantilever suspension involves only very small frictional forces, the mechanical power is almost completely invested into the kinetic energy of the cantilever arm and may be easily described based on the predefined kinematic scenario in Figure 5. It is assumed, that a constant share of the consumed electrical power is transformed into heat energy and conducted to the surface of the machine base. Then, the localised heat supply $q = q(y)$ is completely defined by (16).

Figure 6 presents the mechanical power consumption versus the relative position of the cantilever arm on the magnet liner with the time as curve parameter. The start and stop positions are denoted by the relative position 0 and 1 respectively.

The power consumption, specified by the instantaneous product of mass m, velocity $\dot{\bar{y}}$ and acceleration $\ddot{\bar{y}}$ from Figure 5, has distinct maxima in the neighbourhood of start and stop positions. Therefore, the quasi-stationary heat flux accumulates at

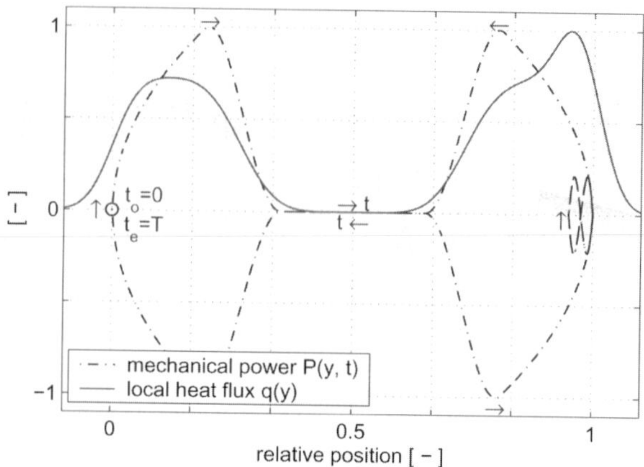

Fig. 6. Mechanical power consumption and quasi-stationary heat flux.

specific positions on the magnet liner. On the other hand, there are no heat loads at those parts of the magnet liner at which the cantilever arm moves with constant velocity.

4.2 Finite element analysis

The thermal finite element model of the machine base consists of 20 641 tetrahedral shaped elements of type Solid90 [30] with 40 471 nodes or thermal degrees of freedom, respectively.

A steady state heat transfer FE analysis has been performed using the analytical heat source introduced above as quasi-stationary load. The solid curve in Figure 6 is taken as heat flux distribution in y-direction along the upper surface of the magnet liner, which is visualised in Figure 7. In the ξ-direction on the upper surface, the heat flux is modelled to be constant. Robin boundary conditions are defined on the complete surface of the machine base with two different film coefficients to reflect different cooling conditions due to the air-stream forced by the moving cantilever.

The heat transfer analysis is performed separately for each cantilever drive at both sides of the machine. Since the thermal as well as the subsequent mechanical structural analysis are linear, the solutions may be superimposed. That way the model definition is open to consider any linear load combination caused by the two cantilever drives. For ease of interpretation the results to present from now on refer to a single load scenario, i.e., the second cantilever drive on the backside of the machine is assumed to be out of use.

Figure 7 presents the results of one heat transfer FE analysis. Since the temperatures mainly vary on the magnet liner while the other elements of the machine base show only small temperature differences, Figure 7 only visualises the temperature

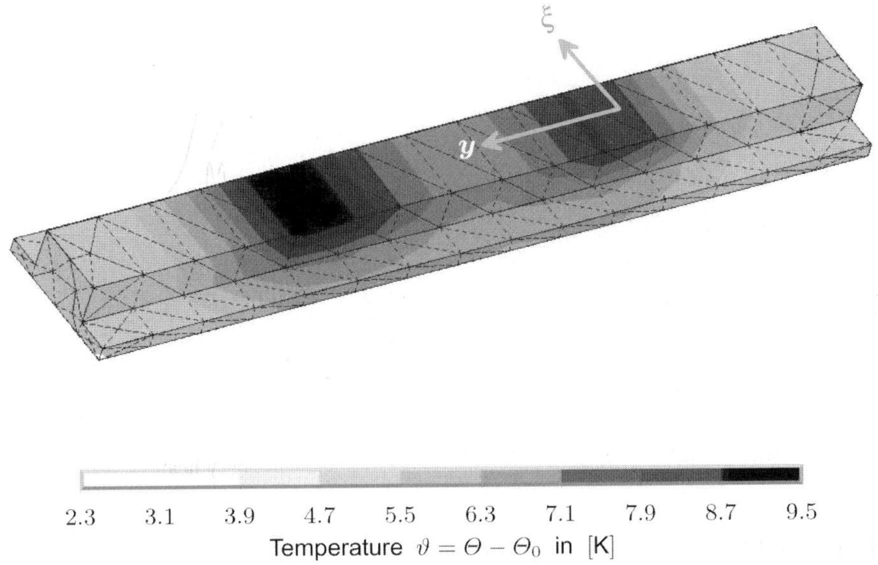

Fig. 7. Temperature field at the magnet liner, obtained by finite element analysis.

field of the magnet liner. Two distinct temperature maxima on the magnet liner are clearly visible.

The two FE temperature field solutions of the complete machine base have been applied as separate thermal loads on the mechanical finite element model of the base structure, which uses the same mesh as the thermal FE-model. However, the mesh now specifies 20 641 elements of type Solid95 [30] with 121 413 mechanical degrees of freedom. The solutions of these steady state FE analysis yield the displacement fields of the machine base caused by the temperature fields and are interpreted as thermal response modes according to Section 3.

Figure 8 plots the thermal displacements of a reference point on the deformed surface of the magnet liner, which moves along the motion path of the cantilever arm with constant velocity from its start to its stop position. The FE results along the motion path are compared with the corresponding multibody deformations modelled by thermal response modes.

In addition to the steady state analysis, a finite element eigenvalue analysis is performed and 27 eigenmodes of the machine base are obtained. That way, the dynamical properties of the mechanical structure up to the frequency of 400 Hz are considered.

A unified set of modes consisting of 27 mechanical eigenmodes and two thermal response modes has been used to reduce the finite element description of the machine base by the modal multifield representation according to Section 3. Eigenmodes and

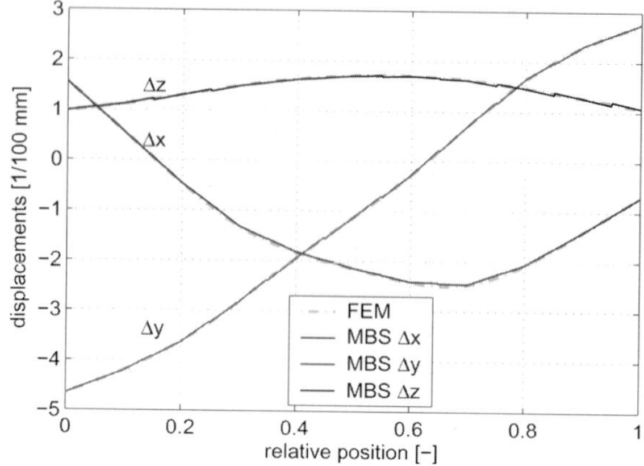

Fig. 8. Kinematic comparison of the thermal displacements as evaluated by FEM and the corresponding thermal response mode used for multibody simulation (MBS).

thermal response modes are only weakly coupled, but the unified set of modes is not orthogonal with respect to the mass and the stiffness matrix. In view of the fact that both groups of modes contain a different physical information which is worth to be retained, the unified set of modes has not been orthogonalised for the multibody simulation.

On a trial basis a supplementary orthogonalisation of the 29 modes has been performed yielding 1 694 Hz and 2 368 Hz as additional frequencies due to the thermal response modes.

4.3 Multibody simulation

Figure 4 shows the principle structure of the multibody model and the FE mesh of the magnet liner. Also the complete machine base originates from the FE model and is mechanically represented as flexible body in modal representation.

Since the machine base rests on six feets, which are not fixed to the foundation, its reference frame has three degrees of freedom that allow a plane motion of the machine base frame w.r.t. the inertial frame. Six stick-slip force elements reflect the dry friction conditions between machine base and ground.

The suspension of the cantilever arm is modelled by spring-damper elements, which connect the deformed magnet liner and the base of the cantilever arm. The cantilever arm itself is assumed to be rigid. Since the arm moves along the liner, so called moved markers [31] have been defined that represent the working points of the suspension forces at the machine base. A moved marker is used as well to model the reference point for Figure 8.

In order to simulate the working task of the machine tool, a controller for the y-drive is modelled. The kinematic scenario from Figure 5 serves as target specific-

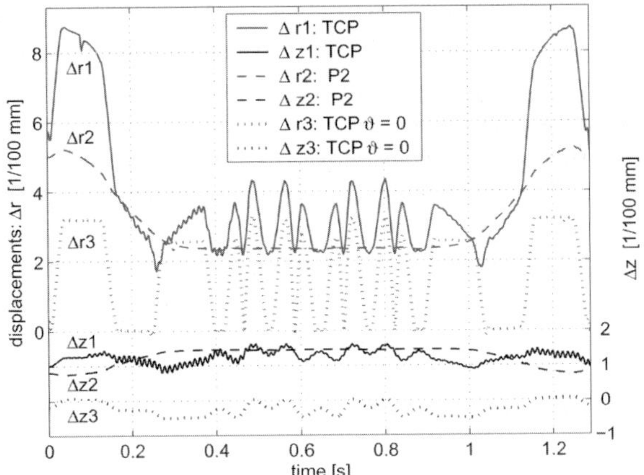

Fig. 9. Displacements at the tool center point (TCP) in solid lines and at the reference point P2 on the magnet liner in dashed lines. $\Delta r_{\ldots}$ denotes the absolute displacement parallel to the work plane, $\Delta z_{\ldots}$ is measured normal to the work plane. The third couple of curves in dotted lines visualises the TCP displacements in a simulation without any thermal loads ($\vartheta = 0$).

ation of the control loop, which is adjusted in such a way that the positioning error induced by the drive control is at least one order of magnitude smaller than the other displacements and cannot falsify the results.

In Figure 9 three different measurements are compared. All results represent displacements w.r.t. the workpiece on the table of the machine tool.

The dotted curves give the tool centre point (TCP) displacements of a multibody simulation without any thermal loads and serve as a reference. These displacements are only caused by the response of the machine base structure and the cantilever suspension to the dynamical loads given by the predefined kinematic scenario.

The other two measurement types in Figure 9 additionally involve the displacements which are induced by the temperature field of the machine base.

The dashed curves plot the displacements of the reference point P2 on the magnet liner, which moves with the cantilever arm. The solid curves again give the displacements of the TCP at the tip of the cantilever arm. The difference between the P2- and TCP-displacements are caused by the kinematic amplification of the cantilever arm.

The thermally induced displacements influence the motion of the TCP mostly in the neighbourhood of the start position at the beginning and at the end of the simulation. However, since the working task consists of a point-to-point job, these deviations are not crucial in this case. More important are the deflections at the stop position, which is reached several times in the time interval between 0.4 s and 0.9 s. The dotted and the solid curves differ here by about 10 μm to 20 μm, which is a relevant error concerning the accuracy requirements of machine tools.

Besides the thermal deflection, the response of the machine base structure corresponds mainly to the acceleration curve in Figure 5 and is of static nature. The vibrations in Figure 9 primarily originate from the compliance of the cantilever suspension. This statement could be verified by an accompanying simulation, for which the compliance of the cantilever suspension is neglected. The structural damping of the machine base has also no significant influence (Lehr's damping coefficient $d = 0.004$).

The time integration has spent 3 580 CPU-s on a HP 9000/785 workstation with 3 GB memory. This high computational effort is caused by the high frequency band width of the multibody model. Since the inertia terms that correspond to the thermal response modes have not been neglected for this simulation, frequencies up to 2 400 Hz are present, which leads to a very stiff system [5].

5 Summary

The rapidly growing interest in multidisciplinary applications is motivated by the increasing importance of coupled physical effects and strongly coupled technical components in the design and optimization of advanced engineering systems. The positive experience with methods and tools of multibody dynamics shows that multidisciplinary problems may often be analysed efficiently by suitable extensions of classical monodisciplinary simulation tools or by the combination of two or more monodisciplinary tools in a block-oriented simulation environment.

Furthermore, modal reduction methods that are well approved in the analysis of flexible multibody systems have been generalized to a modal multifield approach that is successfully used in the analysis of thermoelastic problems. Its practical use has been illustrated by an advanced case study that considers positioning errors of a high-performance machine tool being caused by thermal loads.

References

1. M.M. Tiller. *Introduction to Physical Modeling with Modelica.* Kluwer Academic Publishers, Boston/Dordrecht/London, 2001.
2. O.C. Zienkiewicz and R.L. Taylor. *The Finite Element Method.* Butterworth Heinemann, Oxford, 5th edition, 2000.
3. A. Heckmann and M. Arnold. Flexible bodies with thermoelastic properties in multibody dynamics. In J.M. Goicolea, J. Cuadrado, and J.C. García Orden (eds), *Proc. of Multibody Dynamics 2005 (ECCOMAS Thematic Conference)*, Madrid, Spain, 2005.
4. R.E. Roberson and R. Schwertassek. *Dynamics of Multibody Systems.* Springer-Verlag, Berlin/Heidelberg/New York, 1988.
5. E. Hairer and G. Wanner. *Solving Ordinary Differential Equations. II. Stiff and Differential-Algebraic Problems.* Springer-Verlag, Berlin/Heidelberg/New York, 2nd edition, 1996.

6. W.O. Schiehlen (ed.). *Multibody Systems Handbook.* Springer-Verlag, Berlin/Heidelberg/New York, 1990.
7. W. Kortüm, W.O. Schiehlen, and M. Arnold. Software tools: From multibody system analysis to vehicle system dynamics. In H. Aref and J.W. Phillips (eds), *Mechanics for a New Millennium*, pages 225–238, Kluwer Academic Publishers, Dordrecht, 2001.
8. E. Eich-Soellner and C. Führer. *Numerical Methods in Multibody Dynamics.* Teubner-Verlag, Stuttgart, 1998.
9. M. Arnold. Simulation algorithms and software tools. Accepted for publication in G. Mastinu and M. Plöchl (eds), *Road and Off-Road Vehicle System Dynamics Handbook.* Taylor & Francis, London, 2006.
10. A.A. Shabana. *Dynamics of Multibody Systems.* Cambridge University Press, Cambridge, 2nd edition, 1998.
11. W. Kortüm, R.M. Goodall, and J.K. Hedrick. Mechatronics in ground transportation – Current trends and future possibilities. *Annual Reviews in Control*, 22:133–144, 1998.
12. G. Hippmann. An algorithm for compliant contact between complexly shaped bodies. *Multibody System Dynamics*, 12:345–362, 2004.
13. M. Hoschek, P. Rentrop, and Y. Wagner. Network approach and differential-algebraic systems in technical applications. *Surveys on Math. in Industry*, 9:49–75, 1999.
14. O. Vaculín, M. Valášek, and W.R. Krüger. Overview of coupling of multibody and control engineering tools. *Vehicle System Dynamics*, 41:415–429, 2004.
15. O. Vaculín, M. Valášek, and W. Kortüm. Multi-objective semi-active truck suspension by spatial decomposition. In H. True (ed.), *The Dynamics of Vehicles on Roads and on Tracks, Proc. of the 17th IAVSD Symposium, Denmark, 20-24 August 2001*, pages 432–440. Supplement to Vehicle System Dynamics, Vol. 37, Swets & Zeitlinger, 2003.
16. R. Kübler and W. Schiehlen. Modular simulation in multibody system dynamics. *Multibody System Dynamics*, 4:107–127, 2000.
17. M. Arnold and M. Günther. Preconditioned dynamic iteration for coupled differential-algebraic systems. *BIT Numerical Mathematics*, 41:1–25, 2001.
18. O.C. Zienkiewicz. Coupled problems and their numerical solution. In R.W. Lewis, P. Bettess, and E. Hinton (eds), *Numerical Methods in Coupled Systems.* John Wiley & Sons, 1984.
19. B. Simeon. On Lagrange multipliers in flexible multibody dynamics. *Comp. Meth. Appl. Mech. Eng.*, 2006 (in press). DOI: doi:10.1016/j.cma.2005.04.015 .
20. O. Brüls, P. Duysinx, and J.C. Golinval. A unified finite element framework for the dynamic analysis of controlled flexible mechanisms. In J.M. Goicolea, J. Cuadrado, and J.C. García Orden (eds), *Proc. of Multibody Dynamics 2005 (ECCOMAS Thematic Conference)*, Madrid, Spain, 2005.
21. A. Heckmann. *The modal multifield approach in multibody dynamics.* Fortschritt-Berichte VDI Reihe 20, Nr. 398. VDI-Verlag, Düsseldorf, 2005.
22. J.L.H. Nowinski. *Theory of Thermoelasticity with Applications.* Sijthof & Noordhoff International Publishers B.V., Alphen aan den Rijn, Netherlands, 1978.
23. B. Schweizer and J. Wauer. Atomistic explanation of the Gough–Joule effect. *The European Physical Journal*, B23:383–390, 2001.
24. H.J. Bathe. *Finite Element Procedures.* Prentice Hall, New Jersey, 1996.
25. M.A. Biot. *Variational Principles in Heat Transfer.* Oxford University Press, Oxford, UK, 1970.

26. B.A. Boley and J.H. Weiner. *Theory of Thermal Stresses*. Dover Publications, Mineola, New York, 1997.
27. R. Schwertassek and O. Wallrapp. *Dynamik flexibler Mehrkörpersysteme*. Vieweg, 1999.
28. W. Schiehlen. Multibody system dynamics: roots and perspectives. *Multibody System Dynamics*, 1:149–188, 1997.
29. S. Dietz. *Vibration and Fatigue Analysis of Vehicle Systems using Component Modes*. Fortschritt-Berichte VDI Reihe 12, Nr. 401. VDI-Verlag, Düsseldorf, 1999.
30. ANSYS© Release 7.1 Theory Reference. ANSYS, Inc., 2003.
31. SIMPACK Reference Manual, Release 8.5. INTEC GmbH, Wessling, Germany, 2002.

Computational Methods in Applied Sciences

1. J. Holnicki-Szulc and C.A.M. Soares (eds.): *Advances in Smart Technologies in Structural Engineering*. 2004 ISBN 3-540-22331-2
2. J.A.C. Ambrósio (ed.): *Advances in Computational Multibody Systems*. 2005 ISBN 1-4020-3392-3
3. E. Oñate and B. Kröplin (eds.): *Textile Composites and Inflatable Structures*. 2005 1-4020-3316-8
4. J.C. Garcia Orden, J.M. Goicolea and J. Cuadrado (eds.): *Multibody Dynamics*. Computational Methods and Applications. 2007 ISBN 1-4020-5683-4